Advances in Aquatic Ecology

— Volume 6 —

Advances in Aquatic Ecology

— Volume 6 —

Editors
Dr. Vishwas B. Sakhare
Head,
Post Graduate Department of Zoology
Yogeshwari Mahavidyalaya,
Ambajogai – 431 517
Maharashtra
INDIA

&

Dr. Patricio René De los Ríos Escalante
School of Environmental Sciences,
Catholic University of Temuco
CHILE

2012
DAYA PUBLISHING HOUSE®
New Delhi - 110 002

ISBN 978-81-7035-782-7

Published by : **Daya Publishing House®**
4760-61/23, Ansari Road, Darya Ganj,
New Delhi - 110 002
Phone: 23245578, 23244987
Fax: (011) 23260116
E-mail: dayabooks@vsnl.com
website: www.dayabooks.com

Laser Typesetting : **Classic Computer Services**
Delhi - 110 035

Printed at : **Chawla Offset Printers**
Delhi - 110 052

PRINTED IN INDIA

Preface

The present book entitled *'Advances in Aquatic Ecology Volume 6'* is the compilation of esteemed articles of internationally acknowledged experts in the field of aquatic ecology with the intention of providing a sufficient depth of the subject to satisfy the needs at a level which will e comprehensive and interesting. With its application oriented and interdisciplinary approach, we hope that the students, teachers, researchers, scientists, policy makers and environmental lawyers will find this volume much more useful.

The authors express their gratitude to the Director, Research and Post Graduate Section, University of Temuco, Chile for the financial assistance for the publication of this volume.

Our special thanks and appreciation goes to experts and research workers whose contributions have enriched this volume. We express our sincere gratitude to Hon'ble Dr. S.T. Khursale, President, Yogeshwari Mahavidyalaya, Ambajogai who has been source of constant inspiration. We are especially thankful to Dr. R.V. Kirdak, Principal and Dr. R.D. Joshi, Vice Principal of Yogeshwari Mahavidayalaya, Ambajogai for encouragements.

We are indebted to publisher Shri Anil Mittal of Daya Publishing House, Delhi for taking pains in bringing out the book. We take this opportunity to thank everybody who had helped us in preparation of this volume.

Editors

Contents

List of Contributors

Begum, Nafeesa
Department of Zoology, Shayadri Science College (Autonomous), Shivamogga – 577 203

Benarjee, G.
Fisheries Research Lab, Department of Zoology, Kakatiya University, Warangal – 506 009

Bharti, P.K.
Department of Zoology and Environmental Science, Gurukula Kangri, University, Hardwar – 249 404

Bharti, Umesh
Department of Zoology and Environmental Science, Gurukul Kangri University, Hardwar – 249 404

Bhivgade, S.W.
Department of Botany, Yogeshwari Mahavidyalaya, Ambajogai –431 517

Chavan, B.R.
Associate Research Officer, Marine Biological Research Station (Dr. BSKKV), Ratnagiri

Chavan, S.P.
P.G. Department of Zoology, Dnyanopasak College, Parbhani – 431 401

Ekhande, A.P.
Department of Zoology The M.S. University of Baroda, Vadodara – 390 002

Gaurav
Shriram Institute for Industrial Research, Delhi – 110 007

Girkar, M.M.
College of Fishery Science, Udgir, Maharashtra

Gore, S.B.
College of Fishery Science, Udgir, Maharashtra

Goudar, Mahesh Anand
Department of Chemistry, D.V.S. College of Arts and Science, Shivamogga – 577 203

Gowri, P.
Fisheries Research Lab, Department of Zoology, Kakatiya University, Warangal – 506 009

Haregaonkar, R.B.
Post Graduate Department of Zoology, Yogeshwari Mahavidyalaya, Ambajogai – 431 517

Hosmani, S.P.
Department of Studies in Zoology, University of Mysore, Manasagangotri, Mysore – 570 006, Karnataka

Injal, A.S.
Department of Zoology, R.P. Gogate College of Arts and Science and R.V. Jogalekar College of Commerce, Ratnagiri – 415 612

Jalilzadeh, Koorosh
Fresh Water Microbial Ecology Laboratory, Department of Zoology, University of Mysore, Manasagangothri, Mysore – 570 006

Jindal, Meenakshi
Department of Zoology and Aquaculture, CCS HAU, Hisar – 125 004

Jothi Lakshmi, S.
Department of Zoology, The Madura College (Autonomous), Madurai – 625 011

Khabade, S.A.
Department of Zoology, D.K.A.S.C. College, Ichalkaranji, Maharashtra

Khade, S.K.
Department of Botany, D.K.A.S.C. College, Ichalkaranji, Maharashtra

Khajure, P.V.
Department of Marine Biology, Karnataka University Post Graduate Centre, Karwar – 581 303

Krishan Ram, H.
Department of Studies in Zoology, University of Mysore, Manasagangotri, Mysore

Kulkarni, A.S.
Department of Zoology, R.P. Gogate college of Arts and Science and R.V. Jogalekar College of Commerce, Ratnagiri – 415 612

Kumar, Pawan
Department of Zoology and Environmental Science, Gurukul Kangri University, Hardwar – 249 404, Uttarakhand

Latha, K. Sree
Department of Zoology, Dr. S.R.K. Government Arts College, Yanam – 533 464

Malik, D.S.
Department of Zoology and Environmental Science, Gurukula Kangri, University, Hardwar – 249 404

Manjare, S.A.
Department of Zoology, Shivaji University, Kolhapur

Manjunath, R.
Department of Studies in Zoology, University of Mysore, Manasagangotri, Mysore

Mohan, M. Ramachandra
Department of Zoology, Bangalore University, Bangalore – 560 056

Mruthunjaya, T.B.
Department of Studies in Zoology, University of Mysore, Manasagangotri, Mysore – 570 006

Mule, M.B.
Department of Environmental Science, Dr. Babasaheb Ambedkar Marathwada University, Aurangabad, Maharashtra

Muley, D.V.
Department of Zoology, Shivaji University, Kolhapur

Murugesan, P.
Centre of Advanced Study in Marine Biology, Faculty of Marine Sciences, Annamalai University, Parangipettai – 608 502

Muthuvelu, S.
Centre of Advanced Study in Marine Biology, Faculty of Marine Sciences, Annamalai University, Parangipettai – 608 502

Naik, K.L.
Department of Zoology, Sahayadri Science College (Autonomous), Shivamogga – 577 203

Naik, U.G.
Department of Studies in Marine Biology, Karnataka University Post Graduate Centre, Kodibag, Karwar – 581 303

Nikam, S.M.
Department of Zoology, R.P. Gogate college of Arts and Science and R.V. Jogalekar College of Commerce, Ratnagiri – 415 612

Niture, S.D.
Department of Zoology, JES College, Jalna – 431 203

Padate, G.S.
Department of Zoology, The M.S. University of Baroda, Vadodara – 390 002

Pandure, N.B.
Department of Botany, Dr. Babasaheb Ambedkar Marathwada University, Aurangabad

Patil, J.V.
Department of Zoology, The M.S. University of Baroda, Vadodara – 390 002

Patil, V.G.
Department of Botany, D.K.A.S.C. College, Ichalkaranji, Kolhapur, Maharashtra

Patil, Y.T.
College of Fishery Science, Udgir, Maharashtra

Patole, V.M.
Dr. Balasaheb Khardekar College, Vengurla, Maharashtra

Rathod, J.L.
Department of Studies in Marine Biology, Karnataka University Post Graduate Centre, Kodibag, Karwar – 581 303

Ravi, V.
Centre of Advanced Study in Marine Biology, Faculty of Marine Sciences, Annamalai University, Parangipettai – 608 502

Reddy, B. Laxma
Fisheries Research Lab, Department of Zoology, Kakatiya University, Warangal – 506 009

Roopa, S.V.
Department of Studies in Marine Biology, Karnataka University Post Graduate Centre, Kodibag, Karwar – 581 303

Salve, U.S.
Department of Botany, Sawarkar College, Beed, Maharashtra

Sanap, G.B.
Post Graduate Department of Zoology, Yogeshwari Mahavidyalaya, Ambajogai – 431 517

Sathe, S.S.
Department of Botany, P.D.V.P. College, Tasgaon, Maharashtra

Sayeswara, H.A.
Department of Zoology, Sahayadri Science College (Autonomous), Shivamogga – 577 203

Sharma, Kavita
Department of Zoology and Aquaculture, CCS HAU, Hisar – 125 004

Shivabasavaiah
Department of studies in Zoology, University of Mysore, Manasagangotri, Mysore

Singh, Vijender
Shriram Institute for Industrial Research, Delhi – 110 007

Sitre, S.R.
Department of Zoology, N.S. Science and Arts College, Bhadrawati – 442 902

Tandale, A.T.
College of Fishery Science, Udgir, Maharashtra

Taware, A.S.
Department of Botany, Yogeshwari Mahavidyalaya, Ambajogai – 431 517

Tendulkar, M.V.
Department of Zoology, R.P. Gogate College of Arts and Science and R.V. Jogalekar College of Commerce, Ratnagiri – 415 612

Thamarai Selvan, R.
Department of Zoology, Thiagarajar College (Autonomous), Madurai – 625 009

Thirumurugan, R.
Department of Zoology, The Madura College (Autonomous), Madurai – 625 011

Todkari, S.S.
College of Fishery Science, Udgir, Maharashtra

Vasanthkumar, B.
Department of Zoology, Government Arts and Science College, Karwar – 581 303

Vhanalkar, S.A.
Department of Zoology, Shivaji University, Kolhapur

Vignesh, V.
Department of Zoology, The Madura College (Autonomous), Madurai – 625 011

Vijaykumar, K.
Environmental Biology Research Unit, Department of Zoology, Gulbarga University, Gulbarga – 585 106

Yakupitiyage, A.
Aquaculture and Aquatic Resources Management (AARM), SERD, Asian Institute of Technology, Pathumthani, Thailand

Yamakanamardi, Sadanand M.
Fresh Water Microbial Ecology Laboratory, Department of Zoology, University of Mysore, Manasagangothri, Mysore – 570 006

Yeragi, S.G.
K.J. Somaiya College of Commerce and Science, Vidyavihar (E), Mumbai

Yeragi, S.S.
K.J. Somaiya College of Commerce and Science, Vidyavihar (E), Mumbai

Zade, S.B.
PG Department of Zoology, RTM Naagpur University Campus, Amravati Road, Nagpur – 440 033

Chapter 1

A Preliminary Study on Biotic Community and Water Quality of Umaim River in East Khasi Hills, Meghalaya

☆ *Umesh Bharti, P.K. Bharti, D.S. Malik, Pawan Kumar and Vijender Singh*

ABSTRACT

Umaim river is a major river of East Khasi Hills district in Meghalaya along with many small tributaries near Bangladesh Border. It has a great importance as a natural habitat among the various ecosystems of the region, whereas there is a large scale limestone mining area and no more industrial and agricultural pollution. Anthropogenic factors mainly activities of local people and catchments runoff may influence the index of nutrients in river water and may alter the physico-chemical characteristics and also the whole water quality and the structure of biotic community. Physico-chemical parameters play an important role into niche restoration maintenance, self-regulation of water quality. Location variation in nutrients concentration of the river was studied with special reference to physico-chemical parameters and heavy metals in the river water.

Calcium and magnesium were observed in very low range in Umaim river water at upstream as well as downstream in comparison to rivers and hill-streams of North India. Dissolved oxygen was found 9.2 to 11.0 mg/l at upstream and downstream during the study period. Heavy metals were found almost nil or in very low concentrations at both selected site. The present study deals with the preliminary physico-chemical characteristics of the river water and exhibits the natural quality of water with minimum anthropogenic activities, which contribute in pollution load of an ecosystem.

Introduction

Water is a resource circulated throughout all ecosystems and is one of the most important factors of them (Odum 1971). From the point of view of ecohydrology, for the main elements of an ecosystem including inorganic and organic compound, producers and consumers both the quantity and the chemistry of water have a decisive impact on its biomass and biotic composition. Information obtained from current hydro-chemical analyses of water specimens is very important but it is sometimes difficult to gauge the overall situation especially the occurrence of nutrient, which may be very important to an ecosystem (Malik and Bharti, 2005a). Ecological, geo-chemical and hydrological research has been carried out in various ecosystems to understand the factors controlling the chemistry of natural water (Baron and Bricker, 1990; Malik and Bharti, 2005a). Many of such studies have been under taken during the last two decades to understand the processes that control the hydro-chemistry of alpine and sub-alpine systems of North America and Europe (William *et al.*, 1993 and Psenner, 1989). Similar studies were also done in China and central Asia (Xue and Schnoor, 1994).

Indian mountains are the cradle of a large number of streams and mighty rivers. In North-East region and very close to Bangladesh border, a hill-stream Umaim, 90 Km. far away from Shillong city is the study site for the accounting of physico-chemical parameters and heavy metals in natural water and fluctuations at the different locations along with the stream. Adequate understanding of the North-East regional streams is extremely important for the development of a realistic program for utilizing the potential of water that exist in the form of hidden water resource in the area. The impact of anthropogenic pollution from industrial, agricultural, sources, quarrying and tourists activity on water quality has concerned environmentalists and scientists for the past three decades.

The present study reveals to characteristics the water nutrient chemistry, influenced by anthropogenic activity and quarrying of the geologically sedimental environments and to determine the nature and degree of anthropogenic impacts on qualitative and quantitative variations occurred in nutrients in relation to physico-chemical parameters and heavy metals of stream water.

Study Area

The seven states of North-East India are quite famous as Seven Sisters, and Meghalaya is very well known for the record of rainfall, the state consists of the two places namely Cherrepunjee and Mawsinram for maximum rainfall throughout the year. So, the maximum water resources are depending chiefly on total precipitations of the region. The Meghalaya has been surrounded by the natural beauties with lovely trees and cool climate, which is not only a pleasant place to live in but relaxing for holiday also. This is not only a popular state but also valuable or important place for tourism, Shillong is the capital of Meghalaya state. Mowlong Cherra Cement Ltd and Lafarge Umaim Mining Ltd are the major industries of the region.

Umaim river originated from the hills of Assam-Meghalaya basically. Upstream of Umaim is near village Dissong and Downstream near village Pyrkan. The water of Umaim is very useful for the local people. Population of the region is completely depending upon the river water for drinking, bathing, and other activities. Geologically, North-East hills are enriched with various minerals and the hills near Bangladesh are rich in limestone. Geographically, the study area is situated in the globe on a Latitude 25° 11′ 40.8″ N and Longitude 91° 38′ 16.4″ E for Upstream site and Latitude 25° 10′ 43.8″ N and Longitude 91° 38′ 19.2″ E for downstream site. Meteorologically region has a cool and pleasant climate.

Materials and methods

The water samples were collected from Umaim River upstream (U/S sampling site A) near Disong village and downstream (D/S sampling site B) near Shella Bazar according to the analytical requirement in morning period 9:00 Hrs. to 10:00 Hrs.

The samples for physico-chemical parameters and heavy metals were collected by using rinsed Borosil glassware, and analyzed with the help of the procedure described by APHA (1995) and Trivedi and Goel (1984). Colour, odour, turbidity, velocity, temperature and Dissolved Oxygen were analyzed on spot at sampling sites. Samples were collected from selected sites and immediately preserved in ice boxes, and transfer to the lab for further analysis. Water samples were digested and heavy metals were detected using Atomic Absorption Spectrophotometer. The study of biotic community of Umaim river was carried out with the help of Schwoerbel (1991), Santhanam, *et al.* (1989). The macro-benthic biota and plankton were identified with the help of Edmondson (1959).

Results and Discussion

Umaim river is flowing throughout a valley of North-east hills chain near the Bangladesh Border, enriched with limestone and lignite rocks, which affect the water quality of stream according to the locations. Nutrients concentration, heavy metals and related physico-chemical parameters from selected sites are depicted in Tables 1.1–1.3.

Table 1.1: Physical Characteristics of Umaim River Water

Sl.No.	*Parameters*	*Unit*	*Umaim Upstream*			*Umaim Downstream*			*Desirable Limit*
			Winter	*Summer*	*Monsoon*	*Winter*	*Summer*	*Monsoon*	
1.	Temperature	°C	13	17	16	12	18	15	–
2.	Colour	–	Clear	Clear	Clear	Clear	Clear	Clear	–
3.	Odour	–	Nil	Nil	Nil	Nil	Nil	Nil	–
4.	Turbidity	NTU	2	3	8	3	1	5	5
5.	Velocity	m/s	0.2	0.2	0.3	0.1	0.1	0.3	–
6.	TDS	mg/l	41	43	48	46	48	55	500

Table 1.2: Chemical Characteristics of Umaim River Water

Sl.No.	*Parameters*	*Unit*	*Umaim Upstream*			*Umaim Downstream*			*Desirable Limit*
			Winter	*Summer*	*Monsoon*	*Winter*	*Summer*	*Monsoon*	
1.	pH	–	7.1	6.9	7.4	6.8	7.5	7.3	6.5-8.5
2.	Alkalinity	mg/l	22	25	30	28	32	28	200
3.	Total Hardness	mg/l	22	34	26	42	32	24	300
4.	Calcium	mg/l	8	10	8	12	14	8	75
5.	Magnesium	mg/l	1	3	4	1	3	2	30
6.	Chlorides	mg/l	7	6	6	10	5	7	250
7.	DO	mg/l	10.4	11.0	9.5	10.5	10.6	9.2	–
8.	BOD	mg/l	Nil	Nil	Nil	Nil	Nil	Nil	–
9.	COD	mg/l	4	3	5	6	5	6	–

Umaim river has the spatio-temporal variations of water temperature, which plays a vital role in all physico-biochemical reactions and self-purification power of aquatic system (Badola and Singh, 1981). Higher value of temperature was found 18 °C in summer at downstream and minimum 12 °C in winter season at downstream. Turbidity is striking characteristic of the physical status of the water bodies. Although in Umaim river water is clear because there is no more pollution, siltation was the main source of turbidity in tributaries. Detritus and other non-organic material being added to water mass due to rainfall and anthropogenic activities (Camron, 1996). Maximum turbidity was recorded 8 NTU during rainy season at upstream and minimum 1 NTU in summer season at downstream. The maximum depth of photic zone provides the better biological production for all aquatic organisms especially phytoplankton (Malik and Bharti, 2005b).

Table 1.3: Heavy Metals in Umaim River Water

Sl.No.	Parameters	Unit	Umaim Upstream			Umaim Downstream			Desirable Limit
			Winter	Summer	Monsoon	Winter	Summer	Monsoon	
1.	Cadmium	mg/l	BDL	BDL	BDL	BDL	BDL	BDL	0.01
2.	Copper	mg/l	BDL	BDL	BDL	0.01	BDL	0.01	0.05
3.	Iron	mg/l	0.06	0.08	0.09	0.06	0.02	BDL	0.3
4.	Lead	mg/l	0.01	0.02	BDL	BDL	BDL	BDL	0.05
5.	Manganese	mg/l	BDL	0.03	0.01	BDL	BDL	BDL	0.1
6.	Zinc	mg/l	BDL	0.03	0.03	0.07	0.05	BDL	5.0

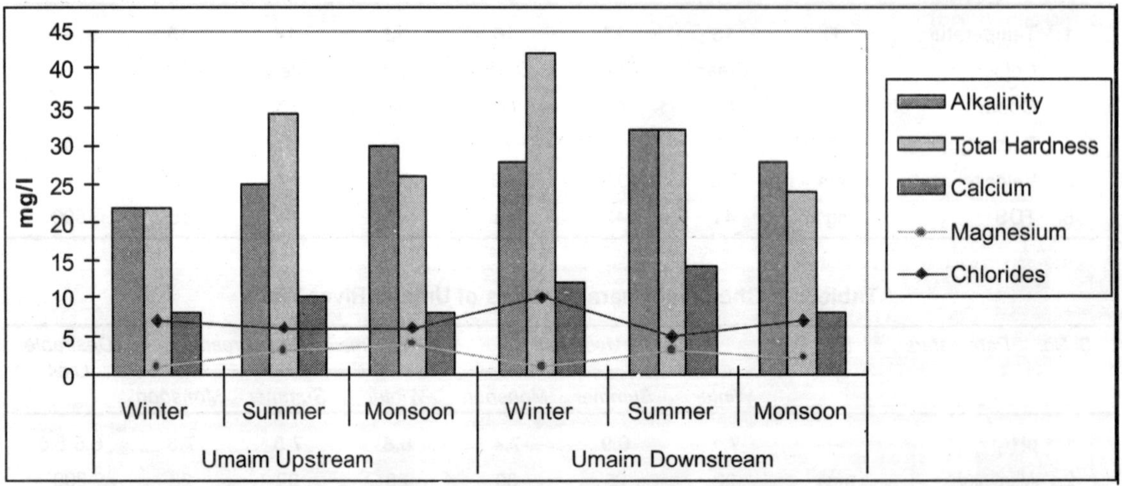

Figure 1.1: Showing the Nutrients Concentration in Umaim River during the Study Period

Total dissolved solids were found in the range of 41 mg/l in to 55 mg/l in monsoon season, due to the gradual increases in velocity of river which favoured effective sedimentation (Subramanian, 1979). Chemical oxygen demand (3-6 mg/l) represents chemically oxidizable organic matter load in water, while biochemical oxygen demand is only biodegradable materials (Malik and Bharti, 2005c). In the present study the values observed during monsoon months may be attributed maximum biological activities and high temperature, stimulate the growth of microorganisms (William *et al.*, 1993).

The pH of natural water was controlled in a great extent by the interaction of hydroxyl ions arising from the hydrolysis of bicarbonate (Sharma, 1986). The pH of Umaim river was recorded slightly alkaline (6.8–7.5). Total hardness is mainly due to percentage of calcium and magnesium salts of bicarbonates, carbonates, sulphates and chlorides, while the value of alkalinity occured due to presence of bicarbonates. The highest concentration of hardness was analyzed 22-32 mg/l during the study. Alkalinity was also found 22-42 mg/l with a small fluctuation. A positive relationship between hardness and alkalinity was recorded in river Ganga at Rishikesh (Chopra and Patric, 1994).

Maximum chloride concentration was recorded (10 mg/l) in winter and minimum in summer (5 mg/l). Chloride and hardness showed a positive relationship to one another (Chopra and Patric, 1994). Chloride was found in the form of chloride ion, and one of the major inorganic anion present in natural water (Malik and Bharti, 2009).

Calcium and magnesium the dominant cations, and these represent the main weathering products, but significant hydro-chemical differences between the two sampling sites associated with the bedrock geology exist (Jenkins *et al.*, 1995). Calcium is one of the essential nutrients, which plays an important role in biological system. Maximum calcium concentration was recorded (14 mg/l) in summer and minimum in monsoon (8 mg/l). Positive relationship between, calcium and temperature was also reported by Khanna and Singh (2000) in river Suswa, Dehradun. Magnesium is also an essential element but it is toxic at higher concentration. The concentration of magnesium in Umaim river was found maximum (4 mg/l) and minimum (1 mg/l) and it was very low in comparison to Hill-streams of Uttarakhand (Bharti, 2004).

During the summer season nutrients concentration in rivers and hill-streams became more. Miller *et al.*(1997) described the nutrients availability in selected environmental settings of the Potomac River and Cameron (1996) showed the similar type of fluctuation in Fraser river of British Columbia. Bond (1979) described similar nutrients concentration pattern in a stream draining a mountain ecosystem in Utah.

Heavy metals like cadmium, copper, iron, manganese, lead and zinc were not found in high concentration at both sites during any season. Cadmium was absolutely absent in all seasons, while copper, lead and manganese concentrations were also found below detection limit in maximum samples. The concentration of iron was maximum observed 0.09 mg/l during monsoon season. Manganese was found maximum 0.03 mg/l in summer and zinc concentration was 0.07 mg/l in monsoon season. Malik *et al.* (2009) described the role of heavy metals in the surface water of north India. The results of Bharti *et al.*(2010) were also indicated the relation of heavy metals and phytoplankton in a north Indian water body.

The concentrations of calcium and magnesium were observed in very low range in Umaim river water at upstream as well as downstream in comparison to rivers and hill-streams of North India. Dissolved oxygen was found 9.2 to 11.0 mg/l at upstream and downstream during the study period. Heavy metals were found almost nil or in very low concentrations at both selected site. The present study deals with the preliminary physico-chemical characteristics of the river water and exhibits the natural quality of water with minimum anthropogenic activities, which contribute in pollution load of an ecosystem.

The present results conclude that significant differences in river water nutrient concentrations exist among different environmental settings within the two subunits. The environmental setting with the highest potential by more soluble nutrients, fluctuations in nutrient concentrations were the land use and carbonate bedrock that was predominated in the North-East valley. The spatial variations in TDS are attributed to climatic and lithological control over the ionic concentrations. Absence or low concentration of heavy metals shows that the water is still industrial pollution free. However, the Indian Standards are not so strict (Bharti, 2007), but heavy metals never cross the limits during the study period. On the basis of nutrients and heavy metals, the water may be considered for drinking and other recreational purposes.

The preliminary study of biotic community of Umaim ecosystem was also carried out in a lengthy stretch of Umaim basin. The genera, species and their related category are given in Table 1.4 with the scientific names. Rundle *et al.* (1993) also observed the similar species of biota in the river of Nepal. Plankton of chlorophyceae family was found dominantly in Umaim river ecosystem. Planktons are the primary producers of and aquatic ecosystem (Hynes, 1970). So, ecosystem always balanced by the presence of each and every necessary species of flora and fauna.

Table 1.4: Preliminary Study of Biotic Communities in Umaim River

Category	*Family/Phylum*	*Genus/Species*
Neuston	**Arthopods**	*Mayfly nymph, Water spider*
Plankton	**Phytoplankton**:	
	Chorophyceae–	*Volvox, Chlorella, Ulothrix, Vaucheria, Zygnema*
	Bacillariophyceae–	Diatoms
	Rhodophyceae–	*Batrachospermun*
	Cyanophyceae–	*Rivularia, Nostoc*
	Zooplankton:	
	Protozoans–	*Vorticalla, Paramoecium, Amoeba, Chrysamocha, Volvox, Cyclopes*
	Coelentrata–	*Hydra, Ceratella*
Necton	Amphebia–	*Rana*
	Pisces–	*Schizothorax richardsonii (Gray)- Asella, Barilius barila (Ham.)*
Benthos	Macroinvertebrate–	*Snails, Beetles, Sepia*
	Annelida–	*Pheretima, Leeches, Flatworms,*
Aquatic macrophytes	Bryophyta–	*Moss, Fern, Weedy rooted aquatic herbs*
	Rannunculaceae–	*Rannunculus*

References

APHA, 1995. *Standard Methods for Examination of Water and Wastewater, 19th edn.* American Public Health Association Inc., New York, pp. 1970.

Badola, S.P. and Singh, H.R., 1981. Hydrobiology of the river Alaknanda of Garhwal Himalaya. *Indian J. Ecol.*, 8(2): 269–276.

Baron, J. and Bricker, O.P., 1990. Hydrological and chemical flux in Loch Vale watershed, Rocky Mountain National Park. In: *Biogeochemistry of Major Rivers*, SCOPE 42. Wiley and Sons, New York, USA.

Bharti, P.K., 2004. Limnobiological study of Sahastradhara hill-stream at Dehradun. *M.Sc. Dissertation*, Gurukula Kangri University, Hardwar, pp. 102.

Bharti, P.K., 2007. Why Indian Standards are not so strict? *Current Science*, 93(9): 1202.

Bharti, P.K., Malik, D.S. and Rashmi Yadav, 2010. Influence of heavy metals on abundance of cyanophyceae members in three spring-fed lake in Kempty, Dehradun. In: *Advances in Aquatic Ecology, Vol. 3*, (Ed.) V.B. Sakhare, Daya Publishing House, New Delhi, pp. 107–111.

Bond, H.B., 1979. Nutrient concentrations patterns in a stream draining a montane ecosystem in Utah. *Ecology*, 60(6): 1184–1196.

Cameron, E.M., 1996. Hydrogeo-chemistry of the Fraser River British Columbia: Seasonal variation in major and minor components. *J. Hydrol.*, 182(1–4): 209–255.

Chopra, A.K. and Patrick, N.J., 1994. Effect of domestic sewage on self-purification of Ganga water at Rishikesh. *A. Bio. Science*, 13(11): 75–82.

Edmondson, W.T., 1959. *Freshwater Biology – Rotifera*, 2nd edn. Wiley and Sons, New York, pp. 420–484.

Hynes, H.B.N., 1970. *The Ecology of Running Waters*. Liverpool University Press, Liverpool, p. 1–555.

Jenkins, A., Sloan, W.T. and Cosby, B.J., 1995. Stream chemistry is the middle hills and high mountain of the Himalaya, Nepal. *Journal of Hydrology*, 166(1–4): 61–79.

Khanna, D.R. and Singh, R.K., 2000. Seasonal fluctuations in the plankton of Suswa River at Raiwala Dehradun. *Env. Conservations J.*, 1(2 and 3): 89–92.

Malik, D.S. and Bharti, P.K., 2005b. Fluctuation in planktonic population of Sahastradhara hill-stream at Dehradun (Uttaranchal). *Aquacult.*, 6(2): 191–198.

Malik, D.S. and Bharti, P.K., 2005c. Primary production efficiency of Sahstradhara hill-stream, Dehradun, *Env. Cons. J.*, 6(3): 117–121.

Malik, D.S. and Bharti, P.K., 2009. Ecology of Sahastradhara hill-stream at Dehradun (Uttaranchal). In: *Advances in Aquatic Ecology*, Vol. 1, (Ed.) V.B. Sakhare. Daya Publishing House, New Delhi, pp. 1–11.

Malik, D.S. and Bharti, P.K., Negi, K.S. and Rashmi Yadav, 2009. Distribution of metals in water of an artificial lake at Mussoorie, Uttarakhand. In: *Aquatic Biology and Aquaculture*, (Ed.) V.B. Sakhare. Manglam Publication, New Delhi, pp. 77–95.

Malik, D.S. and Bharti, Pawan K., 2005a. Nutrient dynamics in Rhithron zone of Shivalik Himalayan stream Sahastradhara, Dehradun (Uttaranchal). *Env. Cons. J.*, 6(2): 63–68.

Miller, C.V., Denis, J.M., Ator, S.W. and Brakebill, J.W., 1997. Nutrients in stream during baseflow in selected environmental settings of the Potomac river basin. *J. American Wat. Resources Association*, 33(6): 1155–1171.

Odum, E.P., 1971. *Fundamentals of Ecology*. W.B. Saunders and Co., New York, pp. 1–344.

Psenner, R., 1989. Chemistry of high mountain lakes in siliceous catchments of central Alps. *Aquatic Sci.*, 51: 108–128.

Rundle, S.D., Jenkins, A. and Ormerod, S.J., 1993. Macroinvertebrate communities in streams in the Himalyas, Nepal. *Freshwater Biol.*, 30: 169–180.

Santhanam, R., Velayntham, P. and Jegatheesan, G., 1989. *A Manual of Freshwater Ecology*. Daya Publishing House Delhi, p. 3–17, 53–92.

Schwoerbel, J., 1991. *Handbook of Limnology*. Ellis Horwood Limited, Chichester, England, pp. 95–114.

Shrama, R.C., 1986. Effect of physico-chemical factors on benthic fauna of Bhagirathi River Garhwal Himalaya. *Indian. J. Ecol.*, 13(1): 133–137.

Subramanian, V., 1979. Chemical and suspended sediment characteristics of river of India. *J. Hydrol.*, 44: 37–55.

Trivedi, R.K. and Goel, P.K., 1984. *Chemical and Biological Methods for Water Pollution Studies.* Environmental Publication, Karad, pp. 1–25.

William, M.W., Brown, A. and Melack, J.M., 1993. Geochemical and hydrologic controls on the composition of surface water in the high elevation basin, *Sierra Navada. Limnol. Oceanogr.*, 38: 775–797.

Xue, H.B. and Schooner, J.L., 1994. Acid deposition and lake chemistry in south-west China. *Wat. Air, Soil Pollut.*, 75: 61–78.

Chapter 2

Biodiversity of Yashwant Lake of Toranmal: Seasonal Variation in Molluscs Density and Species Richness

☆ *A.P. Ekhande, J.V. Patil and G.S. Padate*

ABSTRACT

The Yashwant Lake is a high altitude (3300 feet) freshwater ecosystem in the Western part of Satpura range and its biocoenosis is ideal sentinel systems to detect environmental changes. In a two year study of molluscs at Yashwant Lake only six species belonging to five families were recorded. However, their densities were high. The results reveal that the maximum and minimum density and species richness are recorded in post-monsoon and winter respectively. Percentage composition shows that all the species recorded are either Eudominant or dominant at any of the three sites. A Pearson correlation is carried out between mollusc density and various physicochemical and biotic parameters like plankton density and bird density. The results confirm that the density, species richness and domination index are not influenced by any single environmental variables.

Introduction

Molluscs, the soft bodied macroinvertebrates with calcarious shell are among the major components of a wetland ecosystems and play an important role in interconverting and transporting nutrients from one trophic level to another influencing nutrient cycle. This process is important as it maintains the general health of water body by converting the organic matter such as leaf litter and detritus into food, and they themselves become the main source of food for higher water dependent organisms forming an important link in the wetland food web (Hart and Newman, 1995; Ramchandra *et al.*, 2002). Research on selection of habitat by waterfowl and their food habitats suggests that aquatic macro-invertebrates are important factors in determining avian use of marsh area (Murkin and Kadlec,

1986). These macro invertebrates respond differently in response to the fluctuations in water level as well as water cover (Tronstad *et al.*, 2005). Classified in the phylum mollusca this large group of animals is having diverse shapes, sizes and habits and occupy different habitats (Subba Rao, 1989).

Agriculture, municipal wastewater and recreational development have direct impact on local water quality in a wetland. These factors in turn are reported to affect species composition and abundance of gastropod and freshwater mussel communities (Pip, 2006). Those water bodies that have been subjected to less anthropogenic pressures have provided a refuge for many freshwater species (Collinson, 1995; Williams, 2004). Thus, the macro invertebrates in general can also be the robust indicators for biological assessment of wetlands (Doherty *et al.*, 2000) as they respond quickly to the changes in their physical, chemical or biological parameters (Stansly *et al.*, 1997). It has been reported that freshwater gastropod assemblages are structured by a multitude of variables at regional and local scales (Lodge *et al.*, 1987) that provide the explanation for the same (Pyron *et al.*, 2009). Further, area effects on gastropod distribution, species richness and abundance in local habitats are common (Quintero, 2007).

The present study of Yashwant Lake attempts to evaluate the seasonal variations in species richness and density of molluscs and its correlation to abiotic factors. Freshwater molluscs have been known to play significant roles in the public and veterinary health and thus need to be explored more extensively and scientifically as the local tribal people use such water bodies for domestic, agriculture and holy purpose.

Materials and Methods

The Yashwant Lake is one of the major reservoirs of Toranmal plateau located in Nandurbar district of Maharashtra, India. It is situated around 21° 52′ 49″ North latitude and 74° 27′ 44″ East longitude at an altitude of 3300 Ft. where biodiversity has not been documented. It is one of the oldest manmade lakes of British time, constructed by damming the deep gorge. It has a circumference of 3.14 Km. The climate is cooler when compared to overall climate of the area with temperature fluctuating between 5°C in winter to 38°C in summer. Three stations of Yashwant Lake were selected for sampling, YLA, YLB and YLC. YLA is situated on East shore of the lake and is having rocky and sandy floor, least vegetation and maximum human interference, while YLB and YLC have somehow similar characters with muddy floor, maximum macrophytes and less human interference. YLB is situated on West shore while YLC is on South-West bank of the Lake.

The present study was carried out during December 2006 to November 2008. Sampling was carried out by biweekly visit and monthly mean is taken for calculation. Representative soil samples were collected on each sampling day from the selected stations (*Viz.* YLA, YLB, YLC) using a standard soil corer. The soil collected was sieved and the molluscs were collected in a separate sample bottle as described by Michael (1986) and Tronstad *et al.*(2005). The collected molluscs were preserved in 4 per cent formalin and carried to the laboratory for quantitative and qualitative estimation. Identification was done as per the key provided by Subba Rao (1989).

The density was calculated using following formula:

Density = Number of mollusc/volume of the corer converted to m^3

The physico-chemical parameters were analyzed by using standard methods of analysis as per APHA (1998) and Michael (1986). The data for three months is pooled into four seasons' Summer (March, April, and May), Monsoon (June, July, and August), Post monsoon (Sept., Oct., and Nov.) and Winter (Dec., Jan., and Feb.). Further the mean, standard error of mean (SEM) and one way ANOVA with no post test for density of molluscs for the four seasons as well as 3 stations was performed using

Prism Graph Pad version 3.00 for window. The P value for ANOVA is non significant if $P > 0.05$(ns), significant if $P < 0.05$(*), significantly significant (**) if $P < 0.001$ and highly significant (***) if $P < 0.0001$.

The Pearson correlation between molluscan density and various biotic and abiotic parameters was carried out using SPSS 7.5 software for window, if ** correlation is significant at the 0.01 level (2-tailed), and if * correlation is significant at the 0.05 level (2-tailed).

The percentage density of each species was calculated as domination index as described by Iga and Adam, 2006, as follow:

$$DO = na/n \times 100.$$

where,

na: The number of individuals of species.

n: The total number of individuals in the sample.

The values of the domination index DO were divided into five classes according to Gorny and Grum (1981): eudominants > 10.0 per cent of sample, dominants 5.1–10 per cent of sample, subdominants 2.1–5.0 per cent of sample, recedents 1.0–2.1 per cent of sample, subrecedents < 1.0 per cent of the sample.

Results

Altogether only five species belonging to four Gastropod families and one belonging to Pelecypoda were found from the three Stations (YLA, YLB and YLC) of Yashwant Lake studied. Out of these *Lymnaea acuminata, Lymnaea luteola, Bellamya bengalensis, Thiara tuberculata, Indoplanorbis exustus* were observed frequently and *Lamellidens marginalis* was observed occasionally at all the three stations.

Seasonal variations in density (Table 2.1, Figure 2.1) and species richness (Table 2.2, Figure 2.1) of mollusc were noted at all the three stations. The percentage (domination index) of mollusc species (Table 2.3) and the Pearson Correlation of the density with physico-chemical parameters are also calculated (Table 2.4).

Table 2.1: Seasonal Variation in Density of Molluscs/m³ at Yashwant Lake during Dec-2006 to Nov-2008

Station	*Winter*	*Summer*	*Monsoon*	*Ptmonsoon*	*F Value*	*df Value*	*P-Value Summary*
YLA	870.3± 85.2	1320± 51.3	1866± 55.4	2405 ± 105	65.3	3, 20	***
YLB	995 ± 64.3	1493± 64.3	1990± 90.7	2944±118.7	90.47	3, 20	***
YLC	953.5± 76.5	1576± 52.5	2156 ± 83	3110±124.3	108.6	3, 20	***

Table 2.2: Seasonal Variation in Species Richness of Molluscs at Yashwant Lake during Dec-2006 to Nov-2008

Station	*Winter*	*Summer*	*Monsoon*	*Post-monsoon*	*F Value*	*df Value*	*P-Value Summary*
YLA	3 ± 0.25	3.8 ± 0.17	4.5 ± 0.22	5.3 ± 0.21	20.7	3, 20	***
YLB	3.3 ± 0.21	4.3 ±.21	5.16 ±.16	5.6 ± 0.21	25.9	3, 20	***
YLC	3.6 ± 0.21	5 ± 0.26	5.3 ± 0.21	6 ± 0.00	24.7	3, 20	***

Table 2.3: Domination Index (D_o per cent) of Molluscs at 3 Stations of Yashwant Lake during December 2006 to November 2008.

Sl.No.	Name of the Species	YLA	YLB	YLC
1.	*Lymnaea accuminata*	7.55	20.7	23.1
2.	*Lymnaea luteola*	9.4	16.2	22.1
3.	*Bellamya bengalensis*	22	17.8	17.7
4.	*Thiara tuberculata*	23.9	19	15.6
5.	*Indoplanorbis exustus*	24.5	16.7	12.9
6.	*Lamellidens marginalis*	12.6	9.5	8.6

Table 2.4: Pearson Correlation of Molluscs Density with Abiotic Parameters, Total Phytoplankton, Total Zooplanktons and Birds, at YLA, YLB and YLC during December 2006 to November 2008

Sl.No.	Parameters	YLA	YLB	YLC
1.	Acidity	–0.100	–0.190	–0.145
2.	Alkalinity	–0.172	–0.275	–0.210
3.	Atmospheric temperature (°C)	0.365	0.252	0.228
4.	Chloride	–0.208	–0.383	–0.347
5.	CO_2 (Carbon dioxide)	0.418*	0.312	0.348
6.	DO (Dissolved Oxygen)	–0.454*	–0.357	–0.394
7.	NO_2 (Nitrite)	0.514*	0.443*	0.538**
8.	NO_3 (Nitrate)	0.525**	0.461*	0.484*
9.	pH	0.126	–0.019	0.004
10.	PO_4 (Phosphate)	0.476*	0.384	0.334
11.	TDS (Total Dissolved Solids)	–0.151	–0.309	–0.197
12.	TH (Total Hardness)	–0.382	–0.482*	–0.559**
13.	Transparency	–0.659**	–0.694**	–0.692**
14.	TS (Total Solids)	0.262	0.063	0.240
15.	TSS (Total Suspended Solids)	0.672**	0.499*	0.703**
16.	Water temperature (°C)	0.085	0.034	–0.040
17.	WC (Water Cover)	0.261	0.429*	0.352
18.	TDP (Total Density of Phytoplankton)	–0.479*	–0.719**	–0.549**
19.	TDZ (Total Density of Zooplankton)	–0.121	–0.279	–0.209
20.	TDB (Total Density of Birds)	–0.444*	–0.341	–0.424*

** Correlation is significant at the 0.01 level (two-tailed).

* Correlation is significant at the 0.05 level (two-tailed).

Molluscan density varied significantly across the season. Minimum density was recorded in winter at all stations YLA (870.3 ± 85.2 no./cu.m), YLB (995 ± 64.2 no./cu.m) and YLC (953.5 ± 76.5 no./cu.m). It increased slightly in summer and monsoon and was maximum in post-monsoon-YLC

Figure 2.1: Seasonal Variation of Molluscs Density and Species Richness at Yaswwant Lake during December 2006 to November 2008

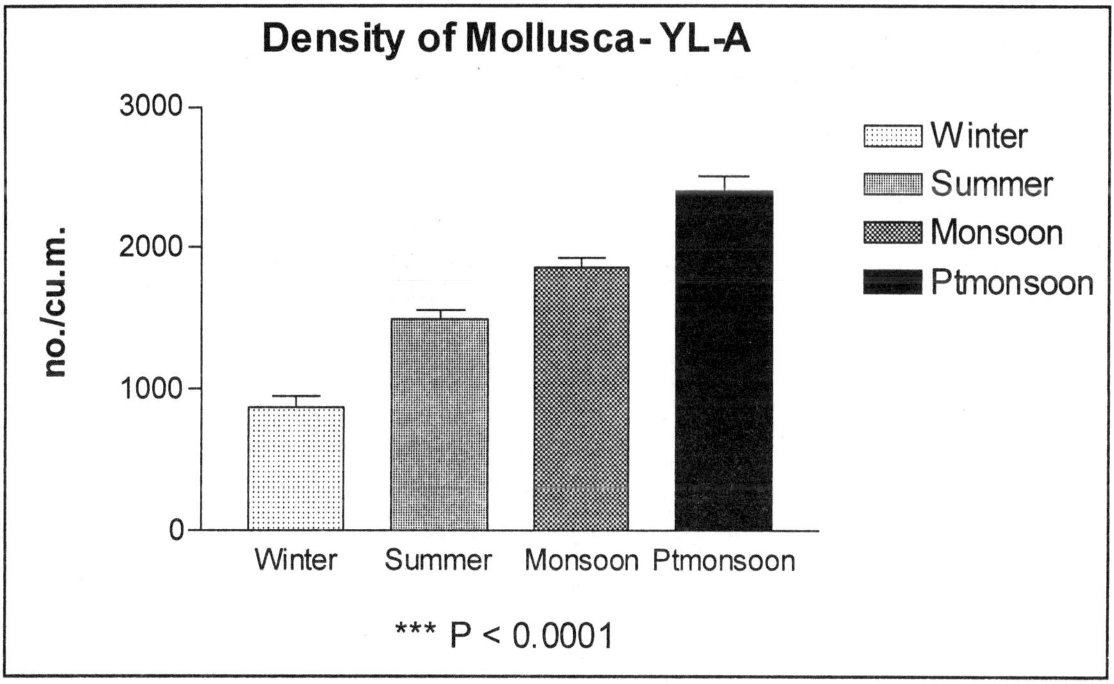

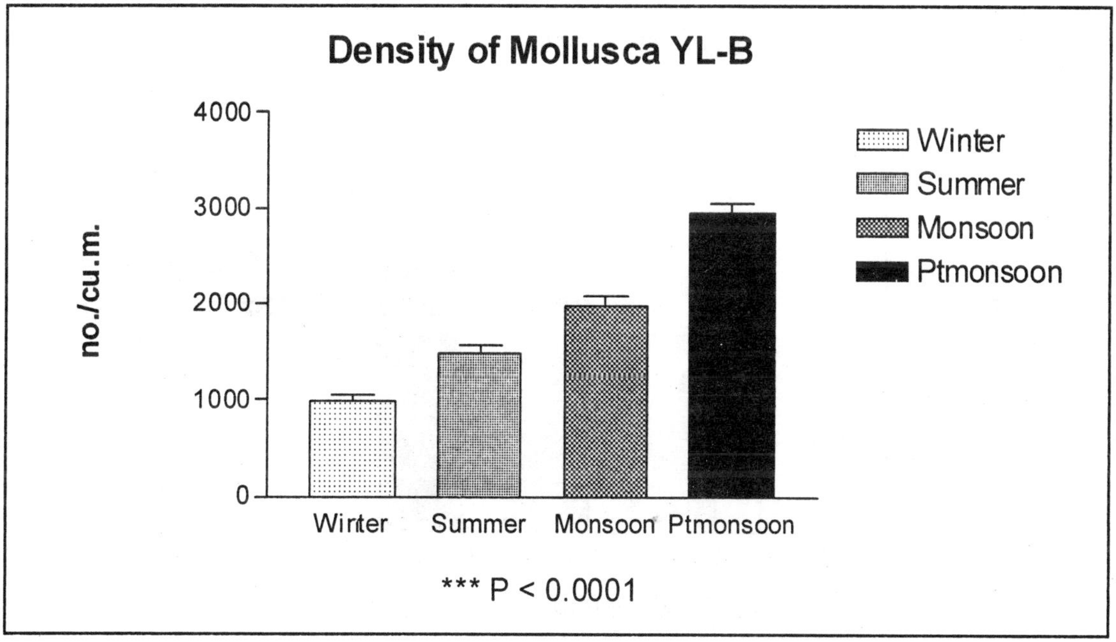

Contd...

Figure 2.1–*Contd...*

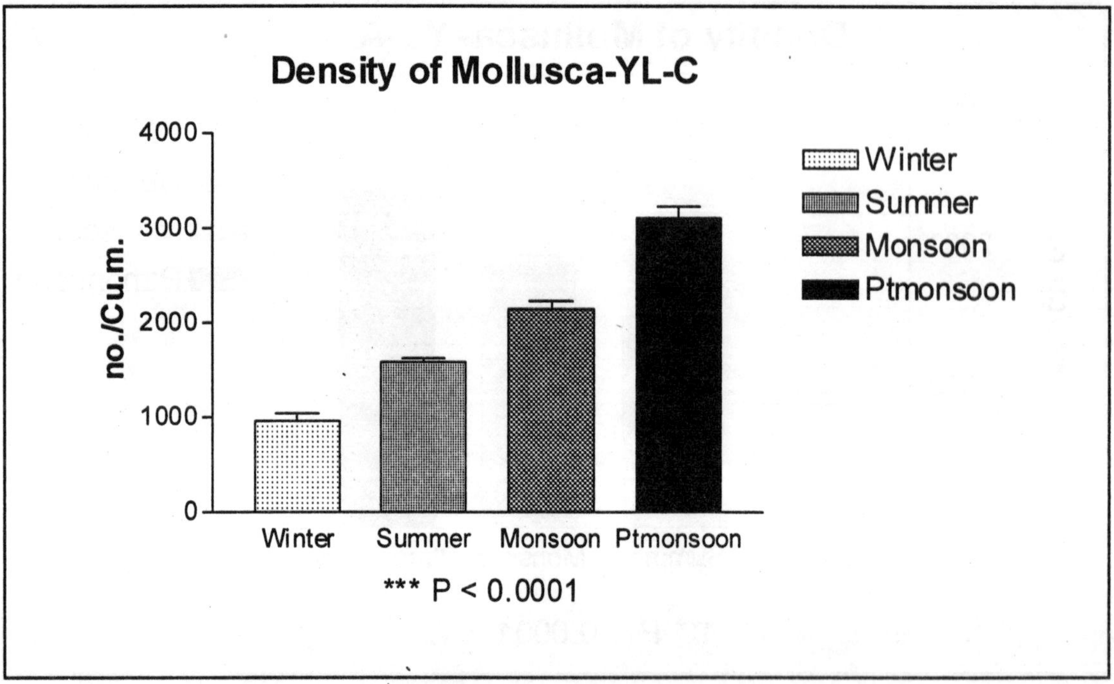

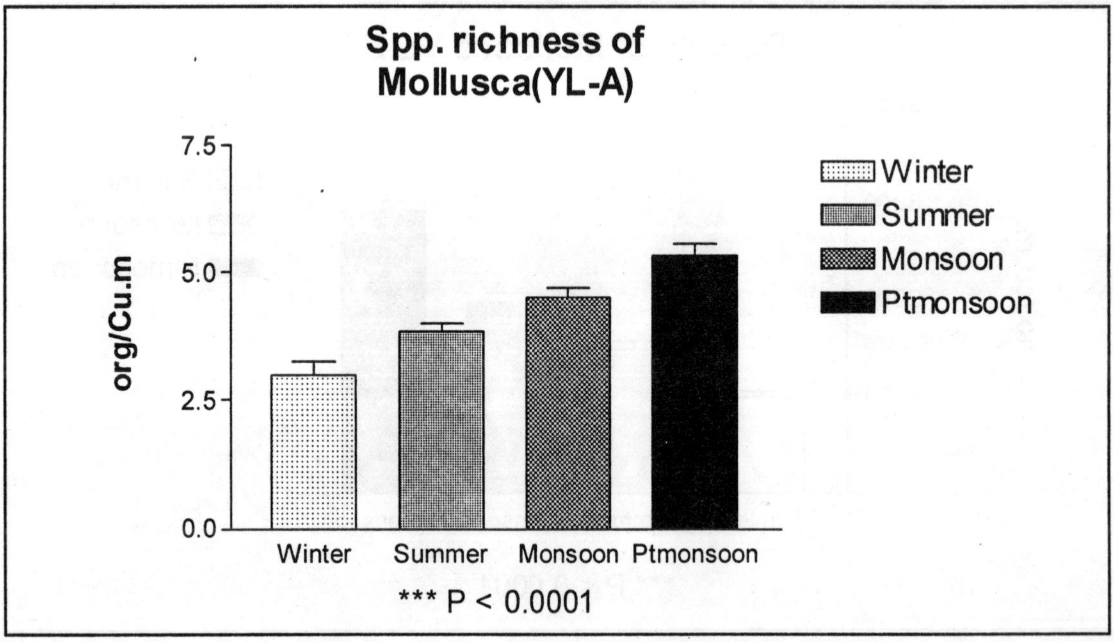

Contd...

Figure 2.1–*Contd...*

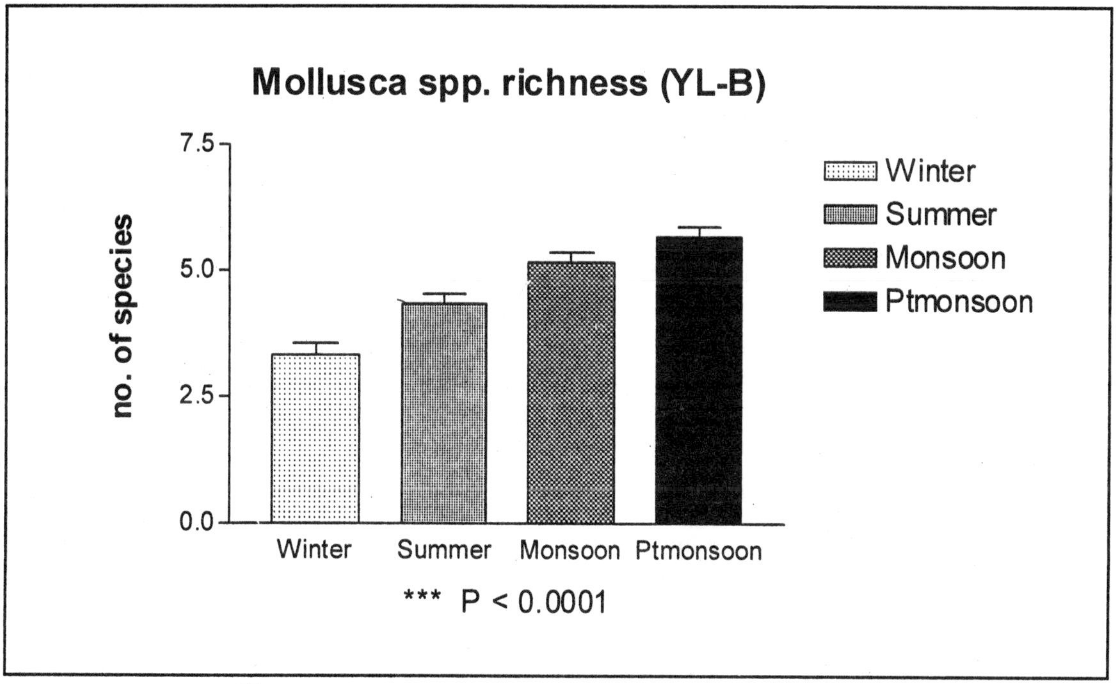

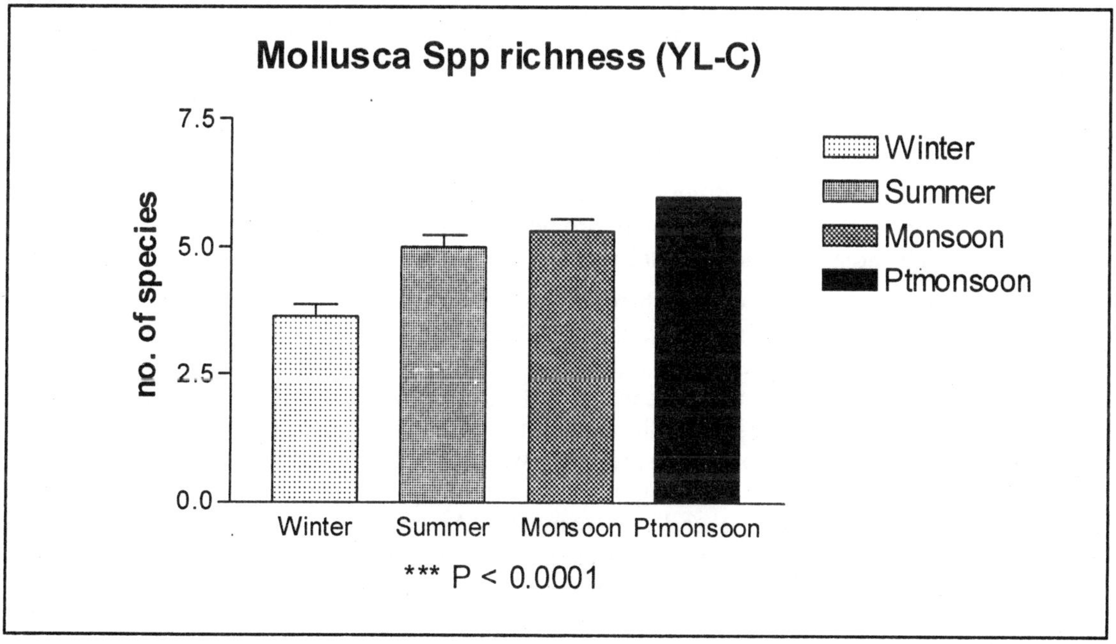

(3110 ± 124.3 no./cu.m), YLB (2944 ± 118.7 no./cu.m) and at YLA (2405 ± 105 no./cu.m). The species richness also varied significantly across the seasons, with minimum in winter with 3 ± 0.2 at YLA, 3.3 ± 0.2 at YLB and 3.6 ± 2 at YLC, it increased nonsignificantly in summer with 3.8 ± 0.17, 4.3 ± 2.1, 5 ± 0.26 at YLA, YLB and YLC respectively and further increased to 4.5 ± 0.22, 5.16 ± 0.16 and 5.3 ± 0.21 at the three stations in monsoon and reached to maximum 5.3 ± 0.21, 5.6 ± 0.21, 6.0 ± 0.0 respectively in post-monsoon.

The domination index (Do- per cent, Table 2.3) shows that species *B. bengalensis, T. tuberculata, I. exustus* were eudominant at all three stations (YLA, YLB and YLC), while *L. accuminata* and *L. luteola* were eudominant at YLB, YLC and dominant at YLA. The *L. marginalis* was dominant at YLB and YLC while it was eudominant at YLA.

When three sites are compared it is noted that maximum density as well as species richness occurred at YLC followed by YLB except in winter and minimum at YLA (Tables 2.1 and 2.2, Figure 2.1).

When Pearson Correlation of mollusc density with physico-chemical parameters and other biotic component are considered (Table 2.4), the molluscan density is positively correlated at the level of 0.01 with Nitrate at YLC, Nitrite at YLA, TSS at YLA and YLC whereas negatively correlated at same level with Total hardness at YLC, Transparency at all three stations and with Total Density of phytoplankton at YLB and YLC, Further, Mollusc density is correlated at 0.05 level positively with CO_2 and PO_4 at YLA, Nitrate, Nitrite, TSS and water cover at YLB, and nitrite at YLC and negatively with DO, Total Density of phytoplankton, and Total Density of birds at YLA; Total Hardness at YLB and Total Density of Birds at YLC.

Discussion

In the present study only six species of mollusc were recorded. However, the densities of the said species were always high. Thus, the species richness of mollusc at 'Yashwant Lake' a lacustrine habitat at higher altitude surrounded by dry deciduous forest in Eastern Satpuda is low. In another part of Western Satpuda where the dry deciduous forest is present with both lacustrine and riverine habitat, only fifteen species of molluscs have been reported (Magare, 2007). In lacustrine and riverine habitats different species of molluscs are known to occur (Subba Rao, 1989). Molluscs are influenced directly by plants as they provide different architecture and periphyton substrate but they are also likely to reflect other local factors such as hydroperiod or water chemistry and physico-chemical variations caused by adjacent land use (Andrew and Michael, 2009).

The highest density of molluscs at all three Stations (YLA, YLB and YLC) of the lake was observed during the post-monsoon. The time when water level is highest and photoperiod as well as temperatures are moderate favoring the growth of the macrophytes and probably the breeding performance of mollusc. In the wetlands of semi arid zone in central Gujarat, India also similar results are reported. (Deshkar, 2008). Among the three stations the maximum density of molluscs is observed at YLC. Adult molluscs require macro-vegetation for the growth as well as attachment (Boycott 1936, Macan, 1950) while these vegetation also serve as hiding places for their larvae (Bronmark, 1985). At YLC macro vegetation is available and hence higher density of mollusc. According to Macan (1950) a good mollusc habitat should have fair but not excessive rooted plant growth. Major part at YLC is covered with submergent and emergent vegetation with broad leaves. Gastropods have preferences for such macrophyte species with broader leaves (Brown, 1997). The local gastropod occurrence patterns are under the influence of multiple variables. The architecture of available habitats is important for freshwater gastropods.

Minimum densities of mollusc in winter have been attributed to lower ambient temperature with corresponding fall in water temperature (Deshkar, 2008). Around Yashwant Lake also temperature was lower, failing up to 8 to 10 °C.

The station YLA is having minimum total density all throughout the year. However, the percentage of 3 species *Thiara tuberculata, Indoplanorbis exustus,and Lamellidens marginalis* were maximum at this station. YLA has rocky shore with few macrophytes. A significant association exists between gastropods, sand and clay sediments, perhaps due to the film of algae on this kind of substratum (Iga and Adam, 2006). Present study confirms this association at YLC where the clay is present and density and species richness of mollusc are maximum while at YLA there is comparatively least clay and hence the total density and species richness of molluscs are also minimum. Out of 6 species *Thiara tuberculata* is associated with rock and silt. Similar results were obtained by Kornijow and Gulati (1992) at Lake Zwemlust. At YLA anthropogenic pressures are also high as it is used for cleaning utensils and washing activities adding soap and other organic matter to water and possibly influencing density of the molluscs. Collinson *et al.*(1995) and Williams *et al.* (2004) reported that those water bodies that are subjected to less anthropopressure provide a refuge for many freshwater species.

In summer densities of molluscs were low compared to monsoon at YLA. However, when YLA and YLC are compared the density was high at YLC, where anthropopressure are least due to muddy shore aiding macrophyte survival. Darby *et al.* (2002) reported that as pond area decreases because of drying, the molluscs (*e.g.* Apple snail) move towards the water. At the Yashwant Lake as the water cover started declining, several dead empty shells were seen accumulating at receding water margins.

Molluscs are consumed by multiple taxa. According to Dillon (2000) predator effects vary among gastropod species depending upon the shape and strength of the shell. Thus, the impacts of predation and composition on local assemblages vary with the system. As per the reports of Lodge *et al.*(1987) presence of predators is a strong control on local gastropod abundance across lake habitats. Many species of water fowls feed on molluscs (Stanczykowska *et al.*, 1990; Nisbet, 1997; Girmmitt *et al.*, 1998) and large number of migratory population of waterfowls visit water bodies during winter (Padate *et al.* 2008). During winter the ambient temperature around Yashwant Lake fall below 10°C, forcing molluscs to hibernate/move to deeper soil. However, the birds like Godwits, Ibises, Spoonbill, Openbill storks, *etc.* known to feed on molluscs (Urban *et al.*, 1982, Ali and Ripley, 1983) are benefited. Some of these species like ibises, storks and spoonbills having specialized beaks are occasionally observed at Yashwant Lake (unpublished data) in winter.

Species Richness and Composition

The fundamental biogeographic principles hold that the larger lakes support more mollusc species. This is confirmed by several studies, as reviewed by Carlson (2001). The data of the survey obtained by Oertli *et al.* (2002), showed a positive correlation between gastropod species richness and pond surface area ranging from 6 to 94,000 m^2. In contrast, Gee *et al.*, 1997 claimed that number of taxa is not correlated with pond surface area, but at the same time, they showed a statistically significant correlation between the number of taxa and the percentage of the pond's cover by macrophytes. Yashwant Lake has surface area of around 0.78 Sq.Km. and six species of gastropod are observed here. At YLC where maximum macrophytes are present density of mollusc is also high followed by YLB. In present study maximum species richness was observed in post-monsoon when the water cover was maximum hence, species richness and water cover is positively correlated.

The species composition at all the three stations is different. *B. bengalensis* is present at all the three stations. This is a common species of India which is abundantly distributed throughout Western zone

(Subba Rao, 1989). *B. bengalensis* population were maximum at YLA the station with rocky shore, few macrophytes and high anthropopressure, while minimum at YLC (Table 3). The next common species *Lymnaea* was maximum where macrophytes are maximum. These results corroborate with Lodge and Kelly (1985) that gastropod population size depends on macrophyte abundance. Maximum population of *Indoplanorbis exustus* was found at YLA where washing and cleaning activities are seen. *I. exustus* is vector for several parasites (Subba Rao, 1989). Thus, this species thrives at YLA where anthropogenic activities are high. A study of this species as vector and its associated parasite and prevailing parasitic diseases in the locals may add interesting information. The population of fourth species *Thiara tuberculata* well adapted to the sandy and rocky habitats is also maximum at YLA. The quality of the bottom sediment and the physical and chemical parameters of water influence gastropod distribution in the reservoir (Savage *et al.*, 1987).

Shell bearing molluscs need calcium for the successful growth and development (Russell-Hunter, 1964; Mc Mahon, 1983, Dillon, 2000; Robert 2003). In present study the density of mollusc and calcium in water in the form of calcium hardness are negatively significantly correlated. Evaluation of calcium in water and soil needs further investigation. The hardness in the form of $CaCO_3$ is thought to be an important factor which could affect growth of mollusc as the shell building depends on the amount of the $CaCO_3$ (Mackie and Flippance, 1983). In the present study when the correlation between the molluscs density and total hardness is carried out high correlation is noted with significant value at YLC. This has been confirmed by studies on not only calcium but local and regional distribution of gastropod species in relation to water chemistry too (Dussart, 1976, Savage *et al.*, 1987). However, according to Deshkar (2008) molluscs density is not correlated to any single physico-chemical parameter. Probably the sample size being smaller correlation with various factors has been established in present study. The periphyton which is dependent on biogenic elements in water constitutes the base of the diet of molluscs and nitrites are important for periphyton. Thus, the positive correlation between the gastropod densities and nitrate concentration in water may explain alimentation relationships (Iga and Adam, 2006).

As far as domination index is concerned all 6 species are either eudominant or dominant. However, three species *B. bengalensis, T. tuberculata* and *I. exustus* are eudominant at all the stations where as *L. accuminata* and *L. luteola* are dominant at YLA where their association with rocky and sandy bank is comparatively low whereas *L. marginalis* associated with this type of bank is eudominate here. Opposite situation is seen at other two stations with this three species.

Several species appear to be restricted in distribution among the ponds by unsuitable water chemistry parameter values, the most significant of these being total dissolved solids, pH and total alkalinity but in present study no significant correlation is established with them.

Today there is no such degradation problem at Yashwant Lake but as the spot is developing as tourist place it may create anthropogenic pressures so for conservation of biodiversity of the lake proper steps needs to be undertaken.

The molluscs are important biota of the inland wetland and may be helpful for determining the habitat quality of wetland. Further detailed study on the similar aspect may give interesting results regarding the molluscan density and diversity and help in determining the best abiotic/biotic predictor of the molluscs density and diversity in the high altitudinal Lake.

References

Ali, S. and Ripley, S.D., 1983. *Handbook of the Birds of India and Pakistan* (Compact Edition). University Press Bombay, Bombay, India.

APHA, 1998. *Standard Methods for the Examination of Water and Wastewater*, 20th edn. American Public Health Association, American Water Works Association, Water Environment Federation. Washington D.C.

Andrew, M.Z. and Michael, J.J., 2009. The distribution of pond snail communities across a landscape: Separating out the influence of spatial position from local habitat quality for ponds in south-east Northumberland, UK. *Hydrobiologia*, 632: 177–187.

Boycott, A.E., 1936. The habitats of the freshwater mollusca in Britain. *J. Animal Eco.*, 5: 116–186.

Bronmark, C., 1985. Freshwater snail diversity: Effects of pond area, habitat heterogeneity and isolation. *Oecologia*, 67(1): 127–131.

Brown, K.M., 1997. Temporal and spatial patterns of abundance in the gastropod assemblage of a macrophyte bed. *Anim. Malacol. Bull.*, 14: 27–33.

Carlson, R., 2001. Species-area relationships, water chemistry and species turnover of freshwater snails on the Aland Islands, Southwestern Finland. *J. Mollus. Stud.*, 67: 17–26.

Collinson, N.H., 1995. Temporary and permanent ponds: An assessment of the effect of drying out on the conservation value of aquatic macroinvertebrates communities. *Biol. Conserv.*, 74: 125–133.

Darby, P.C., Bennetts, R.E., Miller, S.J. and Percival, H.F., 2002. Movements of Florida apple snails in relation to water levels and drying events. *Wetlands*, 22(3): 489–498.

Deshkar, S.L., 2008. Avifaunal diversity and ecology of wetlands in semi arid zone of Central Gujarat with reference to their conservation and categorization. *Ph.D. Thesis*, Maharaja Sayajirao University of Baroda, Vadodara, India.

Dillon, R.T., 2000. *The Ecology of Freshwater Molluscs*. Cambridge University Press. London.

Doherty, S., Cohen, M., Lane, C., Line, L. and Surdick, J., 2000. Biological criteria for inland freshwater wetlands in Florida: A review of technical and scientific literature (1990–1999). *Report to the United States Environmental Protection Agency*, Centre for Wetlands, University of Florida, Gainesville, Florida, USA.

Dussart, G.B.J., 1976. The ecology of freshwater molluscs in North-West England in relation to water chemistry. *J. Mollu. Studies*, 12: 181–198.

Gee, J.H.R., Smith, B.D., Lee, K.M. and Griffiths, S.W., 1997. The ecological basis of freshwater pond management for biodiversity. *Aquatic Conserve. Marine Freshwater Ecosys.*, 7: 91–104.

Gorny, M. and Grum, L., 1981. Metody stosowane w zoologii gleby. Panstwowe Wydawnictwo Naukowe, Warszawa.

Graphpad Software, San Diego, California U.S.A., www.graphpad.com.

Grimmett, R., Inskipp, C. and Inskipp, T., 1998. *Birds of the Indian Sub Continent*. Christopher Helm, London.

Hart, R. and Newman, J., 1995. The importance of isolated wetlands to fish and wildlife in Florida. Project report for non game wildlife program project NG88–102, Florida Game and Freshwater Fish Commission, Tallahassee, FL.

Iga, L. and Adam, S., 2006. Rare, threatened and alien species in the gastropod communities in the clay pit ponds in relation to environmental factors. *Biodiver. and Conserv.*, 15: 3617–3635.

Kornijow, R. and Gulati, R.D., 1992. Macrofauna and its ecology in lake Zwemlust, after biomanipulation, a Bottom fauna. *Archiv fir Hydrobiologie*, 123(3): 337–347.

Lodge, D.M. and Kelly, P., 1985. Habitat disturbance and the stability of freshwater gastropod populations. *Oecologia*, 68: 111–117.

Lodge, D.M., Brown, K.M., Klosiewski, S.P. and Stein, A.P., 1987. Distribution of freshwater snails: spatial scale and the relative importance of physico-chemical and biotic factors. *American Malcological Bulletin*, 5: 73–84.

Macan, T.T., 1950. Ecology of freshwater mollusca in the English lake. *J. Anim. Ecol.*, 19(2): 124–146.

Mackie, G.L. and Flippance, L.A., 1983. Relationships between buffering capacity of water and the size and calcium content of freshwater molluscs. *Fresh. Invert. Biol.*, 2(1): 48–55.

Magare, S.R., 2007. Biodiversity of freshwater molluscs from satpuda mountains and Tapi River with reference to vector snails. *Flora and Fauna*, 13(1): 161–164.

McMohan, R.O.F., 1983. Physiological ecology of freshwater pulmonates. In: *The Mollusca*, (Ed.) W.D. Russel-Hunter. Academic Press, London, pp. 360–430.

Michael, P., 1986. *Ecological Methods for Field and Laboratory Investigations*. Tata McGraw-Hill Publishing Co. Ltd., New Delhi.

Murkin, H.R. and Kadlec, J., 1986. Relationship between water fowl and macroinvertebrate densities in Northern Prairie Marsh. *J. Wild. Manag.*, 50(2): 212–217.

Nisbet, I.C.T., 1997. Female Common Terns (*Sterna hirundo*) eating mollusc shells: Evidence for calcium deficits during egg laying. *Ibis*, 139: 400.

Oertli, B., Jaye, D.A., Castella, E., Juge, R., Cambin, D. and Lachavanne, J.B., 2002. Does size matter? The relationship between pond area and biodiversity. *Biol. Conserv.*, 104: 59–70.

Padate, G., Deshkar, S. and Sapna, S., 2008. The Influence of Narmada water inundation on the duck populations of Wadhwana irrigation reserviour. In: *Proceedings of 'Taal 2007' the World Lake Conference*, Jaipur, (Eds.) M. Sengupta and R. Dalwani, pp. 131–136.

Pip, E., 2006. Littoral mollusc communities and water quality in southern Lake Winnipeg, Manitoba, Canada. *Biodiv. Conservation*, 15: 3637–3652.

Pyron, M., Beugly, J., Spielman, M., Pritchett, J. and Jacquemin, S., 2009. Habitat variation among gastropod assemblages of Indiana, USA. *The Open Zoology Journal*, 2: 8–14.

Quintero, J.C., 2007. Diversity, habitat use and conservation of freshwater molluscs in the lower Guadiana River basin (SW Iberian Peninsula). *Aqua. Conserv.*, 17: 485–501.

Rahmani, A.R. and Shobrak, M.Y., 1992. Glossy ibises (*Plegadis falcinellus*) and black-tailed godwits (*Limosa limosa*) feeding on sorghum in flooded fields in southwestern Saudi Arabia. *Colonial Waterbirds*, 15(2): 239–240.

Ramchandra, T., Kiran, A., Ahylaya, N. and Deepa, R.S., 2002. Status of wetlands of Banglore. Technical Report 86. available at www. wgbis. ces.iisc.ernet. in/energy/TR86/welcome.html.

Robert, A.B., 2003. Range size and environmental calcium requirements of British freshwater gastropods. *Global Ecology and Biogeography*, 12: 47–51.

Russel-Hunter, W., 1964. Physiological aspects of ecology in non-marine molluscs. In: *Physiology of Mollusca*, (Eds.) K.M. Wilbur and C.M. Yonge. Academic Press, London, pp. 83–125.

Savage, A.A. and Gazey, G.M., 1987. Relationships of physical and chemical conditions to species diversity and density of gastropods in English lakes. *Biol. Conserv.*, 42: 95–113.

Stanczykowska, A., Zyska, P., Dombrowski, A., Kot, H. and Zyska, E., 1990. The distribution of waterfowl in relation to mollusc populations in the manmade Lake Zegrzynskie. *Hydrobiologia*, 191: 233–240.

Stansly, P.A., Gore, J.A., Ceilley, D.W. and Main, M.B., 1997. Inventory of freshwater macroinvertebrates. Final Report for the South Florida Water Management District Isolated Wetland Monitoring Program. Contract # C–7949.

Subba Rao, N.V., 1989. *Handbook: Freshwater Molluscs of India*. Zoological Survey of India, Calcutta.

Tronstad, L.M., Tronstad, B.P. and Benke, A.C., 2005. Invertebrate responses to decreasing water levels in a subtropical river floodplain wetland. *Wetlands*, 25(3): 583–593.

Urban, E.K., Fry, C.H. and Keith, S., 1982. The birds of Africa. Volume II London, Great Birtan. Acadamic Press. As cited by Rahmani and Shobark, 1992.

Williams, P., 2004. Comparative biodiversity of rivers, streams, ditches and ponds in an agricultural landscape in southern England. *Biol. Conserv.*, 115: 329–341.

Chapter 3

Study on Ichthyofauna of Waterbodies Around Ambajogai, Maharashtra

☆ *G.B. Sanap, R.B. Haregaokar and V.B. Sakhare*

Introduction

Fishes are the keystone species which determine the distribution and abundance of other organisms in ecosystem they represent and are good indicators of the water quality and health of the ecosystem. Nearly 20 per cent of the world's freshwater fish fauna is already extinct or is on the verge of extinction (Moyle and Leidy, 1992).

Maharashtra is a state richly endowed with inland water resources. Artificial impoundments in the state form an important source of fish production. No correct information is available because some of the reservoirs are under Zilla Parishads, some under Irrigation Department and others are under the control of State Electricity Board. According to Sugunan (1995) the small reservoirs (<1000 ha) in Maharashtra occupy 119515 ha (44 per cent) out of total 2,73,750 ha. Among the remaining 152205 ha, the large reservoirs (>5000) constitute 115054 ha, the other 39181 ha falling under the medium (1000–5000) category. Indian Institute of Management, Ahemadabad (1984) reported 72 reservoirs with water spread area of 1,05,202 hectares.

An appreciable amount of work has been done concerning fisheries of reservoirs in Maharashtra (Valsangkar, 1980, 1987,1993; Bhartiya, 1990; Lonkar 1992; Sakhare 1999,2001,2003; and Sakhare and Joshi 2001,2002,2003). The present investigation was undertaken to gather detailed information on ichthyofauna of waterbodies around Ambajogai.The details of the work done on fisheries and/or ecology of some reservoirs like Bori (Sakhare and Joshi 2001), Palas-Nilegaon (Sakhare and Joshi 2002), Hangarga (Sakhare, 2004) of the neighbour district (Osmanabad) are available. But no such information is available in respect of waterbodies around Ambajogai in Beed district. Hence, it is felt that there is a need of research work on ichthyofauna of waterbodies around Ambajogai.

The Ambajogai Tahsil of Beed district is having several small sized waterbodies in which the fishery is mainly implemented through fisherman's co-operative societies under administration of State Fisheries Department.

Like other rural areas of country fisherman around Ambajogai are not very cautious about use of scientific culture practices and intensive methods of cultures.From ancient time fishes are used as food, providing excellent source of proteins, fats, and vitamins. Fish flesh is good source of different proteins and vitamins like vit.B, vit.A, and vit.D. Minerals like calcium, magnesium, potassium, sodium, phosphorus, iron, copper, manganese etc.

Materials and Methods

For the present study the fishes were collected with help of local fishermen. The survey study was also made by regular visiting to fish market. Identification was done with help of Standard literature such as Day(1878), Datta Munshi and Srivastava (1968) and Talwar and Jhingran (1991).The classification of the fishes on economic importance was done by following the proforma given by Lagler (1956).

Results and Discussion

The fish fauna is an important aspect of fishery potential of a water body. Several workers have studied fish fauna of Indian reservoirs. The distribution of fish species is quite variable because of geographical and geological conditions. In present study of waterbodies around Ambajogai Tahsil, confirms the natural habitat of 23 fish species belonging to 7 orders. Among them order cypriniformes was dominant with 11species to be followed by order perciformes and siluriformes (3 species each).The

Figure 3.1: Wallago attu

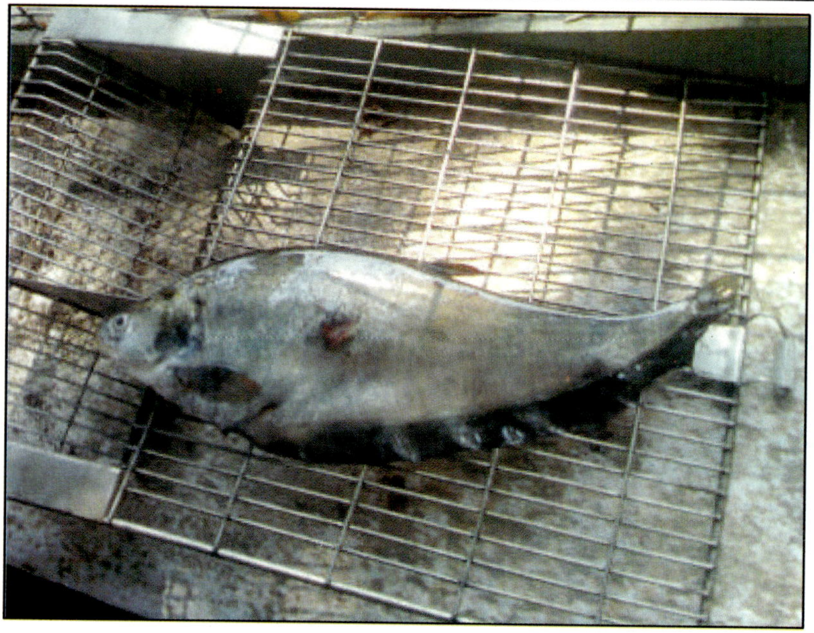

Figure 3.2: ***Notopterus notopterus***

Figure 3.3: ***Channa marulius***

orders osteoglossiformes and channiformes were represented by two species each. Only one species was recorded under order Symbranchiformes and Angulliformes. Valsangkar (1993) recorded 17 indigenous and 5 introduced fish species from Shivajisagar reservoir in Maharashtra. Sakhare (2002)

Figure 3.4: ***Tilapia mossambica***

Figure 3.5: ***Cyprinus carpio***

confirmed the occurrence of 28 fish species in Palas-Nilegaon reservoir of Osmanabad district in Maharashtra. In the present investigation species occurrence is very much similar to Sakhare (2001 and 2006).The observed similarity could be due to the presence of waterbodies in same geographical and climatic regions of Maharashtra.

Figure 3.6: *Catla catla*

Some fish species like *Notopterus notopterus, Notopterus chitala, Puntius sophore* and *Anguilla bengalensis* have been using from long time for medicinal purpose. Species like *Wallago attu, Mystus seenghala, Mystus cavassius, Channa marulius* and *Channa gachua* are carnivorous fish species.

To summarize,the foregoing observations reveal that the waterbodies around Ambajogai supports a good fishery and contributes considerably to the total fish production of the Beed district.

Table 3.1: Ichthyofauna of Waterbodies around Ambajogai

Order: Cypriniformes		
Family: Cyprinidae		
1. *Catla catla*	2. *Cirrhinus mrigala*	3. *Cirrhinus reba*
4. *Labeo rohita*	5. *Labeo calbasu*	6. *Cyprinus carpio*
7. *Hypothalmichthys molitrix*	8. *Puntius sophore*	9. *Puntius sarana sarana*
10. *Puntius ticto ticto*	11. *Puntius kolus*	
Order: Osteoglossiformes		
Family: Notopteridae		
1. *Notopterus notopterus*	2. *Notpterus chitala*	
Order: Anguliformes		
1. *Anguilla bengalensis*		
Order: Perciformes		
Family: Centropomidae		
1. *Chanda nama*	2. *Chanda ranga*	
Family: Cichlidae		
1. *Tilapia mossambica*		

Contd...

Table 3.1–*Contd...*

Order: Siluriformes

Family: Siluridae

1. *Wallago attu*

Family: Bagridae

1. *Mystus seenghala* 2. *Mystus cavassius*

Order: Channiformes

Family: Channidae

1. *Channa marulius* 2. *Channa gachua*

Order: Symbranchiformes

Family: Gobidae

1. *Glossogobius giuris giuris*

Table 3.2: Economic Importance of Fishes Recorded from Waterbodies around Ambajogai

Species	*Commercial*	*Fine food*	*Coarse food*	*Aquarium Fishes*	*Others*
Notopterus notopterus			*		MD
Notopterus chitala			*		MD
Catla catla	*	*			
Cirrhinus mrigala	*	*			
Cirrhinus reba	*	*			
Labeo rohita	*	*			
Labeo calbasu	*	*			
Cyprinus carpio					
(*Var. Scale carp*)	*				
Hypothalmichthys molitrix	*				
Punctius sophore			*	*	MD,Bt,LV
Punctius sarana sarana			*	*	Bt,LV
Punctius ticto ticto			*	*	Bt,LV
Punctius kolius			*	*	Bt, LV
Chanda nama					LV
Chanda ranga					LV
Wallago attu		*			
Mystus seenghala		*			
Mystus cavassius		*			
Channa marulius	*	*			
Channa gachua	*				
Glossogbius giuris					
Tilapia mossambica			*		
Anguilla bengalenasis			*		MD

MD: Medicinal value; LV: Larvivorous fish; Bt: Bait; PH: Public health.

Acknowledgement

We are grateful to trader Shri Kachru and our colleague Shaikh Gaffor for their assistance in field work during course of investigation.

References

Bhartiya, M.T., 1990. Management of reservoir fisheries through fishermen co-operatives in Maharashtra: Constraints and measures. In: *reservoir Fisheries of India*. Proc. of Nat. Workshop on Reservoir Fisheries, 3–4 January, (Eds.) A.G. Jhingran and V.K. Unnithan. Special Publication No. 3. Asian Fisheries Society of India Mangalore, p. 65–70.

Datta Munshi, J.S. and Srivastava, M.P., 1988. *Natural History of Fishes and Systematics Freshwater Fishes of India*. Narendra Publishing House, Delhi.

Day, F.S., 1878. *The Fishes of India*. William Dowson and Sons Ltd., London.

Jayaram, K.C., 1981. *The Freshwater Fishes of India*. Zoological Survey of India, Calcutta.

Lagler, K.F., 1956. *Freshwater Fishery Biology*. WmC Brown and Co., Iowa.

Lonkar, R.L., 1992. Case studies on selected reservoirs of Godavari basin with emphasis to reinforce their fishery. Dissertation submitted to the Extension Training Center of Central Institute of Fisheries Education, Kakinda Center, Andhra Pradesh, India, pp. 75.

Moyle, P.B. and Leidy, R.A., 1992. Loss of biodiversity in aquatic ecosystems: evidence from fish faunas. In: *Conservation Biology: The Theory and Practice of Nature Conservation, Preservation and Management*, (Eds.) P.L. Fiedler and S.K. Jain. Chapman and Hall, New York, pp 127–169.

Sakhare, V.B., 1999. Fisheries of Yeldari reservoir, Maharashtra. *Fishing Chimes*, 19(8): 17–22.

Sakhare,V.B., 2001. Ichthyofauna of Jawalgaon reservoir in Solapur district of Maharshtra. *J. Aqua. Biol.*, 16(1&2): 31–33.

Sakhare, V.B., 2001. Reservoir fisheries in Solapur district of Maharashtra. *Fishing Chimes*, 21(5): 29–30.

Sakhrae, V.B., 2003. Socio-economic status of fishermen around Yeldari Reservoir, Maharashtra. *Aqua Tech.*, 2(5): 77–78.

Sakhare, V.B., 2006. Ecology of Jawalgaon reservoir in Solapur district, Maharashtra. In: *Ecology of Lakes and Reservoirs*, (Ed.) V.B. Sakhare. Daya Publishing House, Delhi, pp. 16–35.

Sakhare, V.B. and Joshi, P.K., 2001. Studies on fish fauna of Bori reservoir, Osmanabad district, Maharashtra, India. Paper presented during 2–day workshop on '*Lake Management in India*', 12–13th December,Hyderabad.

Sakhare,V.B. and Joshi, P.K., 2002. Ecology and ichthyofauna of Bori reservoir in Maharashtra. *Fishing Chimes*, 22(4): 40– 41.

Sakhare, V.B. and Joshi, P.K., 2002. Ecology of Palas–Nilegaon reservoir in Osmanabad district, Maharshtra. *J. Aqua. Biol.*, 18(2): 17–22.

Sakhare,V.B. and Joshi, P.K., 2003.Reservoir fishery potential of Parbhani district of Maharashtra. *Fishing Chimes*, 23(5): 13–16.

Talwar, P.K. and Jhingran, A.G., 1991. *Inland Fishes of India and Adjacent Countries*, Vols. 1 and 2. Oxford and IBH Publishing Co. Pvt. Ltd., New Delhi.

Valsangkar, S.V., 1980. Economic rehabilitation of fishermen in Yeldari reservoir. *India: Today and Tomorrow*, 8(4): 162–163.

Valsangkar, S.V., 1987. Common carp: An asset to Girna reservoir. *Fishing Chimes*, 7(5): 29–39.

Valsangkar, S.V., 1993. Mahseer fisheries of Koyana River (Shivajisagar) in Maharashtra: Scrap to bonanza. *Fishing Chimes*, 12(10): 15–19.

Chapter 4

Seasonal Fluctuation of Phytoplankton in Nagzari Tank Near Ambajogai, Maharashtra

☆ *S.W. Bhivgade, A.S. Taware, U.S. Salve and N.B. Pandure*

ABSTRACT

Phytoplankton diversity of Nagzari tank has been studied from April 2007 to March 2008. The present study deals with standing crop of phytoplankton. It was recorded maximum during the months of July, August, November and December with very large number. During present investigation taxa from cyanophyceae, chlorophyceae, bacillariophyceae and euglenophyceae were identified.

Introduction

Phytoplankton plays an important in the biosynthesis of organic matter in an aquatic ecosystem.

The present study was undertaken to record of phytoplankton diversity in Nagzari tank, which is located at Nagzari area on the outskirts of Shepwadi village near Ambajogai town (3 km North East).

Phytoplankton plays key role in the ecosystem of the environment but over the years the condition of the tank water as well as the surrounding area got deteriorated mainly due to the increase in human usage.The main objective of the present study is to understand the seasonal change of environment in relation to phytoplankton.

Materials and Methods

Nagzari tank is a natural percolation of water from rocks, it flows from the mansoon water, much in rainy season, lessening towards summer season. The water flows from stone cow mouth and is stored in a well built tank.

Water samples were collected monthly from the Nagzari tank for a period of one year (April 2007 to march 2008). For the collection of phytoplankton, 50 L of water was sieved through a ring type terricot net fitted with a wide mouthed glass bottle. The water samples collected were preserved in 5 per cent formaline solution, to which little amount of glycerine was also added. The taxonomic identification of Algae was done qualitatively and quantitatively with the help of standard literature of Smith (1950) and monographs.

Results and Discussion

In the present investigation, algae belong to four families *i.e.* cyanophyceae, chlorophyceae, bacillariophyceae and euglenophyceae were identified in the samples collected from the Ngazari tank.

In the present investigation 19 species of planktonic algae were reported from the samples collected during the period of study from the Nagzari tank (Table 4.1). It was observed that Cyanophyceae was represented by seven genera. Among Cyanophyceae, *Oscillatoria* and *Nostoc* were recorded throughout the investigation period, with maximum numbers occurred during the months of November and December. Cyanophycean growth was also recorded during winter season by Gopal *et al.* (1981), Pandey and Tripati (1984). Barhate (1985) and Zafar (1967) considered that high percentage of dissolved oxygen is favourable for more growth and development of cyanophyceae.

Table 4.1: Phytoplanktonic Algal Farms and their Dominance in the Nagzari Tank, Ambajogai

Phytoplankton	*Apr.*	*May*	*June*	*July*	*Aug.*	*Sep*	*Oct.*	*Nov.*	*Dec.*	*Jan*	*Feb*	*Mar*
Cyanophyceae												
Anabaena	A	A	A	R	R	C	R	C	C	R	A	A
Chroococcus	A	A	R	C	C	R	C	C	R	R	R	A
Nostoc	A	A	R	C	S	R	C	D	D	C	C	A
Oscillatoria	R	A	C	D	D	S	R	C	D	S	S	S
Phormidium	R	A	A	C	R	R	A	A	C	S	A	A
Rivularia	A	A	A	R	R	A	R	R	R	R	R	R
Sytonema	A	A	A	R	R	R	A	R	A	R	S	A
Chlorophyceae												
Chlamydomonas	A	A	R	C	R	C	C	R	D	A	A	A
Cladophora	A	A	A	C	C	A	C	R	C	R	S	A
Closterium	C	A	C	D	D	C	C	D	C	R	A	R
Cosmarium	A	A	R	D	D	S	C	D	D	S	R	R
Mougeotia	R	A	A	A	C	C	C	R	C	C	R	A
Oedogonium	R	R	C	C	D	C	C	C	D	D	C	S
Spirogyra	C	C	D	D	C	C	C	D	D	D	D	S
Stigioclonium	A	A	A	R	R	R	C	R	R	R	R	R

Contd...

Table 4.1–*Contd...*

Phytoplankton	*Apr.*	*May*	*June*	*July*	*Aug.*	*Sep*	*Oct.*	*Nov.*	*Dec.*	*Jan*	*Feb*	*Mar*
Bacillariophyceae												
Gomphonema	A	A	A	R	R	A	A	R	A	R	R	A
Diatom	R	A	A	C	S	R	C	C	D	S	S	R
Navicula	R	A	C	C	C	R	D	C	C	R	C	S
Euglenophyceae												
Euglena	A	A	C	D	C	R	R	R	C	S	C	R

A: Absent; R: Rare; C: Common; S: Sub-dominant; D: Dominant.

The abundance of blue green algae have also been emphasized by Pandey and Verma (1992) and Bairagi and Goswami (1994)

Chlorophyceae was the most abundant group in the Nagzari tank. It is represented by 8 species with *Closterium, Cosmarium, Oedogonium* and *Spirogyra* showing high abundance in the months of July, August, November and December 2007. The density maxima coincided with moderately high temperature. Similar phenomena was reported by Prescott (1939) and Patil *et al.* (1983).

In Bacillariophyceae, diatom and *Navicula* were present in the samples throughout the period of investigation, with its maximum occurrences in the months of July, October and December. Similar observations were made by Venkateswarlu (1969), who observed maximum population of diatoms during winter in the river Moosi in Hyderabad (Andhra Pradesh). Philipose (1960) recorded high density of diatoms in pH range of 7.4 to 8.0 and results of present study are in accordance with the above findings.

Vyas and Kumar (1968) and Barhate (1985) observed Euglenoids in rainy season only in Indrasagar tank, Udaipur, Rajasthan. In the present study density od *Euglena* is maximum in the month of July when water has high level of dissolved oxygen and nutrients in the water.

Conclusion and Recommendation

It can been concluded that, presence of taxa like *Oscillatoria, Chrococcus, Chlamydomonas, Closterium, Cosmarium, Oedogonium, Spirogyra, Diatom, Navicula* and *Euglena* remains dominant in the Nagzari tank.The results could be explained by prevailing the ecological conditions such as continuous flow of domestic load at the Nagzari tank. Our recommendation is to avoid human activities in and around the Nagzari tank.

Acknowledgements

The authors are thankful to Director, B.C.U.D. Dr. Babasaheb Ambedkar Marathwada University, Aurangabad for sanctioning minor research project during 2007-2008.

References

Abbasi, S.A., Bhatiya, K.S., Kunhi, A.V.M. and Soni, R.S., 1996. Studies on the limnology of Kuttadi lake (New Kerala). *Ecol. Env. and Cons.*, 2: 17–27.

Bairagi, S.P. and Goswami, M.M., 1994. Ecology of water blooms in some ponds of N.E. India. *Environ. and Ecol.*, 12(3): 568–571.

Barhate, V.P., 1985. Studies on the algal flora of Vidarbha and Khandesh Maharashtra. *Ph.D. Thesis*, Nagpur University, Nagpur.

Gopal, B., Goel, R.K., Sharma, K.P. and Trivedy, R.K., 1981. Limnological study of a freshwater reservoir Jamwa Ramgarth (Jaipur). *Hydrobiol.*, 83: 283–294.

Jindal, R. and Ghezta, R.K., 1991. Diel variations in eutrophic pond. *Geobios*, 18: 25–28.

Pandey, R.S. and Varma, P.K., 1992. Limnological status of an ancient temple pond, Shivaganga of Deoghar, Bihar. *J. Freshwater Biol.*, 4(3): 163–174.

Pandey, S.N. and Tripathi, A.K., 1984. Algal pollutants of Unnao ponds–II. Qualitative, quantitative and periodical occurrence of Cyanophyceae. *J. Pl. Nature*, 1(1): 83–86.

Patil, S.G., Singh, D.B. and Harshey, D.K., 1983. Ranital (Jabalpur) a sewage polluted water body as evidence by chemical and biological indicators of pollution. *J. Environ. Biol.*, 4(2): 43–49.

Philipose, M.T., 1960. Freshwater phytoplankton of Inland fisheries. *Proc. Symp. Algology*, ICAR, New Delhi, p. 272–291.

Prescott, G.W., 1939. Some relationship of phytoplankton to limnology and aquatic biology in problems of lake biology. *Amer. Assoc. Adv. Sci.*, 10: 65–78.

Sahu, B.K. and Behera, S.K., 1995. Studies on some physico-chemical characteristics of the Ganga river water (Rishikesh–Kanpur) within twenty four hour during winter 1994. *Ecol. Env. Cons.*, 1(1–4): 35–38.

Smith, G.M., 1950. *Freshwater Algae of United States*. McGraw-Hill Book Company, New York, Toronto, London.

Unni, K.S., 1985. Comparative limnology of several reservoirs in central India. *Int. Revue Ges. Hydrobiol.*, 70(6): 845–856.

Venkateswarlu, V., 1969. An ecological study of the river Moosi, Hyderabad (India) with special reference to water pollution. I. Physico-chemical complex. *Hydrobiol.*, 3(1): 117–143.

Yyas, L.N. and Kumar, H.D., 1968. Studies on phytoplankton and other algae of Indrasagar Tank, Udaipur, India. *Hydrobiol.*, 31: 421–434.

Zafar, A.R., 1967. On ecology of algae in certain fish ponds of Hyderabad, India III. The periodicity. *Hydrobiol.*, 30(1): 96–112.

Chapter 5

Total Zooplankton Abundance and Biomass in Three Contrasting Lentic Ecosystems of Mysore, Karnataka State

☆ *Koorosh Jalilzadeh and Sadanand M. Yamakanamardi*

ABSTRACT

A two years study on abundance and biomass of total zooplankton (rotifers, cladocer, copepods and ostracods) as well as some physico-chemical parameters such as pH, CO_2, BOD, DO, calcium, alkalinity, TSS, phosphate, sulphate, chloride, nitrate, TASA, Chlorophyll-*a*, was carried out from February 2007 to January 2009, in three contrasting (highly, moderately and least polluted) lakes, namely, Hebbal (in Industrial area), Lingambuddi (in Domestic/Residential area) and Bannur (in Agriculture area) of Mysore, Karnataka State, India. The seasonal study revealed high abundance of total zooplankton and much variation in Hebbal lake followed by Lingambuddi lake and significantly less abundance and least variations was recorded in Bannur lake. Physico-chemical variables influencing the abundance of total zooplankton were DO in Bannur lake, NO_3, CO_2 and laboratory pH in Lingambudi lake and alkalinity, TSS and BOD in Hebbal lake. The study also revealed low values of wet and dry biomass and much variations in Bannur lake, whereas significantly high values and less variations was recorded in Lingambudi and Hebbal lakes. TASA, Chlorophyll-*a*, NO_3, SO_4 and alkalinity were found to affect the wet biomass of total zooplankton in Bannur lake, NO_3 in Lingambudi lake and PO_4, TASA, Cl_2 and Ca in Hebbal lake. SO_4, field pH and Chlorophyll-*a* was found to affect the dry biomass of total zooplankton in Bannur lake, NO_3, CO_2 and laboratory pH in Lingambudi lake and PO_4, TASA, NO_3 and DO in Hebbal lake.

Introduction

Zooplankton groups are a characteristic indicator of water quality, eutrophication and pollution levels, and are an important source of food chain (Sharma, 1983, Saksena,1987). They play an important role as grazers, suspension feeders and predators within the zooplankton community and also serve as an essential food source for invertebrate and vertebrate predators (Herzig, 1987). Rogozin (2000) stated that it is important to analyze the relationship between the trophic structure of the lake and the zooplankton community. Zooplankton species are cosmopolitan in nature and they inhabit all freshwater habitats of the world, including polluted industrial and municipal wastewaters. Zooplankton are not only useful as bioindicators to help us detect pollution load, but are also helpful for ameliorating polluted waters. Comparisons of size structure, fecundity, and reproductive strategies of zooplanktons can indicate the nature and extent of pollutant loads (Sarma, 1996; Mukhopadhyay *et al.* 2000). Zooplankton populations of tropical freshwater bodies depend on the primary productivity and physico-chemical parameters. Copepods and cladocerans constitute the dominant group of freshwater habitat. They inhabit the ponds, lake, rivers and reservoirs and reported to occur more abundantly in ponds and lakes than in rivers (Reid, 1985 and Sharma, 1991). Various ecological aspects of zooplankton have been a subject of extensive study in India (Zutshi *et al.*, 1980; Babu Roa, 1997; Prasad, 2003; Lendhe and Yeragi, 2004 and Vaishali and Madhuri, 2004). However, these investigations cover only certain period of season. Hence, the present investigation has been carried out to make a systematic analysis of abundance and biomass of total zooplankton along with the physico-chemical parameters with respect to three well marked seasons in Indian climate, in three contrasting lakes of Mysore, Karnataka State, India.

Materials and Methods

Sampling Area

Bannur lake is located at 12° 19′48′ N and 76° 52′12′ E. Water is utilized mainly for agricultural purpose. It is also used for domestic purpose. It is located in rural part and it receives freshwater from its catchment area. The only pollutant it receives is agricultural runoff (Figure 5.1).

Lingambuddi lake is located at 12° 17′N and 75° 27′ E and it is situated at an altitude of 730 m above the sea level. The water spread area is 92.40 ha; the lake independent catchment area is12.41 sq km. Domestic effluent enters this lake (Figure 5.1).

Hebbal lake is one of the polluted tanks. Its geographical location is at 12° 20′00′ N and 76° 36′28′ E at an altitude of 773 m. As it is located within the Hebbal industrial area, the industrial effluent enters this tank (Figure 5.1).

Analysis of Total Zooplankton Abundance

Samples were collected from February 2007 to January 2009 from three contrasting (highly, moderately and least polluted) lakes, namely, Hebbal (in industrial area), Lingambudi (in domestic/residential area) and Bannur (in agriculture area) of Mysore city every month. At each station zooplankton sample were collected by filtering 100 litters of surface water through Nylobolt plankton net (mesh size 60 µm). The samples were stored in 50 ml polyethylene bottles and mixed with ENO-salt (Sodium bi-carbonate 58.8 per cent and anhydrous citric acid 41.2 per cent).

The samples were immediately mixed with buffered (with sodium carbonate) 4 per cent formalin. After fixation, the concentrated zooplankton samples were transferred to a mixture of ethanol (70 per cent), formalin (20 per cent) and glycerol (10 per cent) which acts as a good preservative (Dussart and Defay, 1995). The qualitative analysis of the collected zooplankton samples were carried out in laboratory with the help of phase contrast Olympus microscope model X21, and counting was done using a Sedgewick-Rafter counting cell and calculation were carried out.

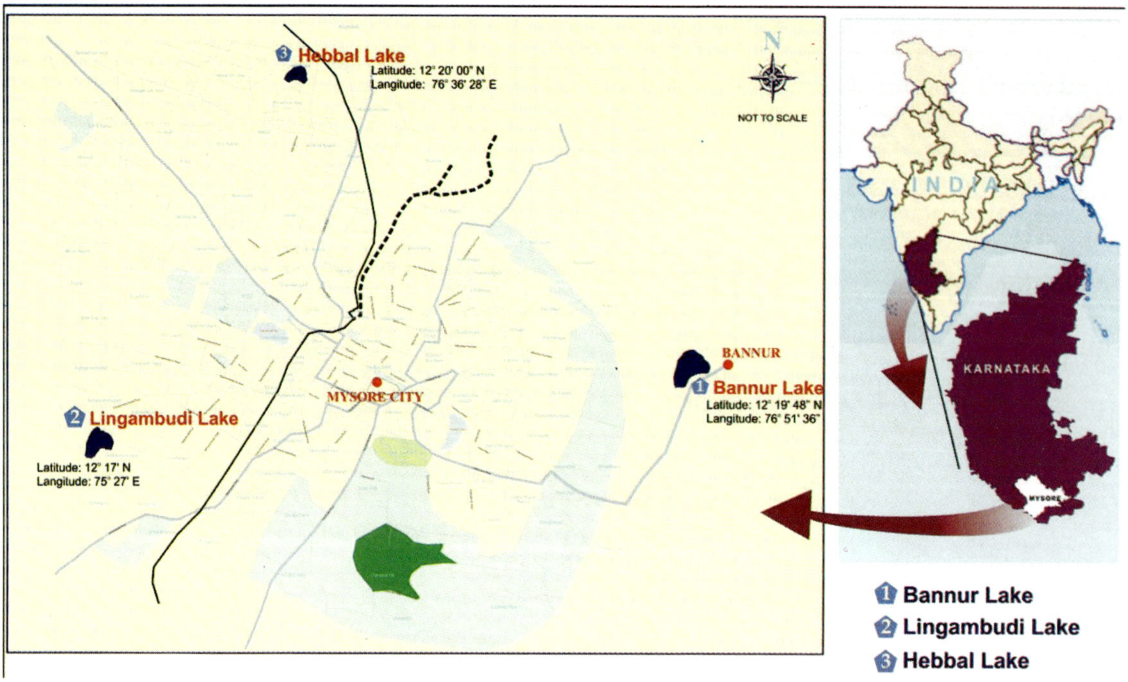

Figure 5.1: Map Showing Samplings Sites of the Freshwater at Bannur, Lingambudi and Hebbal Lakes of Mysore, Karnataka State, India

Analysis of Wet and Dry Biomass of Total Zooplankton

The biomass is estimated by gravimetric method. For determination of wet weight of total zooplankton, the excess water was pipetted out of the 50ml polyethylene container so as to obtain the concentrated zooplanktons in the container. The concentrated zooplanktons were poured into the petridish. The excess water was removed from the petridish using blotting paper. While blotting, due care was taken not to exert too much pressure and cause damage to the zooplankton. The zooplanktons were then transferred onto an aluminum foil which was weighed before and wet weight of total zooplankton was taken. The wet weight is expressed in grams. Dry weight of total zooplankton was determined by drying an aliquot of the zooplankton sample in an electric oven at a constant temperature of 60°C for 48 hours. The dried aliquot was kept in a desiccators until weighed. The dry weight values are expressed in milligram. The dry weight method is dependable as the values indicate the organic contents of the zooplankton.

Analysis of Physico-chemical Parameters

Twenty one physico-chemical parameters namely air temperature, water temperature, field pH, laboratory pH, conductivity, turbidity, dissolved oxygen, BOD, COD, carbon di-oxide, hardness, calcium, alkalinity, TSS, POM, phosphate, sulphate, chloride, nitrate, TASA, and chlorophyll-a were analyzed following the standard methods prescribed by APHA (1992) and Trivedy and Goel (1986).

Statistical Analysis

All the collected data was analyzed for Student-Newman-Keuls, correlation and regression tests. Further, since Indian seasons are mainly controlled by monsoon climate, the study period (February 2007-January 2009) was divided into three well marked seasons, *viz.* Pre-monsoon or summer (February to May), Monsoon or rainy (June to September) and Post-monsoon or winter (October to January). This season wise grouped data was analyzed by ANOVA test employing SPSS 11.0.

Results

Abundance of Total Zooplankton (org/L)

This is the total of the abundance of rotifer, cladocer, copepod and Ostracods. The total zooplankton abundance was similar in Hebbal lake in (560 org/L) and in Lingambuddi lake (493 org/L), but was significantly less in Bannur lake (45 org/L) (Table 5.1).

Table 5.1: Summary of the Abundance and Biomass of Total Zooplankton Groups in the Surface Waters of Bannur, Lingambudi and Hebbal Lakes, February 2007 to January 2009

Sl.No.	*Bannur Lake*			*Lingambudi Lake*			*Hebbal Lake*		
	Mean	*(Range)*	*CV (%)*	*Mean*	*(Range)*	*CV (%)*	*Mean*	*(Range)*	*CV (%)*
1. Abundance of Total zooplankton	45^{a}	(2.3–186.2)	100	493^{b}	(6.32-2148.63)	124	560^{b}	(7.07–5285.92)	191
2. Wet Biomass	9^{a}	(0.24–71.21)	157	48^{b}	(4.82–313.02)	141	58^{b}	(1.47–284.0)	115
3. Dry Biomass	0.36^{a}	(0.01–1.7)	153	2^{b}	(0.06–15.85)	147	2^{b}	(0.1–15.62)	142

n=24, CV: Coefficient of Variation. Mean values with different superscripts are significantly different ($p<0.05$, Student-Newman-Keuls test).

The season wise grouped data revealed no significant seasonal changes in total zooplankton abundance in all the three lakes studied in both first and second year of study (Tables 5.2). The total zooplankton abundance in Bannur lake showed no significant positive correlation with any of zooplankton groups (Table 5.3). However, it showed positive correlation with DO only (Table 5.4). In Lingambuddi lake, the total zooplankton abundance showed only significant positive correlations with Wet and Dry Biomass (Table 5.3). It also showed positive correlation with CO_2 and negative correlations with LpH and Nitrate (Table 5.4). The total zooplankton abundance in Hebbal lake showed no significant correlation with any of zooplankton groups (Table 5.3). However, it showed positive correlations with BOD, TSS and negative correlation with Alkalinity (Table 5.4).

It is noteworthy that the lowest abundance of total zooplankton of 2.3 org/L recorded in Bannur lake and the highest abundance of 5286 org/L in Hebbal lake were the lowest and the highest recorded abundance of total zooplankton during study period (Table 5.1).Thus, along with high abundance of total zooplankton, much variation was observed in Hebbal lake followed by Lingambuddi lake and significantly less abundance and least variations was noted in Bannur lake (Table 5.1 and Figure 5.2).

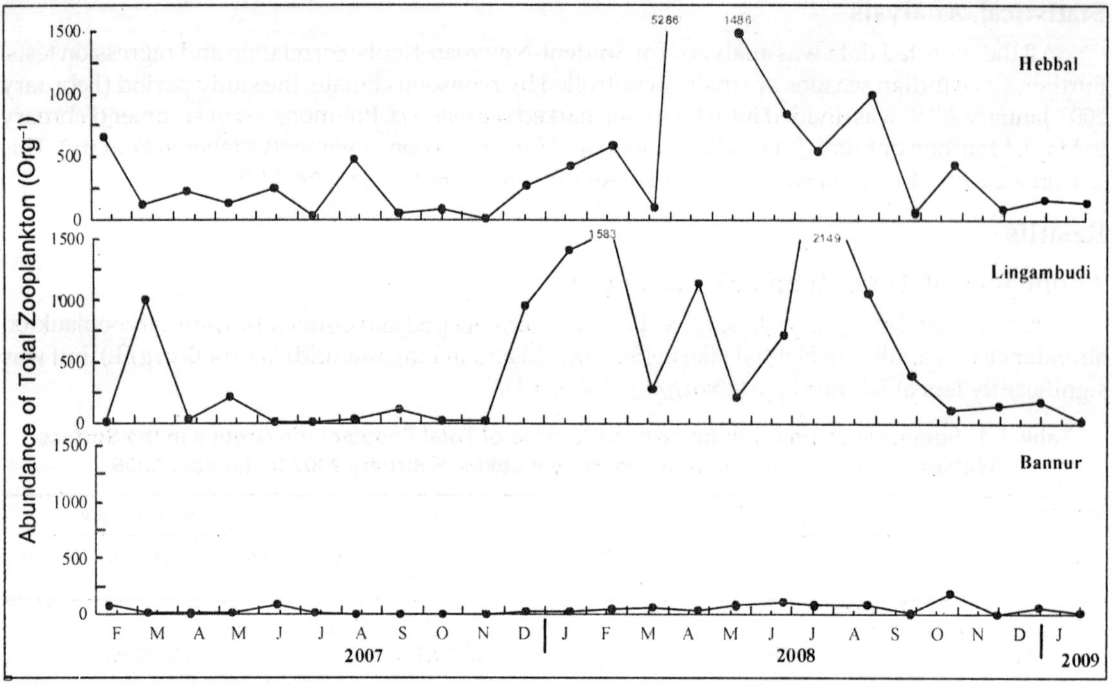

Figure 5.2: Monthly Changes in the Abundance of Total Zooplankton (Org^{-1}) in the Surface Water of Bannur, Lingambudi and Hebbal Lakes, February 2007 to January 2009

Table 5.2: Seasonal Variations in the Abundance and Biomass of Total Zooplankton Groups in the Surface Waters of Bannur, Lingambudi and Hebbal Lakes, February 2007 to January 2009

Sl.No.		Pre-Monsoon (Summer)	Monsoon (Rainy)	Post-Monsoon (Winter)	F-value[1]	P-value[1]
			Bannur Lake			
			1st Year of seasonal study, February 2007-January 2008			
1.	Abundance of Total Zooplankton	32.47^a ± 32.94	27.45^a ± 40.35	16.30^a ± 13.81	0.28	0.760^{NS}
2.	Wet Biomass	9.66^a ± 4.21	27.60^a ± 30.46	7.13^a ± 5.94	1.53	0.269^{NS}
3.	Dry Biomass	0.56^a ± 0.77	0.71^a ± 0.75	0.19^a ± 0.21	0.72	0.512^{NS}
			2nd Year of seasonal study, February 2008-January 2009			
1.	Abundance of Total Zooplankton	56.94^a ± 20.35	71.65^a ± 43.71	67.43^a ± 83.21	0.07	0.929^{NS}
2.	Wet Biomass	2.98^a ± 2.82	1.67^a ± 1.86	6.82^a ± 7.71	1.21	0.342^{NS}
3.	Dry Biomass	0.61^a ± 0.73	0.04^a ± 0.02	0.08^a ± 0.05	2.27	0.159^{NS}

Contd...

Table 5.2–*Contd...*

Sl.No.		*Pre-Monsoon (Summer)*	*Monsoon (Rainy)*	*Post-Monsoon (Winter)*	*F-value*[1]	*P-value*[1]
			Lingambuddi Lake			
			1st Year of seasonal study, February 2007-January 2008			
1.	Abundance of Total Zooplankton	314.05[a]± 466.38	38.91[a]± 50.05	605.16[a] ± 694.14	1.37	0.302 [NS]
2.	Wet Biomass	32.34[a] ± 29.99	11.67[a] ± 8.56	43.05[a] ± 43.69	1.06	0.386 [NS]
3.	Dry Biomass	0.82[a] ± 0.75	0.47[a] ± 0.63	1.95[a] ± 2.04	1.40	0.295 [NS]
			2nd Year of seasonal study, February 2008-January 2009			
1.	Abundance of Total Zooplankton	802.20[a]± 668.94	1081.94[a]±761.51	114.44[a]± 64.63	2.88	0.108 [NS]
2.	Wet Biomass	55.37[a] ± 63.53	97.73[a] ± 144.36	46.41[a] ± 40.38	0.34	0.720 [NS]
3.	Dry Biomass	4.16[a] ± 3.01	4.80[a] ± 7.38	2.01[a] ± 2.20	0.37	0.699 [NS]
			Hebbal Lake			
			1st Year of seasonal study, February 2007-January 2008			
1.	Abundance of Total Zooplankton	281.86[a]± 253.44	200.86[a]±208.45	195.09[a]±188.53	0.20	0.824 [NS]
2.	Wet Biomass	36.47[a] ± 16.62	116.61[a]±116.55	40.81[a] ± 40.27	1.57	0.259 [NS]
3.	Dry Biomass	1.03[a] ± 0.57	5.65[a] ± 6.80	1.38[a] ± 1.10	1.67	0.242 [NS]
			2nd Year of seasonal study, February 2008-January 2009			
1.	Abundance of Total Zooplankton	1864.8[a] ± 2352.89	610.08[a]±414.54	205.17[a]±155.18	1.57	0.261 [NS]
2.	Wet Biomass	109.87[b] ± 66.77	27.91[a] ± 34.38	15.06[a] ± 13.55	5.45	0.028*
3.	Dry Biomass	4.60[b] ± 3.12	1.63[a] ± 1.95	0.37[a] ± 0.10	4.18	0.052*

Values are Mean ± SD, [1]value obtained from ANOVA post hoc nonparametric test. *: Significant, p<0.05, NS: Non Significant, p>0.05. Mean values with different superscripts are significantly different (p<0.05, Student-Newman-Keuls test).

Wet Biomass of Total Zooplankton (g/m^3)

The mean wet biomass of total zooplankton was similar in Lingambuddi (48 g/m^3) and Hebbal (58 g/m^3) lakes, but was significantly less in Bannur lake (9 g/m^3) (Table 5.1).

The season wise grouped data revealed no significant seasonal changes in wet biomass of total zooplankton in Bannur and Lingambuddi lakes in both first and second year of study (Tables 2). In case of Hebbal lake, the season wise grouped data revealed no significant seasonal changes in wet biomass of total zooplankton in the first year of study (Table 5.2). However, in the second year, it revealed significant seasonal changes as the wet biomass of total zooplankton was significantly more during summer season, when compared to rainy and winter seasons both of which had similar values of wet biomass of total zooplankton (Table 5.2).

In Bannur lake, the wet biomass of total zooplankton showed significant positive correlation with dry biomass (Table 5.3) and among the physico-chemical parameters it showed positive correlations with Alkalinity, Sulphate, Nitrate, TASA and Chlorophyll-*a* (Table 5.4). The wet biomass of total zooplankton in Lingambuddi lake showed significant positive correlation only with dry biomass (Table 5.3) and negative correlation with Nitrate (Table 5.4). In case of Hebbal lake, the wet biomass of total zooplankton showed significant positive correlation with only dry biomass (Table 5.3) and significant negative correlations with Calcium, Phosphate, Chloride and TASA (Table 5.4).

Table 5.3: Interrelationships Between the Abundance and Biomass of Total Zooplankton Groups in the Surface Waters of Bannur, Lingambudi and Hebbal Lakes, February 2007 to January 2009

Zooplankton Variables	Dry Biomass	Wet Biomass
	Bannur Lake	
Abundance of Total Zooplankton	NS	NS
Wet Biomass	0.64**	
Dry Biomass		
	Lingambuddi Lake	
Abundance of Total Zooplankton	0.78**	0.65**
Wet Biomass	0.84**	
Dry Biomass		
	Hebbal Lake	
Abundance of Total Zooplankton	NS	NS
Wet Biomass	0.90**	
Dry Biomass		

Values are Pearson correlation coefficient, a 2-tailed test was applied and calculated after $\log_{10}$ transformation of all variables after scaling so that all values were >1, n=24 *P<0.05, **p<0.005 and NS: Non Significant.

It is noteworthy that the lowest wet biomass of total zooplankton value of 0.24 g/m³ recorded in Bannur lake and the highest value of 313.02 g/m³ in Lingambuddi lake were the lowest and the highest recorded values of wet biomass of total zooplankton during study period (Table 5.1).

Thus, along with low values of wet biomass of total zooplankton, much variation was observed in Bannur lake, whereas significantly high values and less variation was noted in Lingambuddi and Hebbal lakes (Table 5.1 and Figure 5.3).

Dry Biomass of Total Zooplankton (mg/m³)

The mean dry biomass of total zooplankton values were similar in Lingambuddi lake (2 mg/m³) and in Hebbal lake (2 mg/m³), but was significantly less in Bannur lake (0.36 mg/m³) (Table 5.1).

The season wise grouped data revealed no significant seasonal changes in values of dry biomass of total zooplankton in Bannur and Lingambuddi lakes in both first and second year of study (Table 5.2). In case of Hebbal lake, The season wise grouped data revealed no significant correlation with any of the zooplankton groups (Table 5.3), but, among the physico-chemical variables, it revealed positive correlation with CO_2 and negative correlations with LpH and Nitrate (Table 5.4). In Hebbal lake, the dry biomass of total zooplankton showed no significant correlation with any of the zooplankton groups (Table 5.3). However, it revealed negative correlations with DO, Phosphate, Chloride and TASA (Table 5.4).

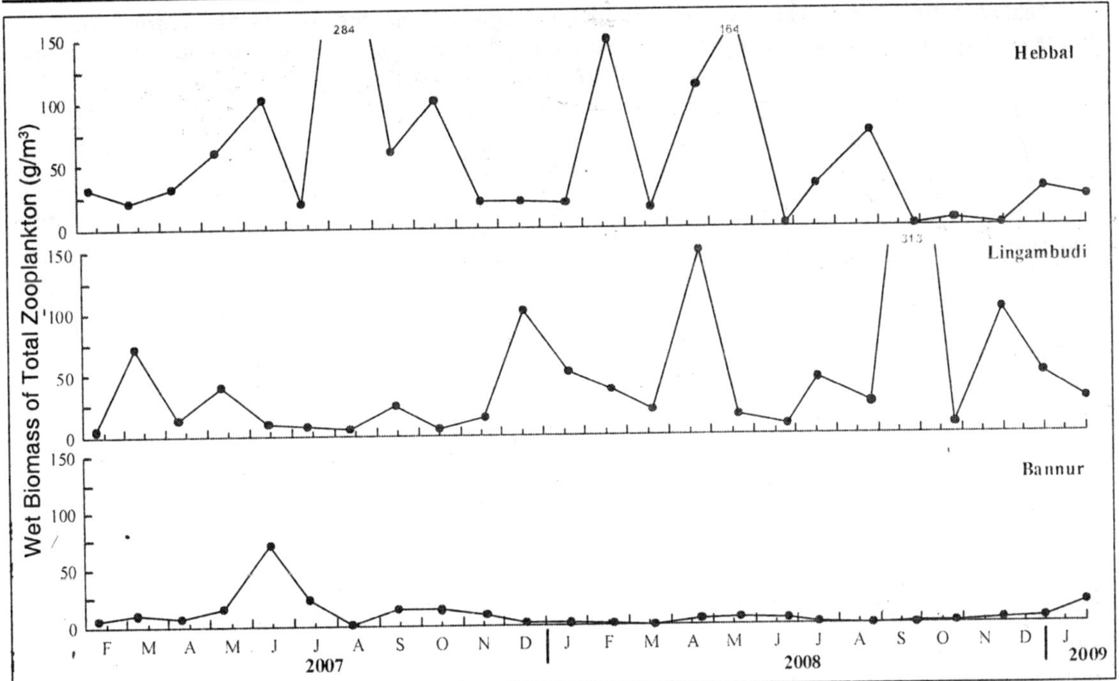

Figure 5.3: Monthly Changes in the Wet Biomass of Total Zooplankton (g/m³) in the Surface Waters of Bannur, Lingambudi and Hebbal Lakes, February 2007 to January 2009

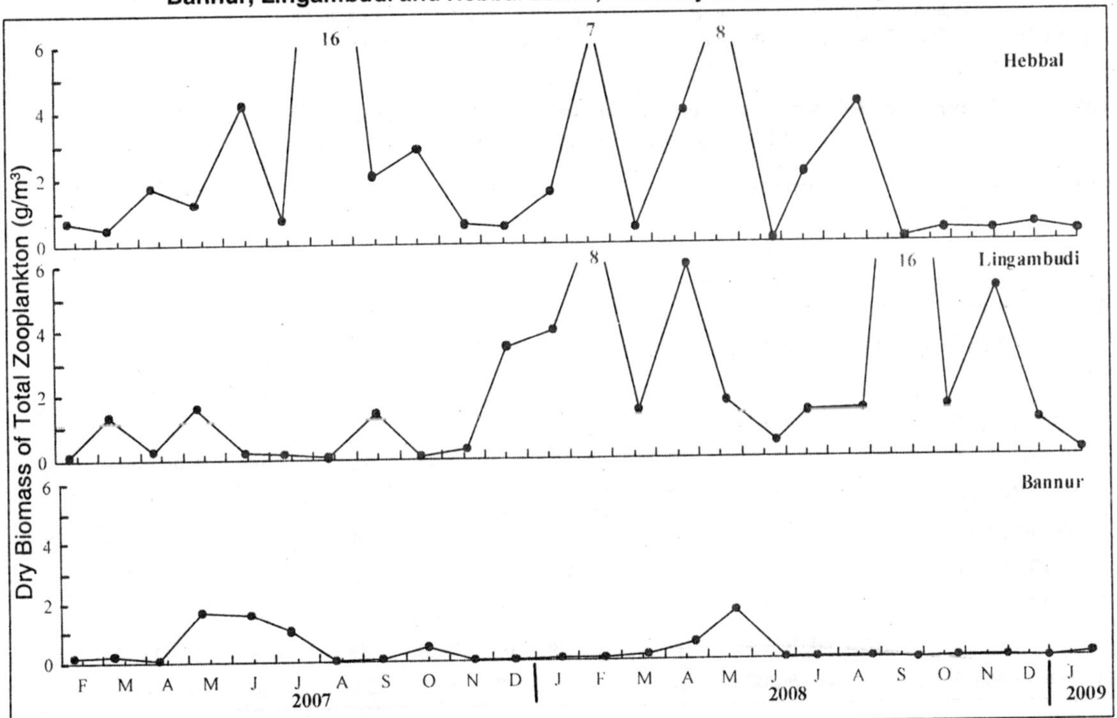

Figure 5.4: Monthly Changes in the Dry Biomass of Total Zooplankton (mg/m³) in the Surface Waters of Bannur, Lingambudi and Hebbal Lakes, February 2007 to January 2009

Table 5.4: Relationships Between the Physico-chemical Variables and the Abundance and Biomass of Total Zooplankton Groups in the Surface Waters Bannur, Lingambudi and Hebbal Lakes, February 2007 to January 2009

Variables	*Dry Biomass*			*Wet Biomass*			*Total Zooplankton*		
	Bannur	*Lingambudi*	*Hebbal*	*Bannur*	*Lingambudi*	*Hebbal*	*Bannur*	*Lingambudi*	*Hebbal*
Field pH	0.49*	NS	NS	NS	NS	NS	NS	NS	NS
Laboratory pH	NS	–0.46**	NS	NS	NS	NS	NS	–0.43*	NS
Dissolved Oxygen	NS	NS	–0.41*	NS	NS	NS	0.45*	NS	NS
BOD	NS	NS	NS	NS	NS	NS	NS	NS	0.46*
Carbon di-Oxide	NS	0.49*	NS	NS	NS	NS	NS	0.43*	NS
Calcium	NS	NS	NS	NS	NS	–0.41*	NS	NS	NS
Alkalinity	NS	NS	NS	0.57**	NS	NS	NS	NS	–0.49*
TSS	NS	NS	NS	NS	NS	NS	NS	NS	0.49*
Phosphate	NS	NS	–0.53**	NS	NS	–0.46*	NS	NS	NS
Sulphate	0.57**	NS	NS	0.63**	NS	NS	NS	NS	NS
Chloride	NS	NS	–0.47*	NS	NS	–0.48*	NS	NS	NS
Nitrate	NS	–0.54*	NS	0.48*	–0.49*	NS	NS	–0.73**	NS
TASA	NS	NS	–0.48*	0.65**	NS	–0.44*	NS	NS	NS
Chlorophyll-a	0.40*	NS	NS	0.47*	NS	NS	NS	NS	NS

Values are Pearson correlation coefficient, a 2-tailed test was applied and calculated after log_{10} transformation of all variables after scaling so that all values were >1, n=24 *P<0.05, **p<0.005, and NS: Non Significant; BOD: Biological Oxygen Demand; TSS: Total Suspended Solid,TASA: Total Anions of Strong Acids.

Table 5.5: Summary of Physico-chemical Variables in the Surface Waters of Bannur, Lingambudi and Hebbal Lakes, February 2007 to January 2009

Sl.No.	*Physico-Chemical Variables*	*Bannur Lake*			*Lingambudi Lake*			*Hebbal Lake*		
		Mean	*(Range)*	*CV (%)*	*Mean*	*(Range)*	*CV (%)*	*Mean*	*(Range)*	*CV (%)*
1.	Air Temp(°C)	23.24[a]	(19-28.8)	11	23.07[a]	(19-28.8)	11	24.1[a]	(21-28.8)	9
2.	Water Temp(°C)	22.11[a]	(18-28)	12	21.29[a]	(15-26)	11	22.43[a]	(18-28)	10
3.	F pH	8.38[a]	(7.4-9.2)	5	08.67[b]	(7.9-9.4)	5	8.29[a]	(7.8-8.8)	3
4.	LpH	8.26[a]	(7.3-9.1)	5	08.33[a]	(7.4-9.4)	7	7.93[b]	(7.4-8.7)	4
5.	Cond.(μS cm^{-1})	810[a]	(460-1200)	27	1463[b]	(610-2550)	31	1866[c]	(990-3600)	36
6.	Tur. (NTU)	6[a]	(1.2-10.7)	56	30[b]	(2.6-99.0)	96	23[b]	(2.4-52.0)	75
7.	DO (mg l^{-1})	3.7[a]	(0.81-6.89)	59	0.75[b]	(0.0-1.63)	68	1.06[b]	(0.0-3.24)	79
8.	BOD (mg l^{-1})	1.21[a]	(0.41-3.65)	79	3.49[b]	(0.41-7.29)	72	3.44[b]	(0.41-9.58)	79
9.	COD (mg l^{-1})	5.13[a]	(1.6-14.4)	56	4.8[a]	(1.6-11.2)	61	5.0[a]	(1.6-19.2)	86
10.	CO_2 (mg l^{-1})	9.48[a]	(0.0-38.4)	123	42.13[b]	(0.0-155.3)	139	43.45[b]	(0.0-118.8)	79
11.	Hard. (mg l^{-1})	131[a]	(41.18-202.0)	28	279[b]	(106.5-532)	33	435[c]	(236.0-886.0)	42

Contd...

Table 5.5–*Contd...*

Sl.No.	*Physico-Chemical Variables*	*Bannur Lake*			*Lingambudi Lake*			*Hebbal Lake*		
		Mean	*(Range)*	*CV (%)*	*Mean*	*(Range)*	*CV (%)*	*Mean*	*(Range)*	*CV (%)*
12.	Cal (mg l^{-1})	28[a]	(16.03-48.1)	26	49[b]	(23.25-76.15)	29	86[c]	(37.67-179.56)	42
13.	Alk.(mg l^{-1})	225[a]	(150.0-350.0)	28	157[b]	(150.0-860)	33	509[c]	(305.0-700.0)	25
14.	TSS (mg l^{-1})	24[b]	(10.0-150.0)	139	63[a]	(10.0-160.0)	65	40[b]	(10.0-110.0)	64
15.	POM (mg l^{-1})	2.67[b]	(1.0-5.0)	50	5.46[a]	(1.0-19.0)	75	3.38[a]	(1.0-6.0)	41
16.	PO_4 (mg l^{-1})	0.07[a]	(0.01-0.41)	157	1.01[b]	(0.01-3.38)	89	0.92[b]	(0.01-3.38)	96
17.	SO_4 (mg l^{-1})	23[a]	(5.0-99.0)	94	101[b]	(11.5-260.0)	83	85[b]	(25.0-210.0)	51
18.	Cl_2 (mg l^{-1})	36[a]	(22.72-49.7)	20	135[b]	(52.54-254.18)	37	240[c]	(100.82-468.6)	50
19.	NO_3 (mg l^{-1})	0.38[a]	(0.01-1.95)	121	0.62[a]	(0.01-4.89)	177	1.06[a]	(0.0-5.15)	138
20.	TASA (mg l^{-1})	62[a]	(33.02-134.9)	42	225[b]	(75.9-402.3)	45	324[c]	(147.06-540.56)	40
21.	Chl-*a* (μg l^{-1})	3.60[a]	(0.0-13.19)	110	17[b]	(2.4-54.0)	81	17[b]	(3.6-45.08)	66

n=24, CV = Coefficient of Variation, Mean values with different superscripts are significantly different ($p<0.05$, Student-Newman-Keuls test).

F pH: pH measured in the field; L pH: pH measured in the laboratory; Temp: Temperature; Cond: Conductivity; Tur: Turbidity; DO: Dissolved Oxygen measured in the Field; BOD: Biological Oxygen Demand; COD: Chemical Oxygen Demand; CO_2: Free Carbon di-Oxide; Hard: Hardness; Alk: Alkalinity, Cl_2: Chloride; NO_3: Nitrate; SO_4: Sulphate; TASA: Total Anions of Strong Acids; Cal: Calcium, PO_4: In-Organic Phosphate; TSS: Total Suspended Solid; Chl-a: Chlorophyll-a.

It is noteworthy that the lowest dry biomass of total zooplankton value of 0.01 mg/m^3 recorded in Bannur lake and the highest value of 15.85 mg/m^3 in Lingambuddi lake were the lowest and the highest recorded values of dry biomass of total zooplankton during study period (Table 5.1). Thus, high values of dry biomass of total zooplankton and moderate variation was observed in Lingambuddi and Hebbal lakes and significantly less values and least variations were noted in Bannur lake (Table 5.1 and Figure 5.4).

The highest number of the abundance of total zooplankton was recorded in Hebbal lake, followed by Lingambuddi and Bannur lakes (Table 5.1).

Discussion

Total Zooplankton Abundance (Rotifer, Cladocer, Copepod and Ostracod)

Zooplankton communities are highly sensitive to environmental variation. They respond to a wide variety of ecological disturbances including nutrient loading (McCauley and Kalff 1981; Pace 1986; Dodson, 1992), acidification (Brett, 1989; Keller and yan, 1991; Marmorek and Kormann 1993), contaminants Yan *et al.*, 1996), fish densities (Carpenter *et al.*, 1985). Zooplanktons are also susceptible to variations in a wide number of environmental factors including water temperature, light, chemistry (particularly pH, oxygen, salinity, toxic contaminants), food availability (algae, bacteria), and predation by fish and invertebrates. As a result, changes in their abundance, species diversity, or community composition can provide important indications of environmental change or ecological disturbance and it is also important in understanding trophodynamics and trophic progression of water bodies

(Mathew, 1977; Verma and Data Munshi, 1987). In the present study, the mean abundance of total zooplankton were similar in Lingambudi and Hebbal lakes but, it was significantly less in Bannur lake (Table 5.1). However, the lowest (2.3 Org l^{-1}) abundance of total zooplankton was recorded in Bannur lake and the highest (586 Org l^{-1}) abundance of total zooplankton of was recorded in Hebbal lake (Table 5.1). The fall in the abundance of zooplankton in Bannur lake may be due to the decrease in the nutrients and phytoplankton population as recorded in the study of Valery Yakovalev (2001) on zooplankton of subarctic Imandra lake in Russia following water quality improvement. The season wise grouped data revealed no significant seasonal changes in the abundance of total zooplankton in all the three lake studied in both first and second year of the study (Table 5.2). The regression analysis revealed that 20 per cent of the variation in the abundance of total zooplankton was due to DO (+) in Bannur lake, 54 per cent of the variations was due to NO_3 (-) in Lingambudi lake and 24 per cent was due to alkalinity (-) in Hebbal lake. Apart from these, in Lingambudi and Hebbal lakes, other physico-chemical variables such as pH, BOD, CO_2 and TSS also entered in the regression equation and thus participated in deciding the abundance of total zooplankton (Table 5.6). High rates of agricultural runoff can cause large quantities of nitrates and other nutrients such as phosphate to enter the water system. When added to a water body, these nutrients can create a large proliferation of algae which is harmful to water quality. The blooms deplete oxygen levels in aquatic ecosystems and thus have a detrimental effect on the zooplankton population and other organisms within the system.

Table 5.6: Results of Stepwise Multiple Regression Analysis Between Abundance and Biomass of Total Zooplankton Groups and Physico-chemical Variables in the Surface Waters of Bannur, Lingambudi and Hebbal Lakes, February 2007 to January 2009

Zooplankton Variables	*Physico-chemical variables*
	Bannur lake
Abundance of Total Zooplankton	DO (+), (R^2=0.20, F=5.60, P<0.05).
Wet Biomass	TASA (+), (R^2=0.42, F=15.63, P<0.005) Chl-*a*(+), NO_3(+), SO_4(+), Alka(+).
Dry Biomass	SO_4 (+), FpH (+), (R^2=0.33, F=10.71, P<0.005) Chl-*a*(+).
	Lingambuddi Lake
Abundance of Total Zooplankton	NO_3 (-), (R^2=0.54, F=25.53, P<0.005)CO_2 (+), LpH(-).
Wet Biomass	NO_3 (-), (R^2=0.24, F=6.98, P<0.05).
Dry Biomass	NO_3 (-), (R^2=0.29, F=9.15, P<0.005)CO_2 (+), LpH(-).
	Hebbal Lake
Abundance of Total Zooplankton	Alka (-), (R^2=0.24, F=6.83, P<0.05) TSS (+), BOD(+).
Wet Biomass	PO_4 (-), (R^2=0.21, F=5.96, P<0.05) TASA (-), Cl_2(-), Cal(-).
Dry Biomass	PO_4 (-), (R^2=0.28, F=8.50, P<0.05) TASA (-), NO_3(-), DO(-).

Environmental (independent) variables in the final regression equation (P in=0.05, P out=0.1) are shown: multiple coefficients of determinations (R^2) and overall F and P values for each equation are given in the parenthesis. Physico-chemical variables which were not in the final equation but which are correlated (P<0.05) with the abundance of Ostracods are then listed in order of decreasing magnitude of correlation coefficient; the sign of the correlation coefficient is indicated in parenthesis.

The Physico-chemical variables are; Wtemp: Water temperature; FpH: pH measured in the field; LpH: pH measured in the laboratory; Turb: Turbidity; Cond: Conductivity; DO: Dissolved Oxygen; BOD: Biological Oxygen Demand; CO_2: Free Carbon di-Oxide; Cal: Calcium; TSS: Total Suspended Solids; Cl_2: Chloride; NO_3: Nitrate; SO_4: Sulphate; PO_4: Inorganic Phosphate; TASA: Total Anions of Strong Acids; Alka: Alkalinity and Chl-*a*: Chlorophyll-*a*.

In this study, it has been proved that Bannur lake is least polluted, Lingambudi lake is moderately polluted and Hebbal lake is highly polluted. The evidence for this conclusion is that Bannur lake which is in agriculture area showed better water quality, that is less conductivity, turbidity, hardness, Ca, TSS, POM, PO_4, SO_4, Cl_2, NO_3 and TASA and least abundance of total zooplankton. As least abundance of zooplankton groups are usually considered to be useful indicators of good water quality, trophic and pollution, status. On contrary, Lingambudi lake which is in domestic/residential area showed moderately polluted status of water quality, that is moderate conductivity, hardness, Ca, TSS, Cl_2, NO_3 and TASA and moderate abundance of total zooplankton. This moderate pollution is because of direct entry of untreated domestic sewage from the catchment area, which probably has organic matter including nutrients and phytoplanktons. In case of Hebbal lake which is in industrial area, highly polluted status of water quality is observed, that is high conductivity, CO_2, hardness, Ca, alkalinity, Cl_2, NO_3 and TASA and highest abundance of total zooplankton. This can be attributed to thick deposits of organic matter due to untreated industrial effluent/sewage carrying large quantity of organic matter. Thus, pollution showed to have positive effect on abundance of zooplankton groups studied.

Wet Biomass of Total Zooplankton (g/m^3)

Zooplankton abundance or density, expressed as number per area or volume units, does not necessarily provide accurate information about community biomass, because zooplankton consists of a great variety of groups or animal species of a large size range (Matsumura-Tundisi *et al.*, 1989). Moreover, the biomass of the zooplankton species is an important and necessary parameter to calculate the secondary production of this community (Melao and Rocha, 2004). In the present study, the mean values of the wet biomass of total zooplankton were similar in Lingambudi and Hebbal lakes but, it was significantly less in Bannur lake (Table 5.1). However, the lowest wet biomass of total zooplankton value of 0.24 g/m^3 was recorded in Bannur lake and the highest wet biomass of total zooplankton value of 313.02 g/m^3 was recorded in Lingambudi lake during study period (Table 5.1). Less wet weight of total zooplankton in Bannur lake as compared to Lingambudi and Hebbal lakes, could probably be a consequence of the lower level of eutophication in which, food limitation for zooplankton could occur as recorded in the study of Gonzalez *et al.* (2008) on size and dry weight of main zooplankton species in Bariri reservoir, Brasil. Significantly high values of wet biomass in Lingambudi and Hebbal lakes may be due to domestic and industrial sewage that affects the zooplankton biomass indirectly by increasing the nutrients availability in water, which leads to an increase in the plankton biomass in these areas as recorded in the study of Al-Najjar and Rasheed (2005) on zooplankton biomass in the most northern tip of the golf of Aqaba, Germany. Additionally chlorophyll-*a* has been shown to affect the zooplankton biomass as recorded in the study of Marine Science Station (MSS, 2006) on National Monitoring Program of Aqaba, Germany. In this study Lingambudi and Hebbal lakes with high values of wet biomass are characterized by higher chlorophyll-*a* content of 17 µg l^{-1} compared to Bannur lake (3.6 µg l^1). Another reason for higher biomass of total zooplankton in Lingambudi and Hebbal lakes could be explained by the dominance of the large sized zooplankton organisms such as copepods and cladocer in these stations. In the second year of this study, in Hebbal lake, the wet biomass of total zooplankton was significantly more during summer season when compared to rainy and winter seasons (Table 5.2). The regression analysis revealed that 42 per cent of the variations in the wet biomass of total zooplankton was due to TASA (+) in Bannur lake, 24 per cent was due to NO_3 (-) in Lingambudi lake and 21 per cent due to PO_4 (-) in Hebbal lake. Apart from these, in Bannur and Hebbal lakes, other physico-chemical variables such as alkalinity, calcium, chloride, SO_4, NO_3, TASA and chlorophyll-*a* also entered the regression equation and thus participated in deciding the wet biomass of total zooplankton (Table 5.6).

Dry Biomass of Total Zooplankton (mg/m^3)

Estimation of zooplankton size and dry weight constitutes an important contribution for the study of trophic-web structure in aquatic ecosystem, considering its relationship with the trophic status of the water bodies (Rocha *et al.*, 1995; Pinto-Coelho *et al.*, 2005). Although there are some indirect methods for calculation of dry weight in zooplankton, as the use of equation, direct measures are more reliable. Pauli (1989) found that transformation of body volumes, which is commonly used as an estimate of fresh weight, with a general conversion factor of 10 per cent dry weight of body volumes, does not reflect the differences among rotifer species which can contain more than 90 per cent of water in their bodies, resulting in unsatisfactory values. However, direct and indirect measure can be combined, in order to obtain accurate estimates of biomass (Culver *et al.*, 1985), but organisms must be collected over a long period of time, in order to represent the seasonal variation of weight per length that occurs in nature. There are many studies which register the dry weight of total zooplankton from temperate ecosystems (Masundire, 1994), but there are few articles concerning the zooplankton biomass of freshwater (Matsumura-Tundisi *et al.*, 1989; Infante *et al.*, 1990; Masundire, 1994; Melao and Rocha, 2004; Sendacz *et al.*, 2006). In the present study, the mean values of dry biomass of total zooplankton were similar in Lingambudi and Hebbal lakes but, it was significantly less in Bannur lake (Table 5.1). The lowest (0.01 mg/m^3) dry biomass of total zooplankton was recorded in Bannur lake and the highest (15.85 mg/m^3) dry biomass of total zooplankton was recorded in Lingambudi lake during study period (Table 5.1). Cladocer and cyclopoid probably contribute in a high degree to the total biomass of the zooplankton community in Lingambudi and Hebbal lakes. In many tropical freshwater ecosystems, copepods could represent the major contribution of zooplankton biomass in water bodies, while rotifer constitute a minor fraction of total biomass (Infante, 1993; Matsumura-Tundisi *et al.*, 1989; Melao and Rocha, 2004). However, there are cases where cladocerans may show the highest values of biomass in other water bodies (Sendacz *et al.*, 2006). There are evidences which show the relationship between the zooplankton biomass and the trophic state of the lake. Eutrophic reservoirs can support the high abundance of zooplankton, which has no food limitation in highly productive ecosystems (Gonzalez *et al.*, 2002). Thus, eutrophication affects structure, size and biomass of the zooplankton community (Pinto-Coelho *et al.*, 2005). In the second year of this study, in Hebbal lake, the dry biomass of total zooplankton was significantly more during summer season when compared to rainy and winter seasons (Table 5.2). This could be the result of the increased food availability as well as the increase of temperature that has a positive effect on their filtering rate as recorded in the studies of Burns (1969) on relation between filtering rate, temperature and body size in four species of Daphnia; and on in situ filtering rates of cladocer (Mourelatos and Lacroix, 1990). Such a coincidence of maximum weight is also reported by Vuille and Maurer (1991). The regression analysis revealed that 32 per cent of the variations in the dry biomass of total zooplankton was due to SO_4 (+) in Bannur lake, 29 per cent was due to NO_3 (-) in Lingambudi lake and 28 per cent was due to PO_4 (-) in Hebbal lake. Apart from these, in all the three lakes studied, other physico-chemical variables such as field pH, laboratory pH, CO_2, DO, NO_3, TASA and chlorophyll-*a* also entered the regression equation and thus participated in deciding the dry biomass of total zooplankton (Table 5.6).

Acknowledgements

First author would like to thank Dr. K. Altaff for providing basic training in his laboratory at New College, Chennai, Tamil Nadu, India. Both the authors are also thankful to the Chairman of the Department of Studies in Zoology, University of Mysore, Mysore for providing the necessary facilities during course of investigation.

References

Al–Najjar, T. and Rasheed, M., 2005. Zooplankton biomass in the most northern tip of the Golf of Aquba, a case study. *Leb. Sci. J.* (In Press).

APHA, 1992. *Standard Methods for Examination of Water and Wastewater*, 18th edn. American Public Health Association, NW, Washington.

Babu Rao, M., 1997. Studies on the ecology and fish fauna of an oligotrophic lake, Himayatsagar, Hyderabad, A.P. In: *Recent Advanced in Freshwater Biology*, Vol. 2, (Ed.) K.S. Rao, p. 73–97.

Brett, M.T., 1989. Zooplankton communities and acidification processes (a review). *Water, Air and Soil Pollution*, 44: 387–414.

Burns, C.W., 1969. Relation between filtering rate, temperature and body size in four species of *Daphnia*. *Limnol. Oceanogr.*, 14: 693–700.

Carpenter, S.R., Kitichell, J.F. and Hodgson, J.R., 1985. Cascading trophic interactions and lake productivity. *Bioscience*, 35: 634–639.

Culver, D.A., Boucherle, M.M., Bean, D.J. and Fletcher, J.W., 1985. Biomass of freshwater crustacean zooplankton from length-weight regressions. *Can. J. Fish. Aq. Sci.*, 42(8): 1380–1390.

Dodson, S., 1992. Predicating crustacean zooplankton species richness. *Limnology and Oceanography*, 37: 848–856.

Dussart, B.H. and Defay, D., 1995. *Copepoda: Introduction to the Copepod*. SPB. Academic Publishing, Netherlands. Vol. 7: 1–253.

Gonzalez, E.J., Matsumura-Tundisi, T., Tundisi, J.G., 2008. Size and dry weight of main zooplankton species in Bariri reservoir (SP, Brazil). *Braz. J. Biol.*, 68(1).

Haberman, J., 1998. Zooplankton of lake Vortsjarv. *Limnologia*, 28(1): 49–65.

Herzig, A., 1987. The analysis of planktonic rotifer populations: a plea for long-term investigations. *Hydrobiologia*, 147: 163–180.

Infante, A., Infante, O., Vegas, T. and Rehl, W., 1990. *Estudio comparative del lago de Valencia (Venezuela) y el lago de Managua (Nicaragua)*. Informe final. Organizacion de los Estados Americanos y Universidad Central de Venezuela. Caracas.

Infante, 1993. Vertyical and horizontal distribution of the zooplankton in lake Valencia. *Acta Limnologica Brasiliensia*, 6(1): 97–105.

Keller, W. and Yan, N.D., 1991. Recovery of crustacean zooplankton species richness in Sudbury area lakes following water quality improvements. *Canadian Journal of Fisheries and Aquatic Sciences*, 48: 1635–1644.

Lendhe, R.S. and Yeragi, S.G., 2004. Physico-chemical parameters and zooplankton diversity of Phirange Kharbav lake, Dist. Thane, Maharashtra. *J. Aqua. Biol.*, 19(1): 49–52.

Marmorek, D.R. and Korman, J., 1993. The use of zooplankton in a biomonitoring program to detect lake acidification and recovery. *Water, Air, and Soil Pollution*, 69: 223–241

Marneffe, Y., Comblin, S. and Thome, J., 1998. Ecological water quality assessment of the Butgenbach lake (Belgium) and its impact on the river. Warche using rotifers as bioindicators. *Hydrobiologia*, 387(388): 459–467.

Mastumura-Tundisi, T., Rietzler, A. and Tundisi, JG., 1989. Biomass (dry weight and carbon content) of plankton crustacea from Broa Reservoir (Sao Carlos, SP. Brazil) and its fluctuation across one year. *Hydrobiologia*, 179(3): 229–236.

Masundire, H.M., 1994. Mean individual dry weight and length-weight regressions of some zooplankton of lake Kariba. *Hydrobiologia*, 272(1–3): 231–238.

Mathew, D.N., 1977. Moult in the Baya Weaver *Ploceus philippinus* Linnaeus. *J. Bomabay. Nat. Hist. Soc.*, 74: 233–245.

Matsunura-Tundisi, T., Rietzler, A. and Tundisi, J.G., 1989. Biomass (dry weight and carbon content) of plankton crustacea from Broa reservoir (Sao Carlos, SP.–Brazil) and its fluctuation across one year. *Hydrobiologia*, 179(3): 229–236.

McCauley, E. and Kalff, J., 1981. Empirical relationships between phytoplankton and zooplankton biomass in lakes. *Canadian Journal of Fisheries and Aquatic Sciences*, 38: 458–463.

Melao, M.G.G. and Rocha, O., 2004. Life history, biomass and production of two planktonic Cyclopoid copeopods in a shallow subtropical reservoir. *J. Plankton Res.*, 26(8): 909–923.

Michael, R.G., 1973. A guide to the study of freshwater organisms, 2. Rotatoria. *J. Madurai. Univ.*, Suppl. 1, 23–36.

Mourelatos, S. and Lacroix, G., 1990. *In Situ* filtering rates of cladocera: Effect of body length, temperature and food concentration. *Limnol. Oceanogr.*, 35(5): 1101–1111.

MSS, 2006. Marine Science Station. National Monitoring Programme of Aqaba, 120 pp.

Mukhopadhyay, S.K., Chatterjee, A., Gupta, R. and Chattopadhyay, B., 2000. Rotiferan community structure of a tannery effluent stabilization pond in east Calcutta wetland ecosystem. *Chem. Env. Res.*, 9(1&2): 85–91.

Pace, M.L., 1986. An empirical analysis of zooplankton size structure across lake trophic gradients. *Limnology and Oceanography*, 31: 45–55.

Pauli, H.R., 1989. A new method to estimate individual dry weights of rotifers. *Hydrobiologia*, 186–187(1): 355–361.

Pinto-Coelho, R.M., Bezerra-Neto, J.F. and Morais-J.R., C.A., 2005. Effects of eutrophication on size and biomass of crustacean zooplankton in a tropical reservoir. *Braz. J. Biol.*, 65(2): 325–338.

Prasad, N.V., 2003. Diversity and richness of zooplankton in Coringa Mangrove Ecosystems: Decedel changes. *J. Aqua. Biol.*, 18(2): 41–46.

Reid, J.W., 1985. Chave de identificacao e lista de referencias bibliograficas para as species continentais sul–americanas de vida livre da Ordem Cyclopoida (Crustacea, Copepoda). *Bolm Zool.*, Sao Paulo. 9: 17: 143.

Rocha, O., Sendacz, S. and Matsumura-Tundisi, T., 1995. Composition, biomass and productivity of zooplankton in natural lakes and reservoir of Brazil. In: *Limnology in Brazil*, (Eds.)J.G. Tundisi, Cem. Bicudo and T. Matsumura-Tundisi. Brazilian Academy of Science and Brazilian Limnological Scociety, Rio de Janeiro, p. 151–165.

Rogozin, A.G., 2000. Specific structural features of zooplankton in lakes differing in tropic status: Species populations. *Russian Journal of Ecology*, 31(6): 405–410.

Saksena, N.D., 1987. Rotifer as indicator of water quality. *Acta. Hydrochim. Hydrobiol.*, 15: 481–485.

Sarma, S.S.S., 1996. Some relationships between size structure and fertility of rotifer populations. In: *Advances in Fish, Wildlife, Ecology and Biology*, (Ed.) B.L. Kaul. Daya Publishing House, Delhi. 1: 37–50.

Sendacz, S., Caleffi, S. and Santos-Soares, J., 2006. Zooplankton biomass of reservoirs in different tropic conditions in the state of Sao Paulo, Brazil. *Braz. J. Biol.*, 66(1b): 337–350.

Sharma, B.K., 1983. The Indian species of the genus *Branchionus* (Eurotatoria: Monogononta: Brachionida). *Hydrobiol.*, 104: 31–39.

Sharma, B.K., 1991. Cladocera. *Animal Resources of India*, pp. 205–233.

Trivedy, R.K. and Goel, P.K., 1986. *Chemical and Biological Methods of Water Pollution Studies.* Environmental Publications, Karad.

Vaishali, Somani and Madhuri Pejavar, 2004. Crustacean zooplankton of lake Masunda, Thane, Maharashtra, India. *J. Aqua. Biol.*, 19(1): 57–60.

Valery Yakovlev, 2001. Zooplankton of subarctic Imandra Lake following water quality improvements, Kola Peninsula, Russia. *Chemosphere*. 42: 85–92.

Verma, P.K. and Datta Munshi, J. S., 1987. Plankton community structure of Badua reservoir of Bhagalpur (Bihar). *Trop. Ecol.*, 28: 200–207.

Vuille, T. and Maurer, V., 1991. Body mass of crustacean plankton in Lake Biel: A comparison between pelagic and littoral communities. *Verh. Int. Verein. Limnol.*, 24: 938–942.

Yan, N.D., Keller, W., Somers, K.M., Pawson, T.W. and Griard, R.E., 1996. Recovery of crustacean zooplankton communities from acid and metal contamination: comparing manipulated and reference lakes. *Canadian Journal of Fisheries and Aquatic Sciences*, 53: 1301–1327.

Zutshi, D.P., Subla, B.A., Khan, M.A. and Wanganee, A., 1980. Comparative limnology of nine lakes of Jammu and Kashmir, Himalayas. *Hydrobiol.*, 72: 101–112.

Chapter 6

Water Quality Status of Kirung Ri River at Nganglam, Pemagatshel (Bhutan)

☆ *Umesh Bharti, P.K. Bharti, D.S. Malik, Pawan Kumar and Gaurav*

ABSTRACT

The project site is surrounded by hills and mountains at high elevation near Nganglam in Eastern Himalayas. The study site is located at about 4.5 km north of Nganglam and 10 km from indo-Bhutan border. The assessment of water quality has been carried out to estimate the effect of any anthropogenic activities on the environment of the study area and to formulate mitigation measures to prevent the surface water bodies and stream water from further pollution.

Heavy metals like cadmium, copper, iron, manganese, lead and zinc were not found in high concentration at both sites during any season. Cadmium, manganese and copper was absolutely absent in all seasons, while lead concentrations were also found below detection limit in maximum samples. The concentration of iron was maximum observed 0.1 mg/l during the entire study. Zinc concentration was also found below the standards limit.

The present study deals with the preliminary physico-chemical characteristics of the river water and exhibits the natural quality of water with minimum anthropogenic activities, which contribute in pollution load of an ecosystem.

Introduction

Bhutan is blessed with enormous water resources as a result of the many glaciers and glacial lakes, vast forest cover and high precipitation. Royal government of Bhutan (RGoB) appointed the National Environment Commission (NEC) as the apex body for coordinating the water resources management in the country. The topography is rugged and characterized by swift flowing of rivers. The climate of Bhutan is dominated by monsoon, which sweeps in from the way of Bengal during

June, is intense during July and August and finally peters out during September. The period from November to January is normally dry. During April and May pre-monsoon showers occur often with substantial rainfall along the southern border with India. The major rivers of the country flow north to south, with their source in the alpine zone and flowing right down to the tropical zone on the border with India. The rivers of Bhutan generally have steep gradients and narrow steep-sided valleys, which occasionally open up to give small areas of flat land for cultivation. Bhutan has abundant water resources, but these resources will face new, complex and pervasive challenges caused by population growth, climate change and socio-economic development.

Water resources in the study area may be classified in to two major categories, (1) stream/spring water, and (2) surface water. The villagers use mostly spring water as their principal source for domestic consumption. In some villages the spring water is collected in a tank and supplied by pipelines to different houses as protected water supply for both drinking and cooking purpose. Assessment of background quality of water in the area was carried out once in every season. Water quality study of water resources with reference to quantity and quality depicts impacts of anthropogenic activities.

To assess the water quality of study area two number of river water samples from different stations were drawn and analyzed for various physico-chemical parameters, heavy metals and biological parameters. The assessment has been carried out to estimate the effect of any anthropogenic activities on the environment of the study area and to formulate mitigation measures to prevent the surface water bodies and stream water from further pollution. There are no lake and ponds and even no well and borewell were found in the study area. In most part of the study area, the water streams are moving from north to south direction. Kirung Ri river flows from Nganglam towards Chenkari entrance bridge and Dungsam Cement Corporation Limited (DCCL) project site. Kirung Ri river passes adjacent to DCCL project site at a distance of 0.1 Km. Kirung Ri river meets the Manas river which is a major tributary of Brahmaputra after traveling around 36 Km after confluence point of Kirung and Kurung rivers.

The present study reveals to characteristics the water nutrient chemistry, influenced by anthropogenic activity and quarrying of the geologically sediment environments and to determine the nature and degree of anthropogenic impacts on quantitative variations in nutrients in relation to physico-chemical parameters and heavy metals of river water.

Study Area

Bhutan is situated in the Eastern Himalayas. The DCCL project site is located at Lower Chenkari village of Nganglam, Pemagatsel Dzonkhag (district) in Eastern Bhutan. The project site is surrounded by hills and mountains at high elevation. The project site is located at about 4.5 Km north of Nganglam and 10 km from indo-Bhutan border. The topography of surrounding area of DCCL project site has an undulating terrain with hills, covered by thick vegetation. Towards west there are some agriculture fields along with some rocks and hilly terrain. The geo-co-ordinates of the study area is 26°52′ to 26°55′ N Latitude and 91°12′ to 91°14′ E Longitude. The average sea level is 1588 meter. The maximum temperature in summer is 32°C and minimum 5°C in winters. There are no industrial activities exist in the area. Nganglam's economy is mainly based on agriculture and animal husbandry. The study area has no transport linkage at present. The villages spread over different hillocks.

Materials and Methods

Grab water samples were collected in 5 litre plastic jerry canes and 250 ml sterilized clean glass/pet bottles for physico-chemical and bacteriological test respectively from selected sampling locations.

Water quality parameters were analyzed as per standard procedures/method given in IS:3025 (revised part) and APHA (1998).

The water samples were collected from Kirung Ri River upstream (U/S sampling site A) Near Military camp and downstream (D/S sampling site B) near DCCL study site according to the analytical requirement in morning period 9:00 Hrs. to 10:00 Hrs. The samples for physico-chemical parameters and Heavy Metals were collected by using rinsed Borosil glassware, and analyzed with the help of the procedure described by APHA (1995) and Trivedi and Goel (1984). Colour, odour, turbidity, velocity, temperature and Dissolved oxygen were analysed on sampling sites. Samples were collected from selected sites and immediately preserved in ice boxes, and transfer to the lab for further analysis. Water samples were digested and heavy metals were detected using Atomic Absorption Spectrophotometer.

Results and Discussion

Kirung Ri river is flowing throughout a valley of hills chain near the Indo-Bhutan Border, enriched with limestone and lignite rocks, which affect the water quality of stream according to the locations. Nutrients concentration and related physico-chemical parameters from selected sites are depicted in Tables 6.1–6.3. Kirung Ri river has the spatio-temporal variations of water parameters, which play a vital role in all physico-biochemical reactions and self-purification of aquatic system (Badola and Singh, 1981). Turbidity is striking characteristic of the physical status of the water bodies. Although in Kirung Ri river water is clear because there is no more pollution, siltation was the main source of turbidity in tributaries. Detritus and other non-organic material being added to water mass due to rainfall and anthropogenic activities (Camron, 1996). Maximum turbidity was recorded 2 NTU during rainy season at upstream and 4 NTU in monsoon season at downstream. Turbidity reduce the transparency and minimize the photic zone and always obstructs into light penetration in a aquatic ecosystem. The maximum depth of photic zone provides the better biological production for all aquatic organisms (Malik and Bharti, 2005b). Total dissolved solids were found in the range of 65 mg/l in to

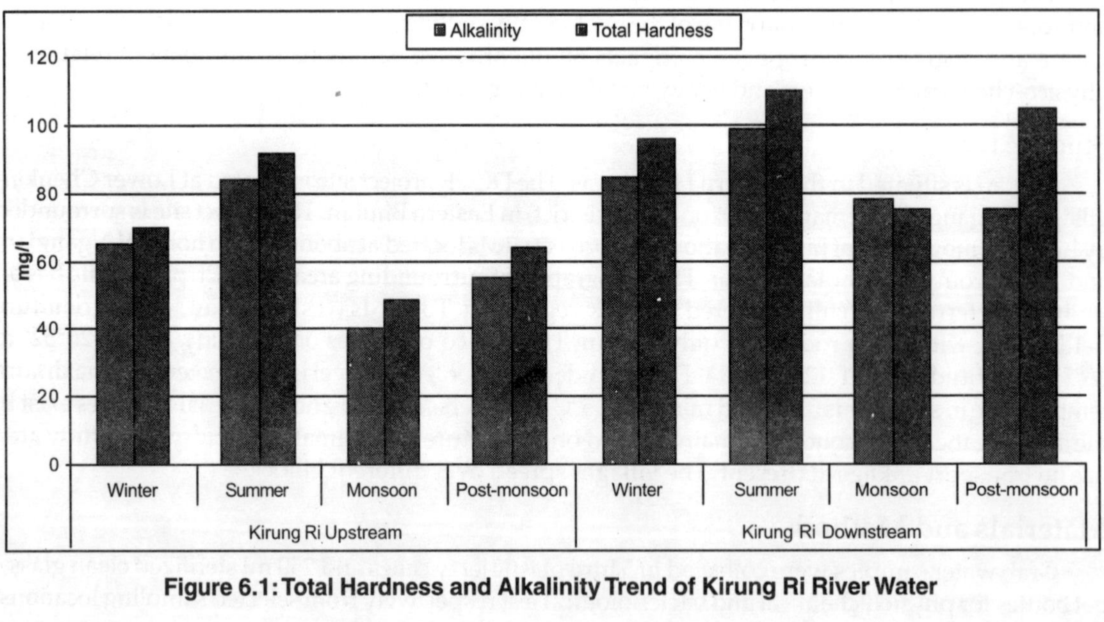

Figure 6.1: Total Hardness and Alkalinity Trend of Kirung Ri River Water

110 mg/l at upstream and 97-135 mg/l at downstream, due to the gradual increases in velocity of river which favoured effective sedimentation (Subramanian, 1979), (Malik and Bharti, 2005c).

Table 6.1: Physical Characteristics of Kirung Ri River Water

Sl.No.	Parameters	Unit	Kirung Ri Upstream				Kirung Ri Downstream				Desirable Limit
			Winter	Summer	Monsoon	Post-monsoon	Winter	Summer	Monsoon	Post-monsoon	
1.	Colour	–	Clear	Clear	Clear	Clear	Clear	Clear	Clear	Clear	–
2.	Odour	–	Nil	Nil	Nil	Nil	Nil	Nil	Nil	Nil	–
3.	Turbidity	NTU	2	1	1	1	4	2	3	1	5
4.	Velocity	m/s	0.9	0.8	1.0	0.8	0.7	0.6	0.8	0.8	–
5.	TDS	mg/l	90	65	110	85	120	97	135	115	500

Table 6.2: Chemical Characteristics of Kirung Ri River Water

Sl.No.	Parameters	Unit	Kirung Ri Upstream				Kirung Ri Downstream				Desirable Limit
			Winter	Summer	Monsoon	Post-monsoon	Winter	Summer	Monsoon	Post-monsoon	
1.	pH	–	8.2	8.1	8.1	7.8	8.3	8.4	7.9	7.9	6.5-8.5
2.	Alkalinity	mg/l	65	84	40	55	85	99	78	75	200
3.	Total Hardness	mg/l	70	92	48	64	96	110	75	105	300
4.	Calcium	mg/l	15	26	11	14	25	28	20	24	75
5.	Magnesium	mg/l	7	8	5	7	8	10	6	11	30
6.	Chlorides	mg/l	3	4	7	6	4	4	5	6	250

Table 6.3: Heavy Metals in Kirung Ri River Water

Sl.No.	Parameters	Unit	Kirung Ri Upstream				Kirung Ri Downstream				Desirable Limit
			Winter	Summer	Monsoon	Post-monsoon	Winter	Summer	Monsoon	Post-monsoon	
1.	Cadmium	mg/l	BDL	BDL	BDL	BDL	BDL	BDL	BDL	BDL	0.01
2.	Copper	mg/l	BDL	BDL	BDL	BDL	BDL	BDL	BDL	BDL	0.05
3.	Iron	mg/l	BDL	0.1	BDL	0.05	0.1	0.1	0.08	BDL	0.3
4.	Lead	mg/l	BDL	0.02	BDL	BDL	BDL	0.02	BDL	BDL	0.05
5.	Manganese	mg/l	BDL	BDL	BDL	BDL	BDL	BDL	BDL	BDL	0.1
6.	Zinc	mg/l	BDL	0.03	BDL	0.04	0.02	0.02	0.02	BDL	5.0

The pH of natural water was controlled in a great extent by the interaction of hydroxyl ions arising from the hydrolysis of bicarbonate (Sharma, 1986). The pH of river water was recorded slightly alkaline (7.8–8.2) at upstream and (7.9-8.4) at downstream. Total hardness is mainly due to percentage

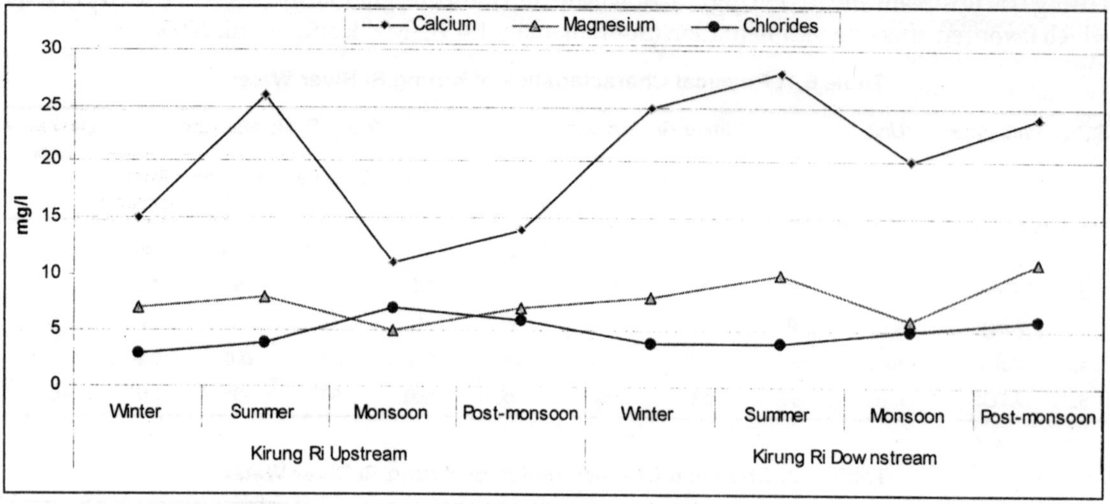

Figure 6.2: Showing the Nutrients Concentration in Kirung Ri River during the Study Period

of calcium and magnesium salts of bicarbonates, carbonates, sulphates and chlorides, while the value of alkalinity occured due to presence of bicarbonates. The concentration of hardness was analyzed 48-92 mg/l at upstream and 75-110 mg/l at downstream during the study. Alkalinity was also found 40-99 mg/l with a small fluctuation. A positive relationship between hardness and alkalinity was recorded in river Ganga at Rishikesh (Chopra and Patric, 1994). Chloride concentration was recorded (3-7 mg/l) at upstream and (4-6 mg/l) at downstream. Chloride and hardness showed a positive relationship (Chopra and Patric, 1994). Chloride was found in the form of chloride ion, and one of the major inorganic anion present in natural water (Malik and Bharti, 2009).

Calcium and magnesium the dominant cations, and these represent the main weathering products, but significant hydro-chemical differences between the two sampling sites associated with the bedrock geology exist (Jenkins *et al.*, 1995). Calcium is one of the essential nutrients, which plays an important role in biological system. Calcium concentration was recorded (11-26 mg/l) at upstream and (20-28 mg/l) at downstream. Magnesium is also an essential element but it is toxic at higher concentration. The concentration of magnesium in river was found maximum (11 mg/l) and minimum (5 mg/l). During the summer season nutrients concentration in rivers and hill-streams became more. Miller *et al.* (1997) described the nutrients availability in selected environmental settings of the Potomac River and Cameron (1996) showed the similar type of fluctuation in Fraser river of British Columbia. Bond (1979) described similar nutrients concentration pattern in a stream draining a mountain ecosystem in Utah.

Heavy metals like cadmium, copper, iron, manganese, lead and zinc were not found in high concentration at both sites during any season. Cadmium, manganese and copper was absolutely absent in all seasons, while lead concentrations were also found below detection limit in maximum samples. The concentration of iron was maximum observed 0.1 mg/l during the entire study. Zinc concentration was also found below the standards limit. Malik *et al.* (2009) described the role of heavy metals in the surface water of north India. The results of Bharti *et al.* (2010) were also indicated the relation of heavy metals and phytoplankton in a north Indian water body.

The present results conclude that significant differences in river water nutrient concentrations exist among different environmental settings within the two subunits. The environmental setting with the highest potential by more soluble nutrients, fluctuations in nutrient concentrations were the land use and carbonate bedrock that was predominated in the Bhutanese valley. The spatial variations in TDS are attributed to climatic and lithological control over the ionic concentrations. Absence or low concentration of heavy metals shows that the water is still industrial pollution free. Heavy metals were found almost nil or in very low concentrations at both selected sites. The present study deals with the preliminary physico-chemical characteristics of the river water and exhibits the natural quality of water with minimum anthropogenic activities, which contribute in pollution load of an ecosystem. However, the Indian Standards are not so strict (Bharti, 2007), but Heavy Metals never cross the limits during the study period. On the basis of nutrients and heavy metals, the water may be considered for drinking and other purposes.

Analysis of the water has been conducted at both sampling sites on the river basins. The natural water quality of Kirung Ri River can be characterized as highly oxygenated, slightly alkaline with low conductivity and no salinity was recorded. The mountain streams and rivers are free of salinity are suitable for irrigation.

References

APHA, 1995. *Standard Methods for Examination of Water and Wastewater*, 19th edn. American Public Health Association, Inc., New York, pp. 1970.

Badola, S.P. and Singh, H.R., 1981. Hydrobiology of the river Alaknanda of Garhwal Himalaya. *Indian J. Ecol.*, 8(2): 269–276.

Baron, J. and Bricker, O.P., 1990. Hydrological and chemical flux in Loch Vale watershed, Rocky Mountain National Park. In: *Biogeochemistry of Major Rivers. SCOPE 42*. Wiley and Sons, New York, USA.

Bharti, P.K., 2007. Why Indian Standards are not so strict? *Current Science*, 93(9): 1202.

Bharti, P.K., Malik, D.S. and Yadav, Rashmi, 2010. Influence of heavy metals on abundance of cyanophyceae members in three spring-fed lake in Kempty, Dehradun. In: *Advances in Aquatic Ecology*, Vol. 3, (Ed.) V.B. Sakhare. Daya Publishing House, New Delhi, pp. 107–111.

Bond, H.B., 1979. Nutrient concentrations patterns in a stream draining a montane ecosystem in Utah. *Ecology*, 60(6): 1184–1196.

Cameron, E.M., 1996.Hydrogeo-chemistry of the Fraser River British Columbia: Seasonal variation in major and minor components. *J. Hydrol.*, 182(1–4): 209–255.

Chopra, A.K. and Patrick, N.J., 1994. Effect of domestic sewage on self-purification of Ganga water at Rishikesh. *A Bio. Science*, 13(11): 75–82.

Jenkins, A., Sloan W.T. and Cosby, B.J., 1995. Stream chemistry is the middle hills and high mountain of the Himalaya, Nepal. *Journal of Hydrology*, 166(1–4): 61–79.

Khanna, D.R. and Singh, R.K., 2000. Seasonal fluctuations in the plankton of Suswa River at Raiwala Dehradun. *Env. Conservations J.*, 1(2 and 3): 89–92.

Malik, D.S. and Bharti, P.K., 2005b. Fluctuation in planktonic population of Sahastradhara hill-stream at Dehradun (Uttaranchal). *Aquacult.*, 6(2): 191–198.

Malik, D.S. and Bharti, P.K., 2005c. Primary production efficiency of Sahstradhara hill-stream, Dehradun, *Env. Cons. J.*, 6(3): 117–121.

Malik, D.S. and Bharti, P.K., 2009. Ecology of Sahastradhara hill-stream at Dehradun (Uttaranchal), In: *Advances in Aquatic Ecology*, Vol. 1, (Ed.) V.B. Sakhare. Daya Publishing House, New Delhi, pp. 1–11.

Malik, D.S. and Bharti, P.K., Negi, K.S. and Yadav, Rashmi, 2009. Distribution of metals in water of an artificial lake at Mussoorie, Uttarakhand. In: *Aquatic Biology and Aquaculture*, (Ed.) V.B. Sakhare. Manglam Publication, New Delhi, pp. 77–95.

Malik, D.S. and Bharti, Pawan K., 2005a. Nutrient dynamics in Rhithron zone of Shivalik Himalayan stream Sahastradhara, Dehradun (Uttaranchal). *Env. Cons. J.*, 6(2): 63–68.

Miller, C.V., Denis, J.M., Ator, S.W. and Brakebill, J.W., 1997. Nutrients in stream during baseflow in selected environmental settings of the Potomac river basin. *J. American Wat. Res. Association*, 33(6): 1155–1171.

Odum, E.P., 1971. *Fundamentals of Ecology*. W.B. Saunders and Co., New York, pp. 1–344.

Psenner, R., 1989. Chemistry of high mountain lakes in siliceous catchments of central Alps. *Aquatic Sci.*, 51: 108–128.

Shrama, R.C., 1986. Effect of physico-chemical factors on benthic fauna of Bhagirathi River Garhwal Himalaya. *Indian J. Ecol.*, 13(1): 133–137.

Subramanian, V., 1979. Chemical and suspended sediment characteristics of river of India. *J. Hydrol.*, 44: 37–55.

Trivedi, R.K. and Goel, P.K., 1984. *Chemical and Biological Methods for Water Pollution Studies*. Environmental Publication, Karad, pp. 1–25.

Xue, H.B. and Schooner, J.L., 1994. Acid deposition and lake chemistry in southwest China. *Wat. Air, Soil Pollut.*, 75: 61–78.

Chapter 7

Food and Feeding Habits of the Mudskipper, *Boleophthalmus boddarti* (Pallas) from Vellar Estuary, Southeast Coast of India

☆ *V. Ravi*

ABSTRACT

The mudskipper, *Boleophthalmus boddarti* (Pallas) (Order Perciformes; Family Gobiidae) are found distributed along the estuarine and mangrove mudflats and they live in burrows. They are amphibious and locomotive on mudflats. During low tide, they feed very actively in the mudflats. In the present study, the food and feeding habits and feeding intensities of the mudskipper from the Vellar estuary were attempted following standard methods. A total of 156 fishes (juveniles, males and females) were randomly caught using cast net and hand nets and they were immediately preserved in formalin. The results revealed that diatoms formed the major food item of *B. boddarti* and found around 55 per cent of its diet throughout the year. Nematodes, polychaetes, algae and fish eggs were also found constituting lower percentage during the study period. Detritus and mud/sand particles were present in moderate percentages in all the seasons. Active feeding was noticed during January [post-monsoon to May (summer)] but it was below 50 per cent during June (summer)to September (pre-monsoon), whereas during October and monsoon the feeding intensities were very low. The poor feeding during these periods may be due to monsoon season.

Introduction

Food is essential for the sustainability of every living organism throughout its life span and feeding is a continuous process to derive energy. The survival, growth and reproduction of a fish depend on the income of energy and the nutrients generated by its feeding activities. The derived energy is utilized in many important ways for the performance of cell repair, muscular contraction, secretory function, nerve impulse conduction etc. although fishes have a considerable capacity to resist starvation (Love, 1980) and many species normally cease to feed at times during their life cycle (Wootton, 1979). Mudskippers (Order Perciformes; Family Gobiidae) are morphologically and physiologically well- adapted amphibious fish capable of living on land and in water. From ecological point of view, mudflat environment is very important for mudskippers as they are active feeders when mudflat exposure during low tide. However they cannot feed during high tide due to submergence of the mudflat. During this time, mudskippers remain inside of their burrow which protects them from predators (Milward, 1974). Investigations on food and feeding of mudskippers reveal them either as herbivores (Ryu *et al.*, 1995) or carnivores (Milward, 1974; Colombini *et al.*, 1996). Information is available on gobiid fishes with reference to food and feeding habits (Stebbins and Kalk, 1961; Gordon *et al.*, 1968; El- Zaidy *et al.*, 1975; Sarkar *et al.*, 1980). More recently, interspecific comparison of diets has been made in relation to competition (Ip *et al.*, 1990). Colombini *et al.* (1995) described the foraging strategy of the mudskipper *Periophthalmus sobrinus*. Study of the diet based upon the analysis of stomach contents is now a standard practice in fish biology. Review of methods and their application on stomach contents analysis was made by Hynes (1950), Pillay (1952), and Hyslop (1980). The morphology of the alimentary tract in relation to diet among gobioid fishes was studied by Geevargheese (1983). Ananda Rao *et al.* (1998) reported the vulnerability of the mudskipper *Boleophthalmus boddarti* (Pallas) while Ravi (2005) reported the loss of mudflat and mudskipper population along the Southeast coast of India. Therefore it is very important to study the food and feeding habits of mudskipper *B. boddarti* to assess its feeding behavior which will certainly help in conserving this important species. The present study was initiated with a view to gather more information concerning the food and feeding habits of this group.

Materials and Methods

To assess the food and feeding habits, a total of 156 fish (juveniles, males and females) specimens of *B. boddarti* were caught randomly from the mudflats of Vellar estuary(Lat. 11° 29′N; Long. 79° 46′E)), Tamil Nadu, Southeast coast of India during July 2008 to June 2009, with the help of cast nets and nylon hand nets. They were immediately preserved in 5 per cent formalin. For each specimen, its total length (TL) in mm, standard length (SL) in mm, weight in g,.sex and maturity stages were noted in fresh condition. During the present study, the gut contents were estimated based on the frequency of occurrence of different food materials and points methods as described by Swynnerton and Worthington (1940) and reviewed and modified Hyslop (1980). By using points method, the points such as 100, 80, 60, 40, 20 and 10 were allotted to the gut contents with due consideration to the size of the organisms as well as their abundance. The points gained by each food items from all the stomach examined were summed up and expressed as percentage of total number of points. This method is essentially a volumetric one and is preferred by many authors since it has the advantage of giving roughly both quantitative and qualitative data without the need for very detailed counts of the food items.

The stomachs were considered gorged when the stomach was found expanded fully with packed food, thin walled and transparent; full when the stomach was packed with food with thick wall and intact; ¾ full when it was partly full with thick wall; ½ full; ¼ full and trace, according to the relative condition of the stomach as indicated above. The empty stomach were either found in contracted state or loosely expanded and appeared full but empty. The later was considered regurgitated stomachs. The feeding intensity based on the degree of fullness of stomach of the fish was determined. The stomach was allotted with points from 0 to 100 in accordance with its fullness (0= empty; 10= trace; 20= ¼ full; 40= ½ full; 60= ¾ full; 80= full; and 100= gorged).

The percentage frequencies of empty, trace, ¼ full, ½ full; ¾ full; full; and gorged stomachs were calculated from the total number of fishes examined in each month, and for the sake of convenience, gorged, full and ¾ full stomachs were clubbed together and designated as actively fed. Under moderately fed and poorly fed heading were included ½ full and ¼ full and trace stomach respectively.

In the present study gut contents of the *B. boddarti* was found to be the diatoms and they were identified up to the genera and species level by following the atlas of diatoms (Desikachary, 1986, 1987 and 1988). Variations in the gut contents in relation to different seasons and size groups of the mudskipper *B. boddarti* were analysed separately for a period of one year.

Results

Various food substances of juvenile, male and female mudskippers (*B. boddarti*) during different seasons are presented in Table 7.1. Diatoms formed the major food item of *B. boddarti* and found around 55 per cent of its diet throughout the year. In juvenile fishes, the nematodes, polychaetes, algae and fish eggs were also found constituting lower percentage while in male and female fishes; they were present slightly in higher percentages than juveniles during the study period. Detritus and mud/sand particles were present in moderate percentages in all the months.

Table 7.1: Percentage Composition of Various Food Items in Juveniles (J), Males (M) and Females (F) of the Mudskipper *B. boddarti*

Food Items/Seasons	*Pre-monsoon (July, Aug, Sep)*			*Monsoon (Oct, Nov, Dec)*			*Post-monsoon (Jan, Feb, Mar)*			*Summer (Apr, May, Jun)*		
	J	*M*	*F*	*J*	*M*	*F*	*J*	*M*	*F*	*J*	*M*	*F*
Diatoms												
Family Coscinodiscaceae												
Coscinodiscus centralis	8.1	6.7	8.4	7.3	8.4	9.3	10.7	10.8	5.9	7.2	1.2	4.3
Family Rhizosolenaceae												
Rhizosolenia sp.	1.5	2.1	2.3	8.9	8.8	10.7	1.7	1.8	1.9	0	0	5
Family Naviculaceae												
Gyrosigma attenuatum	4.5	3.2	4.2	2.1	3.1	1.9	1.7	1.8	1.4	4.3	0.9	1.1
G. balticum	3.2	1.7	3.9	1.9	2.7	2.1	0	4.1	4.2	2.7	1.1	1.2
Navicula gracilis	0	1.8	2.1	14.5	13.2	13.7	3.1	4.2	3.7	3.9	0.9	1.1
Pleurosigma directum	39.2	34	28.8	11.4	9.4	11.1	15.8	7.6	11	33.1	14.7	13.5
P. normanii	5.2	4.7	7.3	0	5.7	4.7	0	0	1.3	13.7	3.8	2.9
Family Nitzschiaceae												
Nitzschia granulata	4.3	1.7	2.3	14.7	13	12.9	4.7	4.5	3.9	15.7	5.3	4.5
N. punctata	2.9	1.2	1.7	8.9	10.3	9	0	3.2	1.7	0	0	0
Family Thalssionemataceae												
Thalassionema sp.	2.1	4.2	3.7	0	0	0	4.2	3.5	4.7	0	0	2.1

Contd...

Table 7.1–*Contd...*

Food Items/Seasons	*Pre-monsoon (July, Aug, Sep)*			*Monsoon (Oct, Nov, Dec)*			*Post-monsoon (Jan, Feb, Mar)*			*Summer (Apr, May, Jun)*		
	J	*M*	*F*	*J*	*M*	*F*	*J*	*M*	*F*	*J*	*M*	*F*
Nematodes												
Family Linhomoeidae												
Terschellingia longicaudata	3.2	5.1	6.2	3.3	2.1	2	5.4	4	4.1	2.9	9.3	10
Paralinhomoeus brevibucca	0	0	0	1.4	3.5	0	4.2	6.7	6	0	5.1	5.1
Family Desmodoridae												
Desmodora luticola	0	2.3	3.1	0	0	0	3.1	3.1	4.1	0	2.9	0
Polychaetes												
Family Nereididae												
Nereis sp.	2.1	6.9	4.2	2.1	2.4	1	5.3	5.1	7.2	2.1	7.1	8
Family Maldanidae												
Euclemene annandalei	0	3.1	3.1	0	2.3	3.3	5.1	6.3	4	0	6.7	6.7
Fish eggs	0	3.7	3.9	0	0	0	8.1	7.1	8.5	0	3	4.1
Algae	5.2	4.2	5.2	4.3	3.1	5.7	7.2	9.2	7	2.3	10.6	9.3
Detritus	6.1	6.3	3.7	7.3	5.7	5.3	10.5	9.7	11.4	4.2	10.3	9
Mud/sand particles	12.4	7.1	5.9	11.9	6.3	8.3	9.2	7.3	9.3	7.9	17.1	12.1

The diatoms were present in higher percentages (83 per cent) in summer (April) and lower percentage (41.9 per cent) in post-monsoon (March) in juvenile fishes while in both males and females, higher percentages (74.7 and 81.5 per cent) and in lower percentages (27.9 per cent and 35.7 per cent) respectively in monsoon(November) and pre-monsoon (July). Nematodes were found to range from 2.1 (Summer-May) to 12.7 per cent (post-monsoon- March) in juveniles, 3.9 (monsoon- October) to 19.3 per cent (summer- June) in males and 2 (monsoon-December) to 15.1 per cent (summer- June) in females. Polychaetes were found in higher percentages of 10.4 per cent (March, 13.8 per cent and 14.17 per cent (June) and in the lower percentages of 2.1 per cent (June), 3.5 per cent (October) and 3.2 per cent (November) in juveniles, males and females respectively. Fish eggs (8.1 per cent) were recorded in March in juveniles. Higher percentages of 7.2 per cent (January) and 8.5 (March) and lower percentages of 2 per cent (August) and 2.5 per cent (April) were recorded in males and females respectively. Algae were present with high percentages of 9.3 per cent, 11.1 per cent and 10.2 per cent (January) and low percentages of 2.3 per cent (June), 3.1 per cent and 1.7 per cent (November) in juveniles, males and females respectively. Detritus were present with low percentages of 2.4 per cent (July), 2.2 per cent (August) and 1.8 per cent (July) and high percentages of 10.5 per cent (March), 11.3 per cent (May) and 11.4 per cent (March) in juveniles, males and females. Mud/sand particles were found with higher percentages of 15.1 per cent (January), 17.1 per cent and 12.1 per cent (June) and low percentages of 4.2 per cent (October), 3.7 per cent (July) and 4.1 per cent (August) in juveniles, males and females respectively.

Variations of Food Items among the Juveniles, Males and Females of *B. boddarti*

The percentage composition of various food items of juveniles, males and females are given Figures 7.1–7.3. It is apparent from the Figure 7.1 that the stomachs of juveniles of *B. boddarti* contained

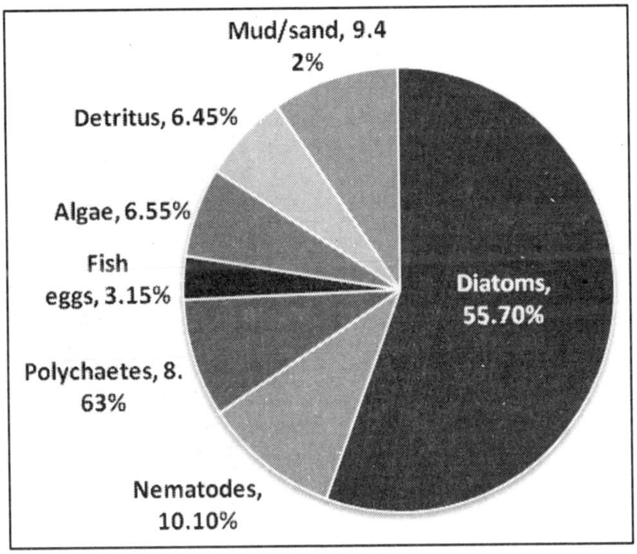

Figure 7.1: Percentage Composition of Food Items in Juvenile, *B. boddarti*

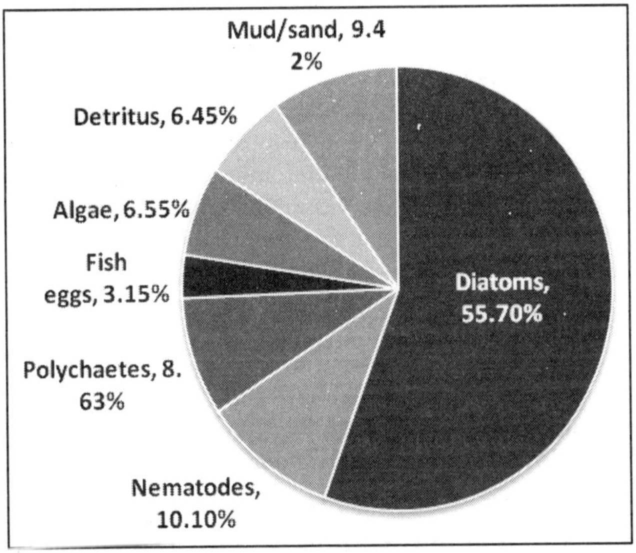

Figure 7.2: Percentage Composition of Food Items in Male, *B. boddarti*

more diatoms (71.5 per cent) when compared to male and female mudskippers. Further, it showed that the juvenile fishes preferred mainly diatoms. Nematodes (4 per cent), polychaetes (3.6) and fish eggs (1.1) were found to be less in the diet of juveniles. Mud/sand particles (8.4 per cent) were present with minimum percentage when compared to male and female fishes. Detritus (6.1 per cent) and algae (5.4 per cent) were recorded with minimum percentage in the food of juveniles. Based on the present study, it can be derived that the juveniles feed on diatoms as a principal diet than the other food items. In the case of males and females, the dietary composition showed a significant preference for other food

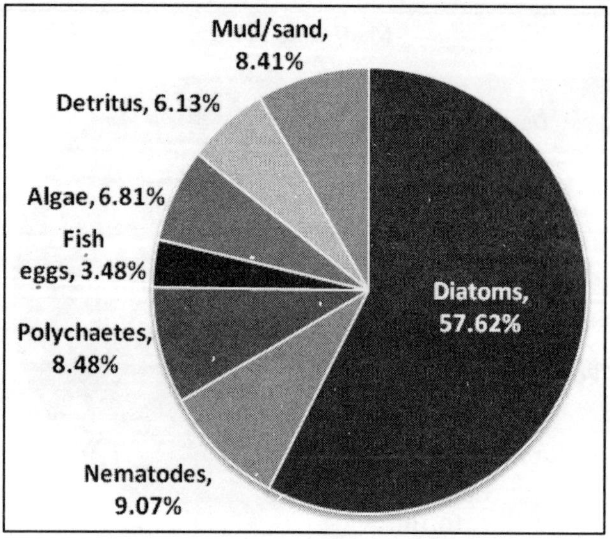

Figure 7.3: Percentage Composition of Food Items in Female, *B. boddarti*

items than the diatoms. The percentage composition of food items of males and females includes diatoms (55.7 and 57.62 per cent), nematodes (10.1 and 9.07 per cent), polychaetes (8.63 and 8.48 per cent), fish eggs (3.15 and 3.48 per cent), algae (6.55 and 6.81 per cent), detritus (6.45 and 6.13 per cent) and mud/sand particles (9.42 and 8.41 per cent) respectively in their diets (Figures 7.2 and 7.3). Various food items present in the males include diatoms, nematodes, polychaetes, fish eggs, algae, detritus and mud/sand particles. All these dietary compositions were also observed in females. Though there was a slight variation in the percentage of food items, the nature of food was always similar in both male and female fishes examined in the present study.

Feeding Intensities in Relation to Different Seasons

The percentage of feeding intensity values for different seasons was calculated and depicted in Table 7.2. Active feeding was noticed during January (post-monsoon to May (summer) but it was below 50 per cent during June (summer)to September (pre-monsoon), whereas during October and monsoon the feeding intensities were very low. The poor feeding during these periods may be due to monsoon season.

Feeding Intensities in Relation to Juveniles, Males, and Females

The percentage of feeding intensities in juveniles, males and females of *B. boddarti* is depicted in Table 7.3. It is clear that the percentage of actively fed mudskipper was more in juveniles than males and females. But it was also been observed that the percentages of moderately fed mudskippers were more or less same among males and females but less in juveniles. While comparing the feeding intensities between males and females, it was observed that the percentage of feeding intensity of actively fed and moderately fed mudskippers was more in males than females. But in the case of fully fed and empty stomachs, the percentage of feeding intensities were more in females than males. The feeding intensity of empty stomachs in juveniles is completely absent in the present study.

Table 7.2: Percentage of Feeding Intensities in Relation to Different Seasons in *B. boddarti*

Feeding Intensity	*Seasons*			
	Pre-monsoon	*Monsoon*	*Post-monsoon*	*Summer*
Gorged	3.92	0	16.32	4.52
Full	11.21	6.15	15.18	11.98
¾ full	30.73	7.13	43.79	48.32
Actively fed	45.86	13.28	75.29	64.82
½ full	21.02	38.27	9.02	13.13
Moderately fed	21.02	38.27	9.02	13.13
¼ full	5.99	22.15	0	5.73
Trace	27.13	26.3	0	6.93
Poorly fed	33.12	48.45	0	12.66
Empty	0	0	15.69	9.39

Table 7.3: Percentage of Feeding Intensities in Relation to Juveniles, Males, and Females of *B. boddarti*

Feeding Intensity	*Juveniles*	*Males*	*Females*
Gorged	19.05	13.91	14.01
Full	23.12	15.73	16.88
¾ full	37.35	24.15	20.72
Actively fed	79.52	53.79	51.61
½ full	12.23	16.02	15.02
Moderately fed	0	16.02	15.02
¼ full	0	8.27	12.21
Trace	8.25	9.13	8.13
Poorly fed	8.25	17.40	20.34
Empty	0	12.79	13.03

Discussion

The mudskipper *B. boddarti* inhabits the mudflats along the Vellar estuary, southeast coast of India. The results of the present study on the food and feeding habits of this species clearly revealed that the fish is a herbivore especially feeding predominantly on diatoms. This is in consonance with the study made by Macnae (1968) describing *B. boddarti* as a herbivore. In general, mudskippers of the genera, *Periophthalmus* and *Periophthalmodon* are carnivores while those of *Boleophththalmus* are herbivores and *Scartelaos* omnivores (Clayton, 1993). The extent of exposure of the mudflats at air depends on the tides, which occur normally twice a day. During this time, the fish emerges from the water-filled burrow and starts to move on the soft mudflats. While doing so, it skims off a thin layer of mud and algae from the surface (Macnae, 1968). The mouthful of such materials is then placed in the mouth, as can be seen by vibrating movements of the lips and opercular region. The mudskipper *B. boddarti* feeds more on diatoms than the other items. Fenchel (1969) stated that the benthic diatoms constitute an important source of food for benthic organisms and other meiobenthos while Heald and Odum (1970) explained that the diatoms play vital role in establishing the food chain relationships between different groups of organisms in the ecosystems.

The stomach content analysis of the present study showed that the diatoms were reprersented by *Coscinodiscus, Gyrosigma, Navicula, Nitzschia, Pleurosigma, Rhizosolenia,* and *Thalassionema.* Other food items were nematodes, polychaetes, fish eggs and algae. Nematodes such as *Desmodora, Paralinhomoeus* and *Terschillingia* constituted the second food items with high percentage following the diatoms. Polychaetes (*Nereis* and *Euclymene*) formed another food item next to nematodes. In addition to the above food items, fish eggs, algae, detritus and mu/sand particles were also present in the stomachs of the mudskipper in moderate percentages. This corroborates with the *B. dussumieri,* which was examined by Mutsaddi and Bal (1969) from Gujarat, West coast of India. They also reported that the mudskipper feeds on algae, diatoms, polychaetes, nematodes, crustaceans and teleost eggs. No crustaceans were recorded in the gut contents of the mudskipper, *B. boddarti.* The dietary preference is probably of great importance from one species to another and also the geographical variations variations play asignificant factor for the food availability. Sarkar *et al.* (1980) studied the food habits of the mudskipper *Pseudopocryptes dentatus* from the mudflats of the Shatt al- Arab estuary as well as along the coastal regions of the Arab Gulf, and reported that the Pennales accounted for 81 per cent and Centrales 19 per cent of the diatoms found in the gut contents of the mudskipper but the study was done in November month alone and the seasonal variations in food preference were not covered by these authors. Krishnamurthy (1971) observed the primary peaks of diatoms and thus the ingestion of large amounts of diatoms during March to June and October to January is in good agreement with the seasonal distribution pattern of phytoplankton in Parangipettai waters (Krishnamurthy, 1971). Seasonal changes in food availability may be caused by changes in the habits available for foraging, changes due to the life- history patterns of food organisms and changes caused by the feeding activities of the fish themselves (Wootton, 1990). Similar observation was made in B. pectinirostris from Korea coast (Ryu *et al.*, 1995). According to them the fish fed principally on diatoms followed by crusctaceans, mysids and delphasids. Nematodes were found in the stomachs of the mudskipper, *B. boddarti* during the study period. Similarly Krishnamurthy *et al.* (1984) and Colombini *et al.* (1996) reported that the mudskippers feed on nematodes. Polychaetes were found in moderate percentage (12 per cent) in the stomach of the mudskipper. Stebbins and Kalk (1961) and Macnae and Kalk (1962) reported that the *Periophthalmus sobrinus* preys on polychaetes and crustaceans. Colombini *et al.* (1996) while studying the foraging strategy of mudskipper, *P. sobrinus* from Kenyan mangroves reported that nearly 39 per cent polychaetes in their diet. In the stomach of the mudskippers, fish eggs were present in good percentage during post monsoon season and it may due to the fact that the majority of estuarine teleosts spawn during this season in Vellar estuary. Algae were common during post monsoon and summer seasons. Detritus and mud/sand particles were also present in high percentages in all seasons. While skimming off the food from the mudflats, these particles may enter into their gut and the hetrotrophic bacteria present in that soil may indirectly be helpful in the digestion. In the present study, there was a slight variation in the percentages of the food items found in males and females. Colombini *et al.* (1996) observed that the males had more varied diet than the females.

Acknowledgements

The author is grateful to the Dean, Faculty of Marine Sciences and authorities of Annamalai University for the facilities provided and to the University Grants Commission, New Delhi for the financial support.

References

Ananda Rao, T., Molur, Sanjay and Salley Walker (Eds.) 1998. Report of the Workshop "*Conservation Assessment and Management Plan for Mangroves of India*" (BCPP– Endangered species project). Zoo Outreach Organization, Conservation Breeding Specialist Group, India, Coimbatore, India, 106p.

Clayton, 1993. Mudskippers. *Oceanogr. Mar. Biol. Annu. Rev.*, 31: 507–577.

Colombini, I., Berti, R., Ercolini, A., Nocita, A. and Chelazzi, I., 1995. Environmental factors influencing the zonation and activity patterns of a population of *Periophthalmus sobrinus* Eggert in a Kenyan mangrove. *J. Exp. Mar. Biol. Ecol.*, 190(2): 135–149.

Colombini, I., Berti, R., Nocita, A. and Chelazzi, I., 1996. Foraging strategy of the mudskipper *Periophthalmus sobrinus* Eggert in a Kenyan mangrove. *J. Exp. Mar. Biol. Ecol.*, 197(2): 219–235.

Desikachary, T.V., 1986. Marine fossil from Indian region. In: *Atlas of Diatoms*, (Ed.) T.V. Desikachary. Madras Science Foundation, Madras, 77p.

Desikachary, T.V., 1987. *Atlas of Diatoms*. Faside III(Diatoms from the Bay of Bengal) pp. 1–10 with plates 222–333. Faside IV (Marine diatoms from the Arabian Sea, Indian Ocean)pp. 1–7 with plates 332–400. Madras Science Foundation, Madras.

Desikachary, T.V., 1988. *Atlas of Diatoms*. Faside V (Marine diatoms of the Indian Ocean) pp. 1–13 with plates 401–621. Madras Science Foundation, Madras.

El-Zaidy, S.S.M. Eissa and Al-Naqi, Z., 1975. Ecological studies on the mudskipper fish, *Periophthalmus chrysospilos* (Bleeker) in Kuwait. *Bull. For. Sci. Cairo University*, 48: 113–133.

Fenchel, T., 1969. The ecology of marine microbenthos. IV. Structure and function of the benthic ecosystem, its chemical and physical factors and the microfaunal communities with special reference to ciliated protozoa. *Ophelia*, 6: 1–182.

Geevarghese, C., 1983. Morphology of the alimentary tract in relation to diet among gobioid fishes. *J. Nat. Hist.*, 17: 731–741.

Gordon, M.S., Boetius, J., Evans, D.H. and Oglesby, 1968. Additional observations on the natural history of the mudskipper, *Periophthalmus sobrinus*. *Copeia*, 1968: 853–857.

Heald, E.J. and Odum, W.E., 1970. The contribution of the mangrove swamps to Florida fishes. *Limnol. Oceanogr.*, 3: 353–361.

Hynes, H.B.N., 1950. The food of freshwater Sticklebacks (*Gastersteus aculeatus* and *Pygosteus pungitus*) with a review of methods used in studies of the food of fishes. *J. Anim. Ecol.*, 19: 36–58.

Hyslop, E.J., 1980. Stomach contents analysis: A review of methods and their application. *J. Fish Biol.*, 17(4): 411–429.

Ip, Y.K., Chew, S.F. and Lim, A.L.L., 1990. Ammoniagenesis in the mudskipper, *Periophthalmus chrysospilos*. *Zool. Sci.*, 7: 187–194.

Krishnamurthy, K., 1971. Phytoplankton pigments in Porto Novo waters (India). *Int. Revue. Ges. Hydrobiol.*, 56(2): 273–282.

Krishnamurthy, K., Sultan Ali, M.A. and Jeyaseelan, M.A. Prince, 1984. Structure and dynamics of the aquatic food web community with special reference to nematodes in mangrove ecosystems. In: *Proc. Asian Symp. Mangrove Env. Res. Management*, (Ed.) E. Soepadmo *et al.* University of Malaysia and UNESCO, Kuala Lampur, p. 429–452.

Love, R.M., 1980. *The Chemical Biology of Fishes*, Vol. 2. Academic Press, London.

Macnae, W., 1968. A general account of the fauna anf flora of mangrove swamps and forests in the Indo-West Pacific region. *Advances in Marine Biology*, 6: 73–270.

Macnae, W. and Kalk, M., 1962. The ecology of the mangrove swamps at Inhaca Island, Macambique. *J. Anim. Ecol.*, 50: 19–34.

Milward, N.E., 1974. Studies on the taxonomy, ecology and physiology of Queensland mudskippers. *Ph.D. Thesis*, University of Queensland, 276 pp.

Mutsaddi, K.B. and Bal, D.V., 1969. Food and feeding of *Boleophthalmus dussumieri* (Cuv. and Val). *J. Univ. Bombay*, 38: 42–55.

Pillay, T.V.R., 1952. A critique of the methods of study of fishes. *J. Zool. Soc., India*, 4: 185–200.

Ravi, V., 2005. Post tsunami studies on the mudskipper *Boleophthalmus boddarti*(Pallas, 1770) from Mudasalodai, Tamil Nadu, southeast coast of India. *Journal of International Goby Society*, 4(1): 9–17.

Ryu, B.S., Kim, S.I. and Choi, Y., 1995. Ecology and life history of *Boleophthalmus pectinirostris* in Korea. *J. Korean Fish Soc.*, 12: 71–78 (In Korean).

Sarkar, A.L., Daham, N.K. Al. and Bhatti, M.N., 1980. Food habits of the mudskipper *Pseudopocryptes dentatus* (Val.). *J. Fish Biol.*, 17: 635–639.

Stebbins, R. and Kalk, M., 1961. Observation on the natural history of the mudskipper, *Periophthalmus sobrinus. Copeia*, 1961: 18–27.

Swynnerton, G.H. and Worthington, E.B., 1940. Note on the food of fish in Hawes water (West Morland). *J. Anim. Ecol.*, 9: 183–187.

Wootton, R.J., 1979. Energy cost of egg production and environmental determinants of fecundity in teleost fishes. *Symp. Zool. Soc. Lond.*, 44: 133–159.

Wootton, R.J., 1990. In: *Ecology of Teleost Fishes*. Chapman and Hall Ltd., 404pp.

Chapter 8

Limnochemistry of Bennithora Dam Near Gulbarga, Karnataka

☆ *B. Vasanthkumar and K. Vijaykumar*

ABSTRACT

Physico-chemical properties of Bennithora dam near Gulbarga city, Karnataka have been studied during June 1999 to May 2000. In the present study the water temperature of Bennithora dam ranges between 19.1-27.3°C, with maximum in April-May and minimum in June and August respectively. The pH of water ranges between 8 to 9. The Dissolved oxygen of water sample ranges between 2.4-8mg/l. The maximum oxygen content during January -February and minimum in August and October. The CO_2 ranged between 0.6-8.6 mg/l, maximum during September. The other parameters like TA, Cl_2, PO_4, and NH_4 studied during the study period.

Introduction

In order to utilize freshwater bodies successfully for fish production it is very important to study the physico-chemical factors, which influence the biological productivity of the water body.

In the recent year remarkable contribution is made in this field. The different trends of studies were observed of them some are important contributions like, Adholia *et al.* (1990 and 1991) More and Gajjar (1990); Vijaykumar and Basalingappa (1992); Subbamma and Sharma (1994); Mukesh *et al.* (1995); Vijaykumar (1995). No information is available o physico-chemical parameters and its related aspects pertaining to reservoirs in the Northerm part of Karnataka especially from the Gulbarga Hence, the present account is an attempt to study the detailed information on some important physico-chemical parameters.

Materials and Methods

The Bennithora Dam is on Bennithora River (Krishna Basin), near Gulbraga, city (76°-04′ to 77°42′ N and 16°-12 to 17° 45 E). This reservoir spreads over area of 45sq. miles. The maximum depth is 25 feet during monsoon and minimum is 15 feet during the dry period. The investigations of physico-chemical parameters were carried out during June 1999 to May 2000. The water samples were collected on monthly basis during 9 am to 11 am and brought to laboratory for the further analysis and analyzed for physico- chemical parameters. Following the Standard Method (APHA, 1985).

Results and Discussion

The physico-chemical parameters of Bennithora Dam are summarized in Table 8.1. It is well known fact that the water temperature directly as well as indirectly influences many abiotic and biotic components of aquatic ecosystem. Many hydrobiological features, parameters such as density, viscosity, conductivity and speciation of nutrient salt etc. undergo change with the temperature. It Also reflects to the dynamics of the living organisms such as metabolic and physiological behaviors of aquatic ecosystem. In the present investigation the minimum water temperature recorded during June (19.1°C) and maximum was recorded in the month of May (27.3°C)

Table 8.1: Monthly Variation in Physico-chemical Parameters in Bennithora Dam

Parameter	*Month/Year*											
	June 1999	*July 1999*	*Aug 1999*	*Sept 1999*	*Oct 1999*	*Nov 1999*	*Dec 1999*	*Jan 2000*	*Feb 2000*	*March 2000*	*April 2000*	*May 2000*
Atmos. temp (°C)	24.6	28.9	24.3	26.8	28.5	28.7	28.9	28.5	30.7	31.7	32.0	33.5
Water temp (°C)	19.9	23.3	23.9	20.2	21.8	23.5	24.5	23.7	24.7	24.6	25.8	27.3
pH	8.5	9.0	8.3	8.4	7.9	8.7	8.0	8.1	8.0	8.1	8.1	8.2
DO mg/l	5.2	4.8	4.4	3.9	3.5	4.0	4.1	4.9	6.8	4.5	4.1	4.0
CO_2 mg/l	+	0.6	6.6	8.6	+	+	1.1	1.1	+	1.1	2.2	1.1
TA mg/l	125	140	117	165	160	185	155	160	175	135	120	125
Cl_2 mg/l	34.60	63.90	62.48	59.64	58.22	48.28	58.22	48.28	35.60	34.60	48.28	34.60
PO_4-P mg/l	0.13	0.14	0.14	0.15	0.16	0.17	0.11	0.12	0.14	0.12	0.11	0.12
NH_4-N mg/l	0.26	0.66	0.61	0.61	0.27	0.66	0.65	0.59	0.69	0.67	0.38	0.30

+ Traces.

Ayyappan and Gupta (1981), Vijaykumar (1992) while working on the limnology of Ramasamundra tank and Jagath tank, Karnataka and many other workers have also observed similar trend in different water bodies (Singh and Swaroop, 1979; Geroge *et al.*, 1986).

It is well-known fact that changes in the pH values of water will bring about slight changes in the structural and functional variations in the organisms of the water body. In the present study the pH values has been recorded from June 1999 to May 2000. During 1999-2000 the pH values ranged from 8.0 to 9.0 and it is observed that the pH of Bennithora Dam followed a specific seasonal trend from June to May. In the study, it is interesting to the note that the pH values were more or less towards alkaline throughout the period of study. Vijaykumar (1992) and Pappa (995) found a direct correlation between water temperature and pH values have recorded the range of variation in pH such a correlation could also be drawn in the present work.

Dissolved oxygen is one of the most important parameters of the water quality, directly effecting surivival and distributing flora and fauna in an ecosystem. The quantity of dissolved oxygen in water is directly or indirectly dependent on water-temperature, partial pressure fo oxygen on the air amount of chlorophyll content etc. (Welch, 1952 and wetzel, 1975) in the present investigation the dissolved oxygen in the Bennithora dam is 2.4 to 7.8 mg/1. The dissolved oxygen concentration recorded higher during February 2000. Singh *et al.*, (1985), Saxena and Mishra (1991), while working on different water bodies, reported low values of dissolved oxygen during the high temperature period. The higher values of dissolved oxygen were observed during summer. In the present investigation substantiates the views of Ayyappan and Guptha (1981), Shukla and Bias (1990) who also reported the values during the summer months. They attributed this to the increased solar radiation and thus leading to a considerable good standing crop of phytoplankton.

The amount of free carbon dioxide depends on the decomposition of topsoil and chemical nature of the underlying rocks. By far, it is relatively abundant in natural waters and its importance in photosynthetic activity requires no further explanation. Since, the seasonal carbon flow in the system forms the very base photo-pyramid (Goldman and Harne, 1983). In the present investigations the free carbon dioxide values were considerably more in the month of August and September (Table 8.1).

Further, it is of interest to note that the observations reveals period of low values of free carbon dioxide, which attributed slight increase in pH values as also, noted by Singhal *et al.* (1985) reported that free carbon dioxide and water temperature varied independently with low values of carbon dioxide when aquatic vegetation was more abundant and high values free carbon dioxide due to high water inflow to the reservoir. In the present study low values of the free carbon dioxide was observed during all seasons.

In the present study it could be observed that the total alkalinity values in the Bennithora dam indicated that similar pattern of fluctuations though with minor differences. The higher values of total alkalinity in the Bennithora dam reached up to 185 mg/$CaCO_3$ during November and February months.

Natural water normally contains low chlorides than carbonated and sulphates. Large contents of chloride in freshwater are an indication of organic pollution (Tharsh *et al.*, 1994). Though chloride levels as high as 250mg/l is safe human consumption, a level above this imparts salty taste to potable water. In the present study chloride concentration was minimum during March, April and June and maximum in July and August 1999.

Table 8.2: Correlation Coefficient between the Various Physico-chemical Parameters in Bennithora Dam

Parameters	*Atmos. Temp °C*	*Water Temp °C*	*pH*	*DO mg/l*	*Free CO_2 mg/l*	*TA mg/l*	*Cl_2 mg/l*	*PO_4-P mg/l*
Atmos. temp °C	–							
Water temp °C	0.7745**	–						
pH	–0.3078	–0.2931	–					
DO mg/l	0.3509	0.2583	–0.3457	–				
Free CO_2 mg/l	–0.4006	–0.2488	0.0520	–0.3552	–			
TA mg/l	0.0124	–0.2371	–0.0116	0.2276	–0.1672	–		
Cl_2 mg/l	–0.4422*	–0.2764	0.2515	–0.6643	0.4698	0.0991	–	
PO_4-P mg/l	–0.3757	–0.4962	0.3806	–0.2515	0.0872	0.5731	0.3001	–
NH_4-N mg/l	–0.0308	0.1028	0.2108	0.1406	0.2046	0.4457	0.2535	0.0997

** Correlation is significant at the 0.01 level; * Correlation is significant at the 0.05 level

The Phosphates (PO_4) in natural water occur in very small quantities and it is an important nutrient for the maintenance of the fertility of the reservoir. During the present investigation the minimum (0.11 mg/l) of the phosphate content was recorded in the month of December and April 1999, 2000 and maximum (0.17 mg/l) was observed in the month of November, though the phosphate content in the reservoir is low during the study period.

Nitrogen is generally known to show no seasonal or depth variations except these resulting from temperature changes, as observed by Pearsall (1930). Ammonical nitrogen estimated in the present investigation remains much variable during the year but with an overall increasing during onset of the summer season (February, 0.66 mg/l).

Simple correlation coefficient analysis in Bennithora dam showed positive correlation of air and water temperature were found to be highly significant at 1 per cent level and atmospheric temperature was positively correlated with Cl_2 at 5 per cent significant level (Table 8.2).

In biological investigation the study of phytoplankton, zooplankton and primary productivity of Bennithora dam have been investigated and studied separately.

References

Adholia, U.N., Chakrabary, A, Srivastava, V. and Vyas, A., 1990. Community studies on macro-benthos with reference to limno-chemistry of Mansarovar, Bhopal. *J. Natcon.*, 3(2): 139–154.

Adholia, U.N., 1991. Phytoplankton community in relation to limno-chemistry of Manasarovar, Bhopal. *J. Natcon.*, 3(2): 155–186.

APHA, AWWA and WPCF, 1985. *Standard Methods for the Examination of Water and Wastewaters*, 16th edn. p. 1268.

Ayyappan, S. and Gupta, T.R.C., 1981. Limnology of Ramasmudra Tank–Hydrography Mysore. *J. Agri. Sci.*, 15: 305–312.

Geroge, J.P., Vengopal, G. and Venkateshvaran, I.P., 1986. Anthropogenic eutrophications in a perennial tank: effect on the growth of *Cyprinus carpio* communities. *Indian J. Environ. Health*. 28(4): 303–313.

Goldman, C.R. and Horne, A.J., 1983. *Limnology*. McGraw Hill Company, New York, USA, pp. 464.

More, P.G. and Gajjar, H.J., 1990. Hydrobiological study of Palan Pond in Valsad, Gujarat. *J. Ecobiol.*, 2(2): 128–135.

Mukesh, S.S., Patil, R.P. and Kulkarni, S.D., 1995. Plankton diversity of Sadatput reservoir. *J. Aqua Biol.*, 10(1): 24–25.

Pappa, A., 1995. Preliminary studies on limnology of Gobbur tank. *M.Phil. Dissertation*, Gulbarga University, Gulbarga.

Pearsall, W.H., 1930. Phytoplankton in the English lakes. I. The water of dissolved substances of biological importance. *J. Ecol.*, 18: 306–320.

Saxena, D.N. and Mishra, S.R., 1991. The water quality index and self-purification capacity of sewage collecting channel, Morar (Kalpi) river at Gwalior (M.P). *J. Freshwater Biol.*, 3(2): 169–175.

Shukla, S.N. and Bais, V.S., 1990. Changes in physico-chemical profile of Bila reservoir during winter season. *Trends on Ecotoxicology*, p. 291–293.

Singh, S.R. and Swaroop, P., 1979. Limnological studies on Suraha Lake (Ballia). II. The periodicity of phytoplankton. *J. Indian Bot. Soc.*, 58: 319–329.

Singh, D.K. and Singh, A.K., 1985. Zooplankton density in relation to physico-chemical factors during diel cycle of river Ganga at Bhagalpur, Bihar. *Environ. Ecol.*, 3(1): 231–234.

Singhal, R.N., Jeet, S. and Devis, R.W., 1985. The relationship among physical, chemical and plankton characteristics of unregulated rural ponds in Haryana, India. *Trop. Ecol.*, 26: 43–53.

Subbamma, D.V. and Sharma, D.V. Rama, 1994. Hydrography of fishpond in the Kolleru Area, A.P. *J. Aqua Biol.*, 9(1&2): 27–29.

Thrash, I.C., Suckling, E.V. and Beal, J.F., 1944. In the examination of water supplies (Ed. E.W. Taylor).

Vijaykumar, K., 1992. Limnological studies of perennial and seasonal standing water bodies of Gulbarga area. *Ph.D. Thesis*, Gulbarga University, Gulbarga, pp. 160.

Vijaykumar, K. and Basalingappa, S., 1992.Effect of physico-chemical factors on seasonal abundance of cladocera in Devikoppa tank. *J. Karnatak Univ. Sci.*, 34: 135–140.

Vijaykumar, K., 1995. Limnology of freshwater pond of Gulbarga. In: *Recent Researchers in Auqatic Environment*, (Ed.) A. Gautam and N.K. Agarwal. Daya Publishing House, New Delhi, p. 88–97.

Welch, P.S., 1952. *Limnology*, 2nd edn. McGraw Hill Book Company, New York, Toronto and London, pp. 538.

Wetzel, R.G., 1975. *Limnology*. W.B. Saunders Company, Toronto, pp. 1–743.

Chapter 9

A Study on Physico-chemical Parameters of Perennial Tank, Laxmiwadi, Kolhapur District, Maharashtra

☆ *S.A. Manjare, S.A. Vhanalakar and D.V. Muley*

ABSTRACT

Various freshwater bodies were sometimes subjected to wastewater discharges originating from different sources. Chemicals such as nitrogen, phosphorus and carbon in certain concentrations might alter and interrupt aquatic ecosystems. This study, purposing to determine water quality characteristics of Laxmiwadi tank in Kolhapur district, Maharashtra began in January, 2009 and was carried out for 12 months by taking monthly water samples from four different selected stations. Water quality parameters of temperature, turbidity, transparency, total dissolved solids, pH, dissolved oxygen, free carbon dioxide, total alkalinity, hardness, chloride, phosphate and nitrates analyses were done. Changes in water quality parameters of Laxmiwadi tank by months were determined. The results of the present study indicate that the tank is non-polluted yet and can be used for domestic, irrigation and pisciculture purposes.

Introduction

The maintenance of water quality standards in lakes and reservoirs is necessary so that they do not become a nuisance to both aquatic life and man. According to Bouck (1977) without high water quality, nothing of quality will live in it. The physical and chemical features of reservoirs are governed by existing hydrological and geomorphic processes. Reservoir water fluctuates in response to surface water runoff, direct precipitation, ground water discharge, rate of evaporation and most importantly human interference. These affect the physicochemical condition of a reservoir and in turn affect the fauna and flora by imposing physiological and behavioural adaptations. Their physiological parameters also act interdependently so as to greatly influence water quality criteria (Kemdirim, 2005). According to Akinbuwa (1992) the physico-chemical factors are limiting, and their presence or absence produce important and vital consequences on the life of aquatic fauna.

Physico-chemical studies on freshwater bodies are important as basis for any ecological work and also in the establishment of water quality standards. In this regard the present work has been carried out to study the physico-chemical parameters of Laxmiwadi tank to determine the monthly variations in water quality.

Materials and Methods

The present study was carried out in a Laxmiwadi tank in Kolhapur district, Maharashtra. It is located in 16°47′ 49.04″N and 74°22′ 57.47″ E. The area of the tank is 100 ha. The water is used for domestic, irrigation and fisheries purpose. The physico-chemical parameters of water were analyzed to know the status of the tank in a period of January, 2009 to December, 2009. Water samples were collected from four different sites of the tank in the morning between 9 to 10 am, in polythene bottle. The temperature and pH were recorded at the time of sample collection, by using thermometer and pocket digital pH meter. Transparency was measured with the help of Secchi disc. For analysis of chemical parameters water was fixed in the field and brought to the laboratory. The estimation of chemical parameters such as dissolved oxygen, total solids, free carbandioxide, hardness, chlorides, alkalinity, phosphate and nitrate was carried out as per the standard methods described by APHA (1985) and Trivedy and Goel (1987).

Results and Discussion

Physical Parameters of Laxmiwadi Tank

The monthly variations in physical parameters of Laxmiwadi tank was shown in Table 9.1.

Table 9.1: Physical Parameters of Laxmiwadi Tank, Kolhapur District, Maharashtra

Months	*Temperature °C*	*Transparency cm*	*Turbidity NTU*	*TDS gm/lit*	*pH*
Jan	24.0	40.0	0.4	0.12	8.4
Feb	25.5	40.0	0.4	0.11	8.3
Mar	26.0	34.0	0.4	0.12	8.3
Apr	23.5	32.5	10.0	0.10	8.1
May	26.5	22.5	11.2	0.20	8.1
Jun	23.5	13.5	10.2	1.10	8.4
Jul	23.5	63.5	1.0	4.55	8.3
Aug	24.5	53.0	2.2	0.30	8.6
Sept	25.0	60.0	3.2	0.50	9.2
Oct	25.0	69.2	1.0	0.22	8.2
Nov	24.5	80.5	1.0	1.70	8.2
Dec	22.5	75.5	2.5	0.30	8.4

Water Temperature

The water temperature ranges from 22.5°C to 26.5°C during the study period. The maximum (26.5°C) temperature was recorded in the month of May and minimum (22.0°C) in the month of December.

It showed that, higher temperature in summer and relatively lowers in winter. Jayabhaye *et al.* (2008), Salve and Hiware (2006) observed that during summer, water temperature was high due to low water level and clear atmosphere. The data of present study was in accordance with the same.

Water Transparency

The transparency of water fluctuates from 13.5 cm to 80.5 cm. The maximum (80.5 cm) was recorded in the month of November and minimum (13.5cm) in the month of June. Higher transparency occurred during winter and summer is due to absence of rain, runoff and flood water as well as gradual settling of suspended particles (Khan and Chowdhury, 1994; Kadam, *et al.* (2007).

Turbidity

The turbidity of water fluctuates from 0.4 NTU to 11.2 NTU. The maximum values (11.2 NTU) was recorded in the month of May and minimum value (0.4NTU) in the month of January, February and March. It might be due to human activities. The reason behind the higher turbidity during summer might be the decreased level of water in the tank (Naik and Purohit, 1996).

Total Dissolved Solids (TDS)

The total dissolved solids fluctuate from 0.10g/l to 4.55g/l. The maximum value (4.55g/l) was recorded in the month of July and minimum value (0.10g/l) in the month of April. Sakhare and Joshi (2001) and Korai *et al.* (2008) reported that the TDS values were higher during monsoon months due to rain and runoff.

pH

The pH values ranges from 8.1 to 9.2. The maximum pH value (9.2) was recorded in the month of September and minimum (8.1) in the month of April. The same results were recorded by Choudhary *et al.* (1979) in Hirakud Dam and Rajshekar *et al.* (2007) from a minor reservoir of Nadergul.

Chemical Parameters of Laxmiwadi Tank

The monthly variations in chemical parameters of Laxmiwadi tank was shown in Table 9.2.

Table 9.2: Chemical Parameters of Laxmiwadi Tank, Kolhapur District, Maharashtra

Months	*Dissolved Oxygen*	*Free CO_2*	*Hardness*	*Chloride*	*Alkalinity*	*Phosphate*	*Nitrate*
Jan	7.41	4.4	101	39.22	127.5	1.68	0.5
Feb	7.03	4.4	100	40.65	140	1.68	0.48
Mar	9.59	4.4	140	45.50	165	2.06	0.08
Apr	9.10	17.6	82.5	52.10	220	4.87	0.20
May	13.00	8.8	140	64.14	225	4.76	0.24
Jun	10.35	4.0	110	57.51	150	8.43	0.80
Jul	8.90	8.8	75	54.66	165	56.62	1.04
Aug	11.05	0.0	50	48.99	150	40.12	1.04
Sept	11.05	0.0	40	49.70	105	36.00	0.67
Oct	11.19	0.0	63	38.34	150	12.37	0.18
Nov	9.05	5.3	65	35.52	150	4.87	0.21
Dec	10.42	14.0	80	40.62	160	4.45	0.45

Value expressed in mg/lit.

Dissolved Oxygen (DO)

The values of DO fluctuate from 7.03 mg/l to 13.0 mg/l. The maximum values (13.0 mg/l) was recorded in the month of May and minimum values (7.03 mg/l) in the month of February. The high DO in summer is due to increase in temperature and duration of bright sunlight has influence on the per cent of soluble gases (O_2 and CO_2). The long days and intense sunlight during summer seem to accelerate rate of photosynthesis by phytoplankton, utilizing CO_2 and giving off oxygen. This possibly accounts for the greater qualities of O_2 recorded during summer. The quality is slightly lesser during winter. Similar results were reported by Ahmed Masood and Krishnamurthy (1990) as the duration of days is shorter and photosynthesis occurs for a relatively shorter time.

Free Carbon Dioxide

The value of free CO_2 ranges from 0.0 mg/l to17.6 mg/l. The maximum value (17.6 mg/l) was recorded in the month of April and minimum value (0.0mg/l) in the month of August, September and October. The value of CO_2 was high in summer. This could be related to the high rate of decomposition in the warmer months and to increased photosynthetic activity during the growing season. Arvind Kumar (1995) and Islam and Pramanik (2009) also reported that high free CO_2 content during summer months were possibly due to the high temperature and heavy rain fall with heavy drainage, which speed up the decomposition of organic matter and low photosynthetic activity.

Hardness

The values of hardness fluctuate from 40 mg/l to 140mg/l. The maximum value (140 mg/l) was recorded in the month of March and May and minimum value (40 mg/l) in the month of September. Hujare (2008) reported total hardness was high during summer than winter and monsoon. High value of hardness during summer can be attributed to decrease in water volume and increase rate of evaporation of water. Similar results were obtained in the present study.

Chlorides

The values of chlorides range from 35.5 mg/l to 64.14 mg/l. The maximum value (64.14 mg/l) was recorded in the month of May and minimum value (35.5 mg/l) in the month of November. In the present study maximum value of chloride reaches in summer and minimum in the winter. Similar results were reported by Swarnalatha and Narsingrao (1998).

Alkalinity

Total alkalinity ranges from 105 mg/l to 225mg/l. The maximum value (225 mg/l) was recorded in the month of May and minimum value (105 mg/l) in the month of September. The maximum alkalinity in the summer month may be due to increase in bicarbonates in the water. Hujare (2008) also reported similar results that alkalinity was maximum in summer and minimum in winter due to high photosynthetic rate.

Phosphate

The value of phosphate fluctuates from 0.08g/l to 1.04 mg/l. The maximum value (1.04 g/l) was recorded in the month of July and August and minimum value in the month of March. The high values of phosphate in July and August months are mainly due to rain, surface runoff, agriculture runoff; washer man activity could have also contributed to the inorganic phosphate content. Similar results were reported by Arvind Kumar (1995).

Nitrates

The values of nitrate ranges from 1.68 g/l to 56.62 mg/l. The maximum value (56.62g/l) was observed in the month of July and minimum in the month of January. Swaranatha and Narsingrao

(1998) reported nitrates are in low concentration in summer and high during monsoon which might be due to surface run off, due to rain. Similar results obtain in the present study.

Acknowledgement

The authors are grateful to University Grants Commission, New Delhi for providing FIP and Prof. G. P. Bhawane, Head, Department of Zoology, Shivaji University, Kolhapur, (M.S.), INDIA, for providing necessary research facilities.

References

Ahmad, M. and Krishnamurty, R., 1990. Hydrobiological studies of Wohar reservoir, Aurangabad (Maharashtra state), India. *J. Environ. Biol.*, 11(3): 335–343.

Akinbuwa, O., 1992. A preliminary study of diurnal vertical distribution of rotifers in Opa Reservoir, Nigeria. *J. Aquat. Sci.*, 7: 19–28.

APHA, 1985. *Standard Methods for Examination of Water and Wastewater*, 20th edn. American Public Health Association, Washington D.C.

Bouck, H.J.M., 1977. The importance of water quality to Columbia river salmon and steelhead. *Trans. Am. Fish. Soc.*, 10: 145–154.

Choudhary, N.K., Pradhan, P.K. and Dash, M.C., 1979.Certain physico-chemical factors and phytoplankton of Hirakud Dam. *Geobios*, 6: 104–106.

Hujare, M.S., 2008. Seasonal variation of physico-chemical parameters in the perennial tank of Talsande, Maharashtra. *J. Ecotoxicol. Environ. Monit.*, 18(3): 233–242.

Islam, M.N. and Pramanik, M.S., 2009. Relationship between physico-chemical and meterological conditions of a fish pond at Rajshahi, Bangladesh. *Res. J. Bio. Sci.*, 4(3): 357–359.

Jayabhaye, U.M., Pentewar, M.S. and Hiware, C.J., 2008. A study on physico-chemical parameters of a minor reservoir, Sawana, Hingoli district, Maharashtra.

Kadam, M.S., Pampatwar, D.V. and Mali, R.P., 2007. Seasonal variations in different physico-chemical characteristics in Masoli reservoir of Parbhani district, Maharashtra. *J. Aqua. Biol.*, 22(1): 110–112.

Kemdirim, E.C., 2005. Studies on the hydrochemistry of Kangimi reservoir, Kaduna State, Nigeria. *Afr. J. Ecol.*, 43: 7-13

Khan, M.A.G. and Choudhary, S.H., 1994. Physical and chemical limnology of lake Kaptai, Bangladesh. *Trop. Eco.*, 35(1): 35–51.

Korai, A.L., Sahato, G.A., Lashari, K.H. and Arbani, S.N., 2008. Biodiversity in relation to physico-chemical properties of Keenjhar Lake, Thatta district, Sindh, Pakistan. *Turk. J. Fish. Aquat. Sci.*, 8: 259–268.

Kumar, Arvind, 1995. Some limnological aspects of the freshwater tropical wetland of Santhal Pargana (Bihar). *Ind. J. Envi. Poll.*, 2(3): 137–141.

Naik, S. and Purohit, K.M., 1996. Physico-chemical analysis of some community ponds of Rourkela. *Ind. J. Environ. Protect.*, 16 (9): 679–684.

Rajshekar, A.V., Lingaiah, M.S., Rao, S. and Piska, R.S., 2007. The study on water quality parameters of a minor Nadergul, Rangareddy district, Andhra, Pradesh. *J. Aqua. Biol.*, 22(1): 118–122.

Sakhare, V.B. and Joshi, P.K., 2003. Studies on some aspect of fisheries management of Yeldari reservoir. *Ph.D. Thesis*, S.R.T.M. University, Nanded.

Salve, V.B. and Hiware, C.J., 2006. Study on water quality of Wanparakalpa reservoir Nagpur, Near Parli Vaijnath, District Beed. Marathwada region. *J. Aqua. Biol.*, 21(2): 113–117.

Swaranlatha, S. and Narsingrao, A., 1998. Ecological studies of Banjara lake with reference to water pollution. *J. Envi. Biol.*, 19(2): 179–186.

Trivedy, R.K. and Goel, P.K., 1986. *Chemical and Biological Methods for Water Pollution Studies.* Environmental Publication, Karad, Maharashtra.

Chapter 10

Monthly Variations in Physico-Chemical Parameters of River Tunga, Shivamogga, Karnataka

☆ *H.A. Sayeswara, K.L. Naik and Mahesh Anand Goudar*

ABSTRACT

Tunga River of Shivamogga is one of the important sources of water supply to agricultural and to urban areas. Water samples were collected from two different sites along the river. Physical and chemical parameters were studied during January 2009 to December 2009. The main aim of the present study was to determine the pollution status of Tunga River and the suitability of the water for domestic and other purposes. The study revealed that there is indication of pollution in the river and hence preventive measures are required to avoid further deterioration of the river water quality.

Introduction

Most of the perennial rivers and their tributaries are being used as a site for disposal of domestic and industrial waste in India which impairs their water quality (Chandanshive *et al.*, 2008). Due to improper and inadequate treatment facilities in a country like India, huge volume of wastewater is being discharged in to the rivers and lakes from various towns and cities (Nataraja *et al.*, 2009). Besides indiscriminate use of fertilizers and pesticides in the irrigated lands has significantly contributed to the non point sources of pollution (prakash *et al.*, 2005). With rapid growth of the Shivamogga city both in urban and industrial areas, the pollution load in the river Tunga has increased.

The River Tunga takes birth in Western Ghats and flows in north east direction and joins the other major river Bhadra, at Kudli about 14.5 kms from Shivamogga. The River Tunga receives copious water supply form the highly wooded and catchment area of Western Ghats. This river can be considered as lifeline of this area, which fulfills the needs of hundreds of villages, situated along the banks of the river. Due to anthropogenic activities, rapid industrial growth, domestic and agricultural activities of the region, the river water is being polluted, which is the case with almost all major rivers of the country (Manjappa *et al.*, 2008). This investigation shows the effects of pollution on the physico-chemical aspects of water of the Tunga River at Shivamogga in different months at two different sampling stations.

Materials and Methods

The sampling is done at two stations.

S_1= Unpolluted station: It is located near Lakshmipura, 5.3 kms away from Shivamogga city.

S_1= Polluted station: It is located near Anjaneya Swamy temple in Shivamogga city.

S_1 and S_2 sites were selected on the basis of varied ecological condition of the exploration zone due to domestic and industrial wastes.

The water samples were collected in properly cleaned plastic bottles. The samples were collected between 8 am to 10 am once in a month for a period of 12 months from January 2009 to December 2009. The Physico- chemcial parameters was analyzed as per APHA (1998).

Results and Discussion

The values of physico-chemical characteristics of Station S_1 and Station S_2 of Tunga River were assessed at monthly interval from January 2009 to December 2009. These have been depicted in Tables 10.1 and 10.2.

Table 10.1: Physico-chemcial Characteristics of Tunga River water at Station S_1 (Unpolluted Station)

Parameters	*Months 2009*											
	Jan	*Feb*	*Mar*	*Apr*	*May*	*Jun*	*Jul*	*Aug*	*Sep*	*Oct*	*Nov*	*Dec*
pH	7.1	7.1	7.3	7.0	6.8	6.8	7.3	7.2	6.9	7.4	7.0	7.2
DO	6.8	7.3	6.7	7.2	7.1	7.0	7.4	7.7	7.9	7.3	6.9	7.0
BOD	2.9	3.1	3.0	2.7	3.6	3.0	2.8	2.7	3.2	2.6	2.9	3.7
Free CO_2	1.3	1.5	1.4	1.9	2.5	2.2	2.6	2.3	2.8	2.9	2.7	2.3
TDS	129.2	139.4	173.9	143.2	151.3	110.3	101.9	123.2	141.2	140.0	139.2	137.2
Sulphates	4.8	5.0	5.3	6.0	5.9	5.3	6.2	6.7	5.7	5.1	6.3	6.0
Phosphates	0.073	0.072	0.068	0.083	0.095	0.090	0.096	0.091	0.083	0.087	0.063	0.079
Nitrates	1.2	0.70	1.0	0.90	1.2	1.3	1.2	1.3	1.2	1.3	1.1	0.90
Ammonia	0.076	0.073	0.079	0.081	0.082	0.081	0.079	0.078	0.087	0.083	0.084	0.082
Chlorides	28.10	23.25	26.15	20.0	22.0	18.30	17.15	18.35	20.15	22.0	22.8	23.10

All values are expressed in mg/L except pH.

pH

Measurement of hydrogen ion concentration which is represented as pH. The pHvalues ranged between 6.8 and 7.3 in station S_1 and 7.2 and 7.9 in station S_2. The average pH of the river water in the present study is well within the permissible limit of 6.5 as recommended by WHO (1971).

Table 10.2: Physico-chemical Characteristics of Tunga River Water at Station S_2 (Unpolluted Station)

Parameters	Months 2009											
	Jan	Feb	Mar	Apr	May	Jun	Jul	Aug	Sep	Oct	Nov	Dec
pH	7.4	7.6	7.2	7.8	7.0	7.3	7.9	7.1	7.7	7.8	7.3	7.9
DO	2.9	2.6	2.7	3.0	2.7	2.8	2.9	2.4	2.3	2.8	3.0	2.2
BOD	6.3	5.9	6.0	6.7	6.1	5.8	7.1	6.9	6.7	7.2	7.0	7.2
Free CO_2	11.9	12.4	12.6	12.0	12.9	12.8	13.8	13.5	14.2	13.5	11.9	12.1
TDS	353.2	372.0	341.3	382.4	300.3	293.2	200.0	280.1	379.8	371.0	383.2	393.2
Sulphates	13.2	11.9	12.6	12.8	13.1	12.5	13.0	12.7	11.6	12.9	12.2	12.6
Phosphates	1.3	1.1	1.7	1.4	1.5	1.3	1.5	1.3	1.7	1.3	1.6	1.6
Nitrates	5.2	4.8	5.9	4.7	5.3	5.2	6.0	5.4	5.2	4.9	5.2	5.8
Ammonia	0.33	0.31	0.34	0.30	0.31	0.34	0.31	0.37	0.32	0.37	0.33	0.32
Chlorides	153.2	151.1	155.1	149.2	147.1	135.3	132.1	133.2	138.1	136.1	142.2	144.2

All values are expressed in mg/L except pH.

Dissolved Oxygen

Dissolved Oxygen concentration in a water body indicates its ability to support aquatic life. The permissible standard of dissolved oxygen is above 5mg/L (Perk and Park, 1980). In the present study the dissolved oxygen level fluctuated between 6.7 mg/L to 7.9 mg/L in station S_1 and 2.2 mg/L to 3.0 mg/l in station S_2. The sampling station S_2 falls under polluted zone because in this zone there is entry of Shivamogga city sewage rich in bacteria. So the bacteria utilize the dissolved oxygen in the process of organic decomposition. Due to the process of biodegradation, the dissolved oxygen has reach lowest level in the station S_2.

Biological Oxygen Demand

BOD is the measure of degradable organic matter present in water. The BOD and other microbial activities are generally increased by the introduction of Sewage (Hynes, 1971). The BOD values ranged between 2.7 mg/L and 3.2 mg/L in station S_1 and 5.8 mg/L and 7.2 mg/L in station S_2. Higher values of BOD in station S_2 indicate the higher consumption of oxygen and higher population load in river water.

Free CO_2

Carbon dioxide is added to aquatic system by directly being mixed from atmosphere. Carbon dioxide in water bodies is contributed by the respiratory activity by organisms. Free CO_2 content was minimum in station S_1 (1.3mg/L to 2.9 mg/L) and maximum in station S_2 (11.9 mg/L to 14.2 mg/L). Free CO_2 helps in buffering the aquatic environment against rapid fluctuations in the acidity or alkalinity and also regulates biological process of aquatic communities (Prasannakumari *et al.*, 2003).

Total Dissolved Solids

The TDS of water samples ranged from 101.9mg/L to 173.9mg/L in station S_1 and 200mg/L to 393.2mg/L in station S_2. High values of TDS may be due to continuous evaporation of water (Agarwal and Agarwal, 1992). The highest desirable limit of TDS is 1500mg/L (WHO, 1994).

Sulphate

Sulphate is naturally occurring anion found in almost all kinds of water bodies. The sulphates are derived from discharge of domestic sewage, surface runoff and agricultural activities near by the water bodies. In the present investigation sulphate level was maximum at station S_2 (11.6 mg/L to 13.2 mg/L). This was due to the dumping of untreated city sewage and waste, which brings sulphate in to the river body at station S_2. In station S_1, the sulphate level ranged between 4.8 mg/L to 6.7 mg/L.

Phosphate

Phosphorus occurs in natural water as various types of phosphates. The most important sources of phosphates are the discharge of domestic sewage, detergents and agricultural runoff (Trivedi *et at.*, 1995). Values of phosphate ranged from 0.063 mg/L to 0.096 mg/L in station S_1 and 1.1 mg/L to 1.7 mg/L in station S_2. Phosphate concentration increases in water bodies that receives domestic waste (Nirmal Kumari., 1984).

Nitrate

The nitrate values ranged between 0.7 mg/L and 1.3 mg/L in station S_1 and 4.7 mg/L and 6.0 mg/L in station S_2. The increase of nitrate in station S_2 indicates the river receives very large amount of organic matter.

Ammonia

The ammonia values ranged between 0.073 mg/L and 0.084 mg/L in station S_1 and 0.30 mg/L and 0.37 mg/L in station S_2.

Chloride

Chloride is one of the important indicators of pollution. Chlorides are present in sewage and farm drainage. Chloride values were fluctuated between 17.15 mg/L to 28.10 mg/L in water samples collected from station S_1 and 132.1 mg/L to 155.1 mg/L in water samples collected from station S_2. The highest desirable limit of chloride is 250 mg/L (ICMR, 1975). High chloride content indicates deterioration of water quality usually linked with increased sewage load (Mini *et al.* 2003).

Conclusion

Moderate pollution is indicated in the study area which can be attributed to the anthropogenic activities. Further pollution parameters revealed that these parameters vary from station to station due to the discharge of domestic sewage and wastes around the study area. Based on the values of parameters, station S_1 can be classified as unpolluted zone. In the present investigation, most of the values of some parameters exceed the desirable limit according to BIS specifications in station S_2. The results indicate the polluted nature of the river at station S_2. Water quality of Tunga River is severely deteriorated due to the discharge of untreated sewage from the shivamogga city at station S_2. The river water quality at the station S_1 is fairly good and the data reveals that river at station S_1 is free from pollution. This water can be used for the human consumption after proper treatment.

Acknowledgements

The authors express their gratitude to Prof. D.B. Sumanthrappa, HOD of Zoology and Prof. B. R. Siddaramappa, Principal, Sahyadri Science College, Shivamogga and Dr. B. M. Hosur, Principal, D.V.S. College of Arts and Science, Shivamogga for providing Lab facility and encouragement.

References

Agarwal, B.L. and Agarwal, S.K., 1992. *Inorganic Chemistry of Water and its Properties*, 5th edn. Ratan Prakashan Mandir Publ., pp. 209–230.

APHA, 1998. *Standard Methods for the Examination of Water and Wastewater*, 20th edn. American Public Health Association, Washington, D.C.

Chandanshive, N.E., Pahade, P.M. and Kumble, S.M., 2008. Physico-chemical aspects of pollution in river Mula-Mutha at Pune, Maharastra. *Aquatic Biol.*, 23(2): 51–55.

Hynes, H.B.N., 1971. *The Biology of Polluted Water*. Univ. Toranto Press, Canada, p. 202.

Manjappa, S., Suresh, B., Aravinda, H.B., Puttaiah, E.T. and Thirumala, S., 2008. Studies on environmental status of Tungabhadra river near Harihara,Karnataka. 23(2): 67–72.

Mini, I., Radhika, C.G. and Ganga Devi, 2003. Hydro-biological studies on a Lotic ecosystem, Vamanpuram river, Thiruvananthapuram, Kerala, South India. *Poll. Res.*, 22(4): 617–626.

Nirmala Kumari, J., 1984. Studies on certain biochemical aspects of Hydrobiology. *Ph.D. Thesis*, Osmania University, Hyderabad.

Perk, J.E. and Park, K., 1980. *Textbook of Preventives and Social Medicine*, 8th edn. Banarsidas Bhanot, Jabalpur (India).

Prakash, K.L., Nataraj, Somashekar, R.K and Monmohan Rao, N., 2005. A model approaches for the water quality: A case study of river Cauvery. *Indian J. Eaviron and Ecoplan.*, 10(3): 557–564.

Prasannakumari, A.A., Ganga Devi, T. and Sukesh Kumar, C.P., 2003. Surface water quality of river Neyyar, Thiruvananthapuram, Kerala, India. *Poll. Res.*, 22(4): 515–525.

Trivedi, R.K., Goel, P.K. and Trisal, C.L., 1995. *Practical Methods in Ecology and Environmental Science.* Environmental Publications, Karad.

WHO, 1984. *International Standards for Drinking Water*, Geneva.

Chapter 11

Studies on Surface Water Quality Evaluation of Gowdana Pond, Shivamogga, Karnataka

☆ *H.A. Sayeswara, Nafeesa Begum and Mahesh Anand Goudar*

ABSTRACT

Studies were carried out to assess the water quality of Gowdana pond at Shivamogga district of Karnataka. An year long study was conducted to measure various physico-chemical parameters like temperature, hydrogen ion concentration, free carbon dioxide, dissolved oxygen, BOD, alkalinity, hardness, chloride, TDS and various radicals. The study revealed that water is polluted as it possess high BOD and free CO_2. The results were compared with the standards given by WHO and ISI for water quality.

Introduction

Gowdana pond is an annual water body as it receives the water from the adjacent paddy fields and Sigge halla dam. Various physico-chemical and biological factors determine the quality of water (Sukunda *et al.*, 2004). Physico-chemical parameters play a vital role in determining the distributional pattern and quantitative abundance of organisms inhabiting a particular aquatic ecosystem (Singh *et al.*, 2009). The normal ranges of physico-chemical characteristics indicate the good water quality (Swaminathan and Manonmani, 1997).

The total area of Gowdana pond is about 175 acres of which water spreads over an area of 135 acres with an average depth of 7 feet. It is located in Ayanur village of Shivamogga district. The Pond water is used for domestic purposes like washing of clothes, vehicles and for domestic animals, etc. The water body is surrounded by paddy and sugarcane fields in all directions. The water has undergone moderate changes in its physico-chemical properties due to ecological degradation, overflowing of water from the adjacent paddy fields and other excessive human activities.

The literature revealed that there is no scientific study carried out with respect to ecological characteristics of this pond. The basis of selection of Gowdana pond was that its water is used by a large population.

Materials and Methods

The study was carried out during the period from January to December 2008. During the study period, the surface water samples were collected in clean plastic cans between 8 am to 10 am once a month. Water temperature was recorded on the spot. The samples for dissolved oxygen fixed immediately on the field itself. The remaining parameters were analyzed as per the Standard methods (APHA, 1998).

Results and Discussion

The results of phyico-chemical parameters of Gowdana Pond water are depicted in Table 11.1 and salient features of the findings are summarized below.

Temperature

The water temperature is influenced by factors such as altitude, season, time and depth of water. Values of temperature ranged from 22 to 27.2°C. The minimum value was recorded in March and maximum in September.

pH

Measurement of hydrogen ion concentration which is represented as pH. The pH of water is slightly acidic to slightly alkaline and found within permissible limit of 6.5 to 8.5 as per the Bureau of Indian Standards (BIS). The values of pH ranged from 6.5 to 7.5. The minimum value was recorded in January and maximum in September. The pH is an important parameter in a water body since aquatic organisms are well adapted to specific pH range and do not withstand abrupt changes in it (George, 1997).

Dissolved Oxygen

Dissolved oxygen is an important gaseous factor that determines the quality of water and inturn regulates the distribution of aquatic organisms. In the present study the dissolved oxygen level fluctuated between 2.7 to 4.2 mg/L. The highest and the lowest values were recorded in October and April, respectively. The variations of dissolved oxygen depend on the primary production and respiration of aquatic organisms.

Biological Oxygen Demand (B.O.D.)

BOD is the measure of degradable organic matter present in water. The B.O.D. and other microbial activities are generally increase by the introduction of Sewage (Hynes, 1971). The values ranged from 7.2 to 10.3 mg/L. The highest and the lowest values were recorded in July and February, respectively. They were found above the permissible limit of 6.5 mg/L as per the WHO (1991).

Free Carbon Dioxide

Carbon dioxide is added to the aquatic system by directly being mixed from atmosphere. The decomposition of organic matter from aquatic ecosystem also adds CO_2 to pond. Free CO_2 values fluctuated between 11.9 to 16.2 mg/L. The highest and the lowest values were recorded in September and May, respectively. The variation of CO_2 was due to the absorption by plants for photosynthesis and activity of other living organisms.

Table 11.1: Physico-chemical Characteristics of Gowdana Pond Water

Parameters	*Months 2008*											
	Jan	*Feb*	*Mar*	*Apr*	*May*	*Jun*	*Jul*	*Aug*	*Sep*	*Oct*	*Nov*	*Dec*
Temperature	23.1	24.3	22	24.9	25	26.5	26.1	26	27.2	26.3	24.8	24.3
pH	6.5	6.7	6.6	6.6	6.8	7.3	7	6.9	7.5	6.8	6.8	6.7
DO	3.7	3.3	4.1	2.7	2.9	2.8	3.1	2.8	2.9	4.2	3.9	3.3
BOD	6.9	7.2	7.9	6.8	7.3	8.4	10.3	10.8	9.9	8.1	9.3	9.7
Free CO_2	13.9	12.6	15.1	14.3	11.9	13.2	14.1	15.3	16.2	15.7	14.2	14.7
Chloride	106	112	110	123	109	97	93	98	73	84	69	72
TH	64	73	69	86	80	96	72	45	49	59	64	102
Alkalinity	62.5	70	81	69	75	63	58	53.5	61	54	67	63.5
TDS	140	155	105	132	144	162	138	133	152	161	195	180
Calcium	22	26	20	33	23	31	25	13	16	23	23	36
Magnesium	40	45	48	50	55	60	45	30	32	35	38	68
Nitrates	4.3	5.1	4.9	5.3	5.2	6.1	6	5.3	5	4.9	5.7	5.7
Phosphates	1.1	1.2	1.2	1.3	1.5	1.3	1.2	2	1.7	1.4	1.6	1.8
Sulphates	8.9	7.3	7.9	7.2	9.9	9.8	8	7.7	7.9	8.2	8.9	7.7

All values in mg/L except pH and temperature

Chloride

Chlorides are important anions found in variable amounts in water bodies. Chlorides present in sewage and form drainage, control the salinity of water and osmotic stress on biotic communities (Benerjee, 1967). Chlorides increase the degree of eutrophication (Goel *et al.*, 2003). In the present study, chloride values fluctuated between 69 mg/L (November) to 123 mg/L (April). The highest desirable limit of chloride is 250 mg/L (ICMR, 1975). High chloride content indicates deterioration of water quality usually linked with increased sewage load (Mini *et al.*, 2003).

Total Hardness

Total hardness of water is not a pollution parameter but indicates water quality mainly in terms of Ca^{2+} and Mg^{2+} contents. Total hardness values observed are 45 to 102 mg/L. The highest and the lowest values were recorded in December and August, respectively. Total hardness above 200 mg/L is not suitable for domestic use in drinking and cleaning.

Total Alkalinity

Alkalinity in the water samples is primarily a function of carbonate, bicarbonate and hydroxide contents. It ranged from 54 mg/L (October) to 81 mg/L (March). It is within permissible limit of 600 mg/L. Surface alkalinity may result from the discharge of domestic wastes.

Total Dissolved Solids (TDS)

TDS ranged from 105 mg/L (March) to 195 mg/L (November). The minimum value may due to the stagnant condition of the water body. The values are within permissible limits of 1500 mg/L (BIS, 1982).

Calcium

The calcium values ranged between 13 mg/L (August) to 33 mg/L (April). Calcium is essential for muscular and cardiac functions and its low level may have adverse effect on human health.

Magnesium

The magnesium values ranged between 30 to 68 mg/L. The highest and the lowest values were recorded in December and August.

Nitrate

The nitrate values ranged between 4.3 to 6.1 mg/L. The highest and the lowest values were recorded in June and January, respectively. The increase of nitrate is associated with rain water runoff and sewage discharge.

Phosphate

Phosphorus occurs in natural water as various types of phosphates. The most important sources of phosphates are the discharge of domestic sewage, detergents and agricultural runoff (Tivedi *et al.*, 1995). Values of phosphate ranged from 1.1 to 1.8 mg/L with the minimum value in January and maximum in December.

Sulphate

It is one of the major anions occurring in natural water bodies. It may enter natural water bodies through weathering of sulphate bearing deposits and agricultural activities. In the present study, sulphate values fluctuated between 7.2 mg/L (April) to 9.9 mg/L (May).

Conclusion

The water samples from Gowdana pond was collected and analyzed for various physico-chemical parameters to study the extent of pollution. The results of physico-chemical analysis have revealed that the Gowdana pond is contaminated due to human disturbances. In the light of standard of water quality recommended by WHO, the pond water should not be used by human beings especially for drinking and cooking. It is recommended that the anthropogenic activities should be prevented by organizing awareness programs.

Acknowledgements

The authors express their gratitude to Prof. D.B. Sumanthrappa, Head of Departmen of Zoology and Prof. B. R. Siddaramappa, Principal, Sahyadri Science College, Shivamogga and Dr. B. M. Hosur, Principal, D.V.S. College of Arts and Science, Shivamogga for providing Laboratory facility and encouragement during study period.

References

APHA, 1998. *Standard Methods for the Examination of Water and Wastewater*, 20th edn. American Public Health Association, Washington, D.C.

Banerjee, S.M., 1967. Water quality and soil conditions of ponds in some states of India in relation to fish production. *Indian J. Fish.*, 14: 115–144.

George, J.P., 1997. Aquatic ecosystem: structure, degradation, strategies for management. In: *Recent Advances in Ecological Research*. APH Publ. House, New Delhi, p. 603.

Goel, P.K., Gopal, B. and Trivedy, R.K., 1980. Impact of sewage on freshwater ecosystem. *Int. J. Ecol and Envir. Sci.*, 6: 97–116.

Hynes, H.B.N., 1971. *The Biology of Polluted Water*. Univ. Toranto Press, Canada, p. 202.

Mini, I., Radhika, C.G. and Ganga Devi, 2003. Hydro-biological studies on a lotic ecosystem, Vamanpuram river, Thiruvananthapuram, Kerala, South India. *Poll. Res.*, 22(4): 617–626.

Nirmala Kumari, J., 1984. Studies on certain biochemical aspects of Hydrobiology. *Ph.D. Thesis*, Osmania University, Hyderabad.

Santhoshkumar Singh, A., Dakua, D. and Biswas, S.P., 2009. Physico-chemical parameters and fish enumeration of Maijan Beel (wetland) of upper Assam. *Geobios*, 36: 184–188.

Sukunda, B.N. and Patil, H.S., 2004. Seasonal dynamics of phytoplankton in relation to physico-chemical factors of fort lake, Belgaum (North Karnataka). *J. Environment and Ecology*, 22(2): 337–347.

Swaminathan, K. and Manonmani, K., 1997. Studies on toxicity of viscose rayon factory effluents. *J. Env. Bio.*, 18(1): 73–78.

Trivedi, R.K., Goel, P.K and Trisal, C.L., 1995. *Practical Methods in Ecology and Environmental Science*. Environmental Publications, Karad.

Chapter 12

Heavy Metals in Water and Sediment of Ichchamati River in East Khasi Hills, Meghalaya

☆ *P.K. Bharti, D.S. Malik, Pawan Kumar and Umesh Bharti*

ABSTRACT

Ichchmati river is a major river of East Khasi Hills district in Meghalaya near Bangladesh Border. This river ecosystem has a great importance as a natural habitat among the various ecosystems of the region, whereas there is a little agricultural practices and no more industrial pollution. Location variation in nutrients concentration of the river was studied with special reference to physico-chemical parameters and heavy metals in the river water and sediment.

Heavy metals were found almost nil or in very low concentrations at selected site. The present study deals with the preliminary physico-chemical characteristics and heavy metals in the river water and river sediment, which exhibits the status of water quality and transfer of heavy metals from river water to sediment.

Introduction

The impact of anthropogenic pollution from industrial, agricultural, sources, quarrying and tourists activity on water quality has concerned environmentalists and scientists for the past three decades. Ecological, geo-chemical and hydrological research has been carried out in various ecosystems to understand the factors controlling the chemistry of natural water (Baron and Bricker, 1990; Malik and Bharti, 2005a). In North-East region, a river Ichchamati, 80 km. far away from Shillong city is the study site near Ichchamati village for the accounting of physico-chemical parameters and heavy metals in natural water and fluctuations at the different locations along with the river. Adequate understanding of the North-East regional rivers is extremely important for the development of a realistic program for utilizing the potential of water that exist in the form of surface water resource in the region.

The dynamic balance in the aquatic ecosystem is upset by human activities, resulting in pollution which meter, is manifested dramatically as fish kill, offensive taste, odour, colour and unchecked aquatic weeds. The over production of higher tropic levels biomass and the subsequent decay of dead plants could lead to oxygen depletion, death of aquatic organisms and development of anaerobic zone where bacteria action produce foul odours and bad tastes (Forstner and Wittman, 1979). Water quality characteristic of aquatic environment arise from a multitude of physical, biological and chemical interactions (Dezuane, 1979). The water bodies, lakes, rivers, dams and estuaries are continuously subject to a dynamic state of change with respect to the geological age and geochemical characteristics. This is demonstrated by continuous circulation, transformation and accumulation of energy and matter through the medium of living thing and their activities (Adefemi, 2007). The present study deals with the characteristics the water nutrient chemistry, influenced by anthropogenic activity and quarrying of the geologically sediment environments and to determine the nature and degree of anthropogenic impacts on qualitative and quantitative variations occurred in nutrients in relation to physico-chemical parameters and heavy metals of river water and sediment.

Study Area

Meghalaya is very well known for the record of rainfall, the state consists of the two places namely Cherrepunjee and Mawsinram for maximum rainfall throughout the year. So, the maximum water resources are depending chiefly on total precipitations of the region. The Meghalaya has been surrounded by the natural beauties with lovely trees and cool climate, which is not only a pleasant place to live in but relaxing for holiday also. This is not only a popular state but also valuable or important place for tourism. Shillong is the capital of Meghalaya state. Mowlong Cherra Cement Ltd and Lafarge Umaim Mining Ltd are the major industries of the region. The all seven states of North East India are quite famous as Seven Sisters.

Basically, Ichchamati river originated from the hills of Assam-Meghalaya. Population of the region is completely depending upon the river water for drinking, bathing, and other activities. *Meteorologically*, region has a cool and pleasant climate. Geologically, North-east hills are enriched with various minerals and the hills near Bangladesh are rich in limestone. Geographically, the study area is situated in the globe on a Latitude 25° 09′ 58.1″ N and Longitude 91° 41′ 26.5″ E.

Materials and Methods

The samples for physico-chemical parameters and heavy metals were collected by using rinsed Borosil glassware, and analyzed with the help of the procedure described by APHA (1995) and Trivedi and Goel (1984). The water samples were collected from Ichchamati river near Ichchamati village according to the analytical requirement in morning period 9:00 Hrs. to 10:00 Hrs.

Colour, Odour, Turbidity, Velocity, Temperature and Dissolved oxygen were analyzed on sampling sites. Samples were collected from selected sites and immediately preserved in ice boxes, and transfer to the lab for further analysis. Water samples were digested and Heavy Metals were detected using Atomic Absorption Spectrophotometer. Transfer Factor (TF) was calculated according to Bharti (2007) to assess the status of heavy metals transfer from river water to river sediment of Ichchamati river.

Results and Discussion

Ichchamati river is flowing through a piece of plain and hills chain near the Ichchamati village, enriched with minerals, which affect the water quality of river according to the locations. Nutrients concentration, heavy metals and related physico-chemical parameters from selected sites are depicted in Tables 12.1–12.3.

Table 12.1: Physical Characteristics of Ichchamati River Water

Sl.No.	Parameters	Unit	Ichchamati River Water		
			Winter	Summer	Monsoon
1.	Temperature	°C	12	16	15
2.	Colour	–	Clear	Clear	Clear
3.	Odour	–	Nil	Nil	Nil
4.	Turbidity	NTU	23	21	27
5.	TDS	mg/l	90	100	147

Table 12.2: Chemical Characteristics of Ichchamati River Water

Sl.No.	Parameters	Unit	Ichchamati River Water			
			Winter	Summer	Monsoon	Mean
1.	pH	–	7.8	7.7	7.4	7.6
2.	Alkalinity	mg/l	58	76	91	75.0
3.	Total Hardness	mg/l	71	86	97	84.7
4.	Calcium	mg/l	22	29	38	29.7
5.	Magnesium	mg/l	4	3	2	3.0
6.	Chlorides	mg/l	6	6	5	5.7
7.	DO	mg/l	8.7	9.2	9.1	9.0
8.	BOD	mg/l	1	1	1	1.0
9.	COD	mg/l	6	6	6	6.0

Table 12.3: Heavy Metals in Ichchamati River Water and Sediment

Sl.No.	Heavy Metals	Unit	Ichchamati River Water				Ichchamati Sediment (mg/kg)
			Winter	Summer	Monsoon	Mean	
1.	Cadmium	mg/l	BDL	BDL	BDL	BDL	BDL
2.	Copper	mg/l	BDL	BDL	BDL	BDL	24
3.	Iron	mg/l	0.8	0.62	0.17	0.53	2000
4.	Lead	mg/l	BDL	0.02	BDL	0.01	59
5.	Manganese	mg/l	0.04	0.06	0.01	0.04	100
6.	Zinc	mg/l	BDL	0.02	0.03	0.02	79

Ichchamati river has the spatio-temporal variations of water temperature, which plays a vital role in all physico-biochemical reactions and self-purification power of aquatic system (Badola and Singh, 1981). Higher value of temperature was found 16 °C in summer and minimum 12 °C in winter season.

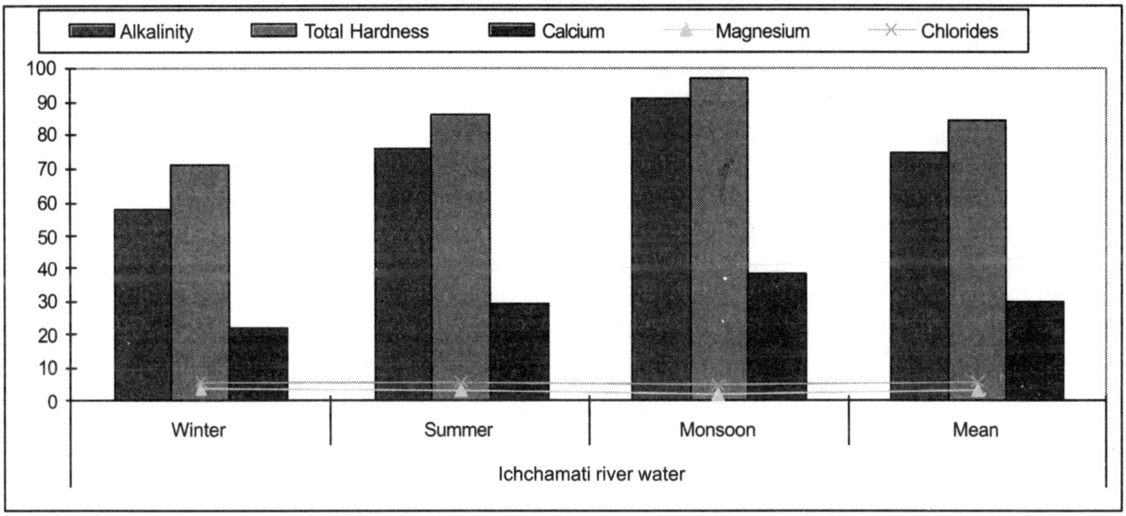

Figure 12.1: Showing the Nutrients Concentration in Ichchamati River during the Study Period

Turbidity is striking characteristic of the physical status of the water bodies. Although in Ichchamati river water is clear because there is no more pollution, siltation was the main source of turbidity in tributaries. Detritus and other non-organic material being added to water mass due to rainfall and anthropogenic activities (Camron, 1996). Maximum turbidity was recorded 27 NTU during monsoon season and minimum 21 NTU in summer season. The maximum depth of photic zone provides the better biological production for all aquatic organisms in a river (Malik and Bharti, 2005b).

Total dissolved solids were found in the range of 90 mg/l in winter to 147 mg/l in monsoon season, due to the gradual increases in velocity of river which favoured effective sedimentation (Subramanian, 1979). Chemical oxygen demand was found 6 mg/l during the study period. Chemical oxygen demand represents chemically oxidizable organic matter load in water, while biochemical oxygen demand is only biodegradable materials (Malik and Bharti, 2005c). In the present study the values observed during monsoon months may be attributed maximum biological activities and high temperature, stimulate the growth of microorganisms (William *et al.*, 1993).

The pH of natural water was controlled in a great extent by the interaction of hydroxyl ions arising from the hydrolysis of bicarbonate (Sharma, 1986). The pH of Ichchamati river was recorded alkaline (7.4-7.8). Total hardness is mainly due to percentage of calcium and magnesium salts of bicarbonates, carbonates, sulphates and chlorides, while the value of alkalinity occured due to presence of bicarbonates. The concentration of hardness was analyzed 26-74 mg/l during the study. Alkalinity was also found 32-63 mg/l with a small fluctuation. A positive relationship between hardness and alkalinity was recorded in river Ganga at Rishikesh (Chopra and Patric, 1994). Maximum chloride concentration was recorded maximum (15 mg/l) in summer and minimum in monsoon (6 mg/l). Chloride and hardness showed a positive relationship to one another (Chopra and Patric, 1994). Chloride was found in the form of chloride ion, and one of the major inorganic anion present in natural water (Malik and Bharti, 2009).

Calcium and magnesium the dominant cations, and these represent the main weathering products, but significant hydro-chemical differences between the two sampling sites associated with the bedrock geology exist (Jenkins *et al.*, 1995) Calcium is one of the essential nutrients, which plays an important role in biological system. Maximum calcium concentration was recorded (28 mg/l) in monsoon and minimum in winter (22 mg/l). Positive relationship between, calcium and temperature was also reported by Khanna and Singh (2000) in river Suswa, Dehradun. Magnesium is also an essential element but it is toxic at higher concentration. The concentration of magnesium in Ichchamati river was found maximum (4 mg/l) and minimum (2 mg/l) and it was very low in comparison to Hill-streams of Uttarakhand (Bharti, 2004). During the summer season nutrients concentration in rivers and hill-streams became more. Miller *et al.* (1997) described the nutrients availability in selected environmental settings of the Potomac River and Cameron (1996) showed the similar type of fluctuation in Fraser river of British Columbia. Bond (1979) described similar nutrients concentration pattern in a stream draining a mountain ecosystem in Utah.

Dissolved oxygen was found 8.7 to 9.2 mg/l during the study period. While biochemical oxygen demand was found 1 mg/l in every season. Heavy metals were found almost nil or in very low concentrations except iron. Heavy metals like cadmium, copper, lead and zinc were not found in high concentration at both sites during any season. Cadmium, copper and lead were absolutely absent in all seasons, while zinc concentrations were also found below detection limit in maximum samples. The concentration of iron was maximum observed 0.8 mg/l during summer season. Zinc concentration was found 0.03 mg/l in monsoon season. Malik *et al.* (2009) described the role of heavy metals in the surface water of north India. The results of Bharti *et al.* (2010) were also indicated the relation of heavy metals and phytoplankton in a north Indian water body.

Transfer factors from river water to sediment for all metals were found quite irregular. Transfer factor for cadmium was found 0, while it was found constant for copper in all seasons. It was found 2500, 3225.8 and 11764.7 in winter, summer and monsoon seasons respectively for Iron. Similarly, transfer factor was calculated 2500, 1666.7 and 10000 in winter, summer and monsoon seasons respectively for manganese. For Lead, transfer factor was calculated from 2950 to 11800. For zinc, transfer factor was calculated from 2633.3 to 15800. All results of transfer factors are depicted in Table 12.5. Physico-chemical parameters of river sediments are also depicted in Table 12.4.

Table 12.4: Physico-chemical Characteristics of Ichchamati River Sediment

Sl.No.	*Parameters*	*Unit*	*Ichchamati Sediment*	*Method*
1.	Texture	–	Silty sand	IS: 2720 p-4
2.	Grain size analysis	Per cent by mass		IS: 2720 p-17
	Sand		71	
	Silt		28	
	Clay		1	
3.	Moisture Content	Per cent by mass	7.4	IS: 2720 p-2
4.	pH	–	8.3	IS: 2720 p-26
5.	Conductivity	µmho/cm	110	Conductivity meter
6.	Calcium	Per cent by mass	0.3	APHA (1998)
7.	Magnesium	Per cent by mass	0.2	APHA (1998)
8.	Chlorides	Per cent by mass	0.1	Volhard's method

Table 12.5: Transfer Factor of Heavy Metals from Ichchamati River Water to Sediment

S.N.	*Heavy Metals*	*Transfer Factor*		
		Winter	*Summer*	*Monsoon*
1.	Cadmium	0.0	0.0	0.0
2.	Copper	4800.0	4800.0	4800.0
3.	Iron	2500.0	3225.8	11764.7
4.	Lead	11800.0	2950.0	11800.0
5.	Manganese	2500.0	1666.7	10000.0
6.	Zinc	15800.0	3950.0	2633.3

The present results conclude that significant differences in river water nutrient concentrations in different seasons. The spatial variations in TDS are attributed to climatic and lithological control over the ionic concentrations. Absence or low concentration of heavy metals shows that the water is still industrial pollution free. Heavy metals never cross the limits during the study period. On the basis of nutrients and heavy metals, the water may be considered for other purposes.

References

APHA, 1995. *Standard Methods for Examination of Water and Wastewater*, 19th edn. American Public Health Association Inc., New York, pp. 1970.

Adefemi, O.S., Asaolu, S.S. and Olaofe, O., 2007. Assessment of the physico-chemical status of water samples from major dams in Ekiti State, Nigeria. *Pakistan Journal of Nutrition*, 6(6): 657–659.

Badola, S.P. and Singh, H.R., 1981. Hydrobiology of the river Alaknanda of Garhwal Himalaya. *Indian J. Ecol.*, 8(2): 269–276.

Baron, J. and Bricker, O.P., 1990. Hydrological and chemical flux in Loch Vale watershed, Rocky Mountain National Park. In: *Biogeochemistry of Major Rivers, SCOPE 42*. Wiley and sons, New York, USA.

Bharti, P.K., 2004. Limnobiological study of Sahastradhara hill-stream at Dehradun. *M.Sc. Dissertation*, Gurukula Kangri University, Hardwar, pp. 102.

Bharti, P.K., 2007. Effect of textile industrial effluents on groundwater and soil quality in Panipat region (Haryana). *Ph.D. Thesis*, Gurukula Kangri University, Hardwar, pp. 191.

Bharti, P.K., Malik, D.S. and Yadav, Rashmi, 2010. Influence of heavy metals on abundance of cyanophyceae members in three spring-fed lake in Kempty, Dehradun. In: *Advances in Aquatic Ecology*, Vol. 3, (Ed.) V.B. Sakhare. Daya Publishing House, New Delhi, pp. 107–111.

Bond, H.B., 1979. Nutrient concentrations patterns in a stream draining a montane ecosystem in Utah. *Ecology*, 60(6): 1184–1196.

Cameron, E.M., 1996. Hydrogeochemistry of the Fraser River British Columbia: Seasonal variation in major and minor components. *J. Hydrol.*, 182(1–4): 209–255.

Chopra, A.K. and Patrick, N.J., 1994. Effect of domestic sewage on self-purification of Ganga water at Rishikesh. *A. Bio Science*, 13(11): 75–82.

Dezuane, J., 1979. *Handbook of Drinking Water Quality*. Indiana University Press, pp. 3–17.

Forstner, U. and Wittman, G.T.W., 1979. *Metal Pollution in the Aquatic Environment*. Springer-Verlag, Berlin, pp. 486.

Jenkins, A., Sloan, W.T. and Cosby, B.J., 1995. Stream chemistry is the middle hills and high mountain of the Himalaya, Nepal. *Journal of Hydrology*, 166(1–4): 61–79.

Khanna, D.R. and Singh, R.K., 2000. Seasonal fluctuations in the plankton of Suswa River at Raiwala Dehradun. *Env. Conservations J.*, 1(2 and 3): 89–92.

Malik, D.S. and Bharti, Pawan K., 2005a. Nutrient dynamics in Rhithron zone of Shivalik Himalayan stream Sahastradhara, Dehradun (Uttaranchal). *Env. Cons. J.*, 6(2): 63–68.

Malik, D.S. and Bharti, P.K., 2005b. Fluctuation in planktonic population of Sahastradhara hill-stream at Dehradun (Uttaranchal). *Aquacult.*, 6(2): 191–198.

Malik, D.S. and Bharti, P.K., 2005c. Primary production efficiency of Sahstradhara hill-stream, Dehradun, *Env. Cons. J.*, 6(3): 117–121.

Malik, D.S. and Bharti, P.K., 2009. Ecology of Sahastradhara Hill-stream at Dehradun (Uttaranchal), In: *Advances in Aquatic Ecology*, Vol. 1, (Ed.) V.B. Sakhare. Daya Publishing House, New Delhi, pp. 1–11.

Malik, D.S. and Bharti, P.K., Negi, K.S. and Yadav, Rashmi, 2009. Distribution of metals in water of an artificial lake at Mussoorie, Uttarakhand. In: *Aquatic Biology and Aquaculture*, (Ed.) V.B. Sakhare. Manglam Publication, New Delhi, pp. 77–95.

Miller, C.V., Denis, J.M., Ator, S.W. and Brakebill, J.W., 1997. Nutrients in stream during baseflow in selected environmental settings of the Potomac river basin. *J. American Wat. Resources Association*, 33(6): 1155–1171.

Psenner, R., 1989. Chemistry of high mountain lakes in siliceous catchments of central Alps. *Aquatic Sci.*, 51: 108–128.

Shrama, R.C., 1986. Effect of physico-chemical factors on benthic fauna of Bhagirathi River Garhwal Himalaya. *Indian J. Ecol.*, 13(1): 133–137.

Subramanian, V., 1979. Chemical and suspended sediment characteristics of river of India. *J. Hydrol.*, 44: 37–55.

Trivedi, R.K. and Goel, P.K., 1984. *Chemical and Biological Methods for Water Pollution Studies*. Environmental Publication, Karad, pp. 1–298.

William, M.W., Brown, A. and Melack, J.M., 1993. Geochemical and hydrologic controls on the composition of surface water in the high elevation basin, *Sierra Navada. Limnol. Oceanogr.*, 38: 775–797.

Xue, H.B. and Schooner, J.L., 1994. Acid deposition and lake chemistry in southwest China. *Wat. Air, Soil Pollut.*, 75: 61–78.

Chapter 13

Assessment of Heavy Metals in Water and Sediment of Hatmawdon River near Bangladesh Border in District East Khasi Hills, Meghalaya

☆ *Umesh Bharti, P.K. Bharti, D.S. Malik, Pawan Kumar, Gaurav and Vijender Singh*

ABSTRACT

Hatmawdon river is a river of East Khasi Hills district in Meghalaya along with the Umaim River near Bangladesh Border. The river ecosystem has a great importance as a natural habitat among the various ecosystems of the region, whereas there is a limestone mining area at very large scale and no more industrial and agricultural pollution. Location variation in nutrients concentration of the river was studied with special reference to physico-chemical parameters and heavy metals in the river water and sediment.

Heavy metals were found almost nil or in very low concentrations at selected site. The present study deals with the preliminary physico-chemical characteristics and heavy metals in the river water and river sediment, which exhibits the status of water quality and transfer of heavy metals from river water to sediment.

Introduction

The impact of anthropogenic pollution from industrial, agricultural, sources, quarrying and tourists activity on water quality has concerned environmentalists and scientists for the past three decades. Ecological, geo-chemical and hydrological research has been carried out in various ecosystems to understand the factors controlling the chemistry of natural water (Baron and Bricker, 1990; Malik and Bharti, 2005a). Many of such studies have been under taken during the last three decades to understand the processes that control the hydrochemistry of alpine and sub-alpine systems of North America and Europe (William *et al.*, 1993 and Psenner, 1989). Indian mountains are the cradle of a large number of streams and mighty rivers (Malik and Bharti, 2005a). In North-East region, a hill-stream Hatmawdon, 120 Km. far away from Shillong city is the study site near Bangladesh border for the accounting of physico-chemical parameters and heavy metals in natural water and fluctuations at the different locations along with the river. Adequate understanding of the North-East regional rivers is extremely important for the development of a realistic program for utilizing the potential of water that exist in the form of hidden water resource in the area.

Water quality characteristic of aquatic environment arise from a multitude of physical, biological and chemical interactions (Dezuane, 1979). The water bodies, lakes, rivers, dams and estuaries are continuously subject to a dynamic state of change with respect to the geological age and geochemical characteristics. This is demonstrated by continuous circulation, transformation and accumulation of energy and matter through the medium of living thing and their activities (Adefemi, 2007). The dynamic balance in the aquatic ecosystem is upset by human activities, resulting in pollution which meter, is manifested dramatically as fish kill, offensive taste, odour, colour and unchecked aquatic weeds. The over production of higher tropic levels biomass and the subsequent decay of dead plants could lead to oxygen depletion, death of aquatic organisms and development of anaerobic zone where bacterial action produce foul odours and bad tastes (Forstner and Wittman, 1979).

The present study deals with the characteristics the water nutrient chemistry, influenced by anthropogenic activity and quarrying of the geologically sediment environments and to determine the nature and degree of anthropogenic impacts on qualitative and quantitative variations occurred in nutrients in relation to physico-chemical parameters and heavy metals of river water and sediment.

Study Area

The Meghalaya has been surrounded by the natural beauties with lovely trees and cool climate, which is not only a pleasant place to live in but relaxing for holiday also. This is not only a popular state but also valuable or important place for tourism. Shillong is the capital of Meghalaya state. Mowlong Cherra Cement Ltd and Lafarge Umaim Mining Ltd are the major industries of the region. The all seven states of North East India are quite famous as Seven Sisters. Meghalaya is very well known for the record of rainfall, the state consists of the two places namely Cherrepunjee and Mawsinram for maximum rainfall throughout the year. So, the maximum water resources are depending chiefly on total precipitations of the region.

Basically, Hatmawdon river originated from the hills of Assam-Meghalaya. Population of the region is completely depending upon the river water for drinking, bathing, and other recreational activities. Meteorologically, region has a cool and pleasant climate. Geologically, North-east hills are enriched with various minerals and the hills near Bangladesh are rich in limestone. Geographically, the study area is situated in the globe on a Latitude 25° 10′ 29.7″ N and Longitude 91° 37′ 22.5″ E.

Materials and Methods

The water samples were collected from Hatmawdon River near Hatmawdon Village according to the analytical requirement in morning period 9:00 Hrs. to 10:00 Hrs. The samples for physico-chemical parameters and heavy metals were collected by using rinsed Borosil glassware, and analyzed with the help of the procedure described by APHA (1995) and Trivedi and Goel (1984).

Colour, odour, turbidity, velocity, temperature and dissolved oxygen were analyzed on spot at sampling sites. Samples were collected from selected sites and immediately preserved in ice boxes, and transfer to the lab for further analysis. Water samples were digested and heavy metals were detected using Atomic Absorption Spectrophotometer. Transfer Factor (TF) was calculated according to Bharti (2007) to assess the status of heavy metals transfer from river water to river sediment of Hatmawdon river.

Results and Discussion

Hatmawdon river is flowing throughout a valley of North-east hills chain near the Bangladesh Border, enriched with limestone and lignite rocks, which affect the water quality of river according to the locations. Nutrients concentration, heavy metals and related physico-chemical parameters from selected sites are depicted in Tables 13.1–13.3.

Table 13.1: Physical Characteristics of Hatmawdon River Water

Sl.No.	*Parameters*	*Unit*	*Hatmawdon River Water*		
			Winter	*Summer*	*Monsoon*
1.	Temperature	°C	12	17	16
2.	Colour	–	Clear	Clear	Clear
3.	Odour	–	Nil	Nil	Nil
4.	Turbidity	NTU	14	8	10
5.	TDS	mg/l	95	105	110

Table 13.2: Chemical Characteristics of Hatmawdon River Water

Sl.No.	*Parameters*	*Unit*	*Hatmawdon River Water*			
			Winter	*Summer*	*Monsoon*	*Mean*
1.	pH	–	7.3	8.1	7.3	7.6
2.	Alkalinity	mg/l	63	63	32	52.7
3.	Total Hardness	mg/l	74	72	26	57.3
4.	Calcium	mg/l	21	23	7	17.0
5.	Magnesium	mg/l	5	4	2	3.7
6.	Chlorides	mg/l	10	15	6	10.3
7.	DO	mg/l	10.9	11.2	8.6	10.2
8.	BOD	mg/l	2	2	1	1.7
9.	COD	mg/l	13	11	8	10.7

Hatmawdon river has the spatio-temporal variations of water temperature, which plays a vital role in all physico-biochemical reactions and self-purification capacity of aquatic system (Badola and Singh, 1981). Higher value of temperature was found 17 °C in summer and minimum 12 °C in winter season. Turbidity is striking characteristic of the physical status of the water bodies. Although in Hatmawdon river water is clear because there is no more pollution, siltation was the main source of turbidity in tributaries. Detritus and other non-organic material being added to water mass due to rainfall and anthropogenic activities (Camron, 1996). Maximum turbidity was recorded 14 NTU during winter season and minimum 8 NTU in summer season. The maximum depth of photic zone provides the better biological production for all aquatic organisms in a river (Malik and Bharti, 2005b).

Table 13.3: Heavy Metals in Hatmawdon River Water and Sediment

Sl.No.	*Heavy Metals*	*Unit*	*Ichchamati River Water*				*Hatmawdon Sediment (mg/kg)*
			Winter	*Summer*	*Monsoon*	*Mean*	
1.	Cadmium	mg/l	BDL	BDL	BDL	BDL	BDL
2.	Copper	mg/l	BDL	BDL	BDL	BDL	32
3.	Iron	mg/l	0.7	0.94	0.15	0.60	5600
4.	Lead	mg/l	BDL	BDL	BDL	BDL	38
5.	Manganese	mg/l	0.05	0.08	0.02	0.05	600
6.	Zinc	mg/l	BDL	BDL	0.08	0.03	121

Total dissolved solids were found in the range of 95 mg/l in winter to 110 mg/l in monsoon season, due to the gradual increases in velocity of river which favoured effective sedimentation (Subramanian, 1979). Chemical oxygen demand was found 8 mg/l to 13 mg/l during the study period. Chemical oxygen demand represents chemically oxidizable organic matter load in water, while biochemical oxygen demand is only biodegradable materials (Malik and Bharti, 2005c). In the present study the values observed during monsoon months may be attributed maximum biological activities and high temperature, stimulate the growth of microorganisms (William *et al.*, 1993).

The pH of natural water was controlled in a great extent by the interaction of hydroxyl ions arising from the hydrolysis of bicarbonate (Sharma, 1986). The pH of Hatmawdon river was recorded alkaline (7.3-8.1). Total hardness is mainly due to percentage of calcium and magnesium salts of bicarbonates, carbonates, sulphates and chlorides, while the value of alkalinity occured due to presence of bicarbonates. The concentration of hardness was analyzed 26-74 mg/l during the study. Alkalinity was also found 32-63 mg/l with a small fluctuation. A positive relationship between hardness and

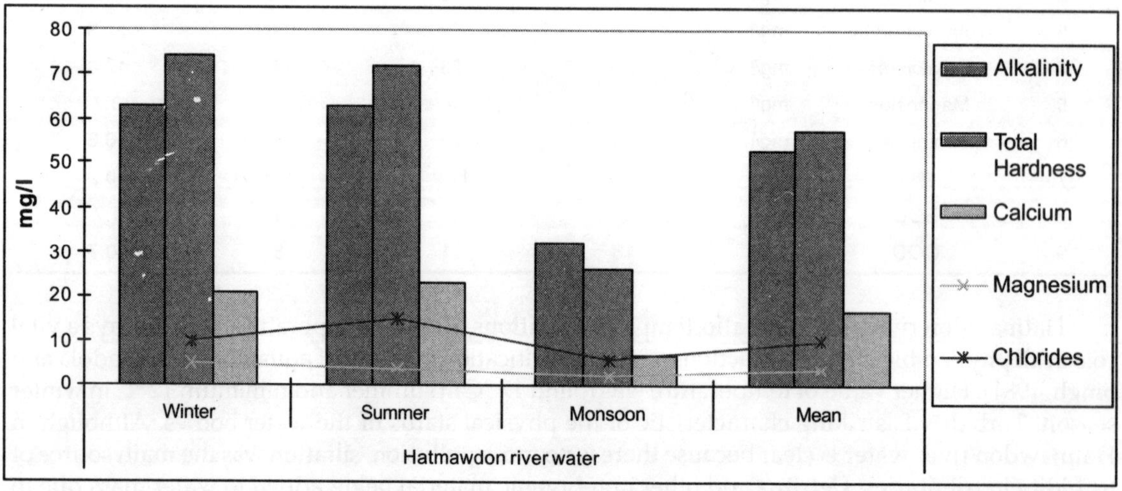

Figure 13.1: Showing the Nutrients Concentration in Hatmawdon River during the Study Period

alkalinity was recorded in river Ganga at Rishikesh (Chopra and Patric, 1994). Maximum chloride concentration was recorded maximum (15 mg/l) in summer and minimum in monsoon (6 mg/l). Chloride and hardness showed a positive relationship to one another (Chopra and Patric, 1994). Chloride was found in the form of chloride ion, and one of the major inorganic anion present in natural water (Malik and Bharti, 2009).

Calcium and magnesium the dominant cations, and these represent the main weathering products, but significant hydro-chemical differences between the two sampling sites associated with the bedrock geology exist (Jenkins *et al.*, 1995). Calcium is one of the essential nutrients, which plays an important role in biological system. Maximum calcium concentration was recorded (23 mg/l) in summer and minimum in monsoon (7 mg/l). Positive relationship between, calcium and temperature was also reported by Khanna and Singh (2000) in river Suswa, Dehradun. Magnesium is also an essential element but it is toxic at higher concentration. The concentration of magnesium in Hatmawdon river was found maximum (5 mg/l) and minimum (2 mg/l) and it was very low in comparison to Hill-streams of Uttarakhand (Bharti, 2004). During the summer season nutrients concentration in rivers and hill-streams became more. Miller *et al.* (1997) described the nutrients availability in selected environmental settings of the Potomac River and Cameron (1996) showed the similar type of fluctuation in Fraser river of British Columbia. Bond (1979) described similar nutrients concentration pattern in a stream draining a mountain ecosystem in Utah.

Dissolved oxygen was found 8.6 to 11.2 mg/l during the study period. While chemical oxygen demand was found 8 mg/l in monsoon and 13 mg/l in winter season. Heavy metals were found almost nil or in very low concentrations except iron. Heavy metals like Cadmium, copper, lead and zinc were not found in high concentration at both sites during any season. Cadmium, copper and lead were absolutely absent in all seasons, while zinc concentrations were also found below detection limit in maximum samples. The concentration of iron was maximum observed 0.94 mg/l during summer season. Zinc concentration was 0.08 mg/l monsoon season. Malik *et al.* (2009) described the role of heavy metals in the surface water of north India. The results of Bharti *et al.* (2010) were also indicated the relation of heavy metals and phytoplankton in a north Indian water body.

Table 13.4: Physico-chemical Characteristics of Hatmawdon River Sediment

Sl.No.	*Parameters*	*Unit*	*Hatmawdon Sediment*	*Method*
1.	Texture	–	Silty sand	IS: 2720 p-4
2.	Grain size analysis	per cent by mass		IS: 2720 p-17
	Sand		69	
	Silt		30	
	Clay		1	
3.	Moisture Content	per cent by mass	6.0	IS: 2720 p-2
4.	pH	–	5.4	IS: 2720 p-26
5.	Conductivity	μmho/cm	27	Conductivity meter
6.	Calcium	per cent by mass	1.1	APHA (1998)
7.	Magnesium	per cent by mass	0.6	APHA (1998)
8.	Chlorides	per cent by mass	0.2	Volhard's method

Transfer factors from river water to sediment for all metals were found quite irregular. Transfer factor for cadmium was found 0, while for copper and lead it was found constant in all seasons. It was found 8000, 5957.4 and 37333.3 in winter, summer and monsoon seasons respectively for Iron. Similarly, transfer factor was calculated 12000, 7500 and 30000 in winter, summer and monsoon seasons respectively for manganese. For zinc, transfer factor was calculated from 1512.5 to 24200. All results of transfer factors are depicted in Table 13.5. Physico-chemical parameters of river sediments are also depicted in Table 13.4.

Table 13.5: Transfer Factor of Heavy Metals from Hatmawdon River Water to Sediment

S.N.	*Heavy Metals*	*Transfer Factor*		
		Winter	*Summer*	*Monsoon*
1.	Cadmium	0.0	0.0	0.0
2.	Copper	6400.0	6400.0	6400.0
3.	Iron	8000.0	5957.4	37333.3
4.	Lead	7600.0	7600.0	7600.0
5.	Manganese	12000.0	7500.0	30000.0
6.	Zinc	24200.0	24200.0	1512.5

The present results conclude that significant differences in river water nutrient concentrations exist among different environmental settings. The spatial variations in TDS are attributed to climatic and lithological control over the ionic concentrations. Absence or low concentration of heavy metals shows that the water is still industrial pollution free. Heavy metals never cross the limits during the study period. On the basis of nutrients and heavy metals, the water may be considered for drinking and other domestic purposes.

References

APHA, 1995. *Standard Methods for Examination of Water and Wastewater*, 19th edn. American Public Health Association Inc., New York, pp. 1970.

Adefemi, O.S., Asaolu, S.S. and Olaofe, O., 2007. Assessment of the physico-chemical status of water samples from major dams in Ekiti state, Nigeria. *Pakistan Journal of Nutrition*, 6(6): 657–659.

Badola, S.P. and Singh, H.R., 1981. Hydrobiology of the river Alaknanda of Garhwal Himalaya. *Indian J. Ecol.*, 8(2): 269–276.

Baron, J. and Bricker, O.P., 1990. Hydrological and chemical flux in Loch Vale watershed, Rocky Mountain National Park. In: *Biogeochemistry of Major Rivers, SCOPE 42*. Wiley and Sons, New York, USA.

Bharti, P.K., 2004. Limnobiological study of Sahastradhara hill-stream at Dehradun. *M.Sc. Dissertation*, Gurukula Kangri University, Hardwar, pp. 102.

Bharti, P.K., 2007. Effect of textile industrial effluents on groundwater and soil quality in Panipat region (Haryana). *Ph.D. Thesis*, Gurukula Kangri University, Hardwar, pp. 191.

Bharti, P.K., Malik, D.S. and Yadav, Rashmi, 2010. Influence of heavy metals on abundance of cyanophyceae members in three spring-fed lake in Kempty, Dehradun. In: *Advances in Aquatic Ecology*, Vol. 3, (Ed.) V.B. Sakhare. Daya Publishing House, New Delhi, pp. 107–111.

Bond, H.B., 1979. Nutrient concentrations patterns in a stream draining a montane ecosystem in Utah. *Ecology*, 60(6): 1184–1196.

Cameron, E.M., 1996. Hydrogeo-chemistry of the Fraser River British Columbia: Seasonal variation in major and minor components. *J. Hydrol.*, 182(1–4): 209–255.

Chopra, A.K. and Patrick, N.J., 1994. Effect of domestic sewage on self-purification of Ganga water at Rishikesh. *A. Bio Science*, 13(11): 75–82.

Dezuane, J., 1979. *Handbook of Drinking Water Quality*. Indiana University Press, pp. 3–17.

Forstner, U. and Wittman, G.T.W., 1979. *Metal Pollution in the Aquatic Environment*. Springer-Verlag, Berlin, pp. 486.

Jenkins, A., Sloan, W.T. and Cosby, B.J., 1995. Stream chemistry is the middle hills and high mountain of the Himalaya, Nepal. *Journal of Hydrology*, 166(1–4): 61–79.

Khanna, D.R. and Singh, R.K., 2000. Seasonal fluctuations in the plankton of Suswa River at Raiwala Dehradun. *Env. Conservations J.*, 1(2 and 3): 89–92.

Malik, D.S. and Bharti, Pawan K., 2005a. Nutrient dynamics in Rhithron zone of Shivalik Himalayan stream Sahastradhara, Dehradun (Uttaranchal). *Env. Cons. J.*, 6(2): 63–68.

Malik, D.S. and Bharti, P.K., 2005b. Fluctuation in planktonic population of Sahastradhara hill-stream at Dehradun (Uttaranchal). *Aquacult.*, 6(2): 191–198.

Malik, D.S. and Bharti, P.K., 2005c. Primary production efficiency of Sahstradhara hill-stream, Dehradun, *Env. Cons. J.*, 6(3): 117–121.

Malik, D.S. and Bharti, P.K., 2009. Ecology of Sahastradhara hill-stream at Dehradun (Uttaranchal), In: *Advances in Aquatic Ecology*, Vol. 1, (Ed.) V.B. Sakhare. Daya Publishing House, New Delhi, pp. 1–11.

Malik, D.S. and Bharti, P.K., Negi, K.S. and Yadav, Rashmi, 2009. Distribution of metals in water of an artificial lake at Mussoorie, Uttarakhand. In: *Aquatic Biology and Aquaculture*, (Ed.) V.B. Sakhare. Manglam Publication, New Delhi, pp. 77–95.

Miller, C.V., Denis, J.M., Ator, S.W. and Brakebill, J.W., 1997. Nutrients in stream during baseflow in selected environmental settings of the Potomac river basin. *J. American Wat. Resources Association*, 33(6): 1155–1171.

Psenner, R., 1989. Chemistry of high mountain lakes in siliceous catchments of central Alps. *Aquatic Sci.*, 51: 108–128.

Shrama, R.C., 1986. Effect of physico-chemical factors on benthic fauna of Bhagirathi River Garhwal Himalaya. *Indian J. Ecol.*, 13(1): 133–137.

Subramanian, V., 1979. Chemical and suspended sediment characteristics of river of India. *J. Hydrol.*, 44: 37–55.

Trivedi, R.K. and Goel, P.K., 1984. *Chemical and Biological Methods for Water Pollution Studies*. Environmental Publication, Karad, pp. 1–298.

William, M.W., Brown, A. and Melack, J.M., 1993. Geochemical and hydrologic controls on the composition of surface water in the high elevation basin, *Sierra Navada. Limnol. Oceanogr.*, 38: 775–797.

Xue, H.B. and Schooner, J.L., 1994. Acid deposition and lake chemistry in southwest China. *Wat. Air, Soil Pollut.*, 75: 61–78.

Chapter 14

Benthic Biodiversity in the Marine Zone of Vellar Estuary, Southeast Coast of India

☆ *P. Murugesan and S. Muthuvelu*

Introduction

Benthic organisms constitute an essential component in the marine environment and play an important role in the ecology both as consumers of plankton and as food for bottom feeding fin and shellfishes (Parulekar *et al.*, 1980). They provide key linkages between primary producers and higher trophic levels in the marine food chains.

Along the coastal regions of India, a number of estuaries, backwaters and mangroves are found and benthos here vouch safe for the extraordinarily fertile nature of these habitats. Studies by Gaufen and Tarzwell (1952) on benthos led to the development of the indicator organism concept that is the presence of a particular species or a group of species at a given locality reflects the state of a particular environment. Benthic macro invertebrates are useful in indicating the degree of contamination in a marine area. Their presence or absence indicates environmental conditions for sometime previous to sampling. Not only do benthic invertebrates serve as indicators of the past conditions but they also serve as indicators of change. The benthic fauna of a clean area is composed of representatives from different feeding groups as herbivores, carnivores, detrital feeders etc. With increasing degrees of pollution, all feeding groups will die and are excluded from recolonisation except the detrital feeders. Benthos have also been used to remove the organic wastes from aquaculture systems (Tenore *et al.*, 1974) and as toxicological test organisms (Reish, 1980). The majority of species are essentially sedentary and therefore changes in their community structure and diversity can be examined in relation to inputs of pollutants (Warwick, 1993; Warwick and Clarke, 1993).

The Vellar estuary is a well-studied estuary of India and comparable to the best studied estuaries of the world. It is highly productive and rich in fishes, prawns, crabs, hermit crabs and bivalves. The marine zone of Vellar estuary, selected for the present study is in the vicinity of mouth of the estuary (Thangaraj *et al.*, 1979; Jeyabalan *et al.*, 1980) and hence this marine zone is supremely important with regard to ecological studies.

Understanding well the importance of benthos in general and polychaetes in particular, the present investigation rather a fact finding mission was made to assess the diversity of benthic fauna and to ascertain the role of physico-chemical parameters on the distribution and relative abundance of benthic organisms using modern statistical tools in the marine zone of Vellar estuary.

Materials and Methods

The sampling methods and procedures were designed in such a way to maximize the usefulness of the data obtained. To achieve this, due attention was paid to obtain specimens in the best possible condition. This helped in sorting, identifying, enumerating and processing of data using the univariate and multivariate techniques. Benthic sampling was done in the marine zone of Vellar estuary for a period of one year from October 1998 to September 1999. The sampling was made in the following five transects (Figure 14.1):

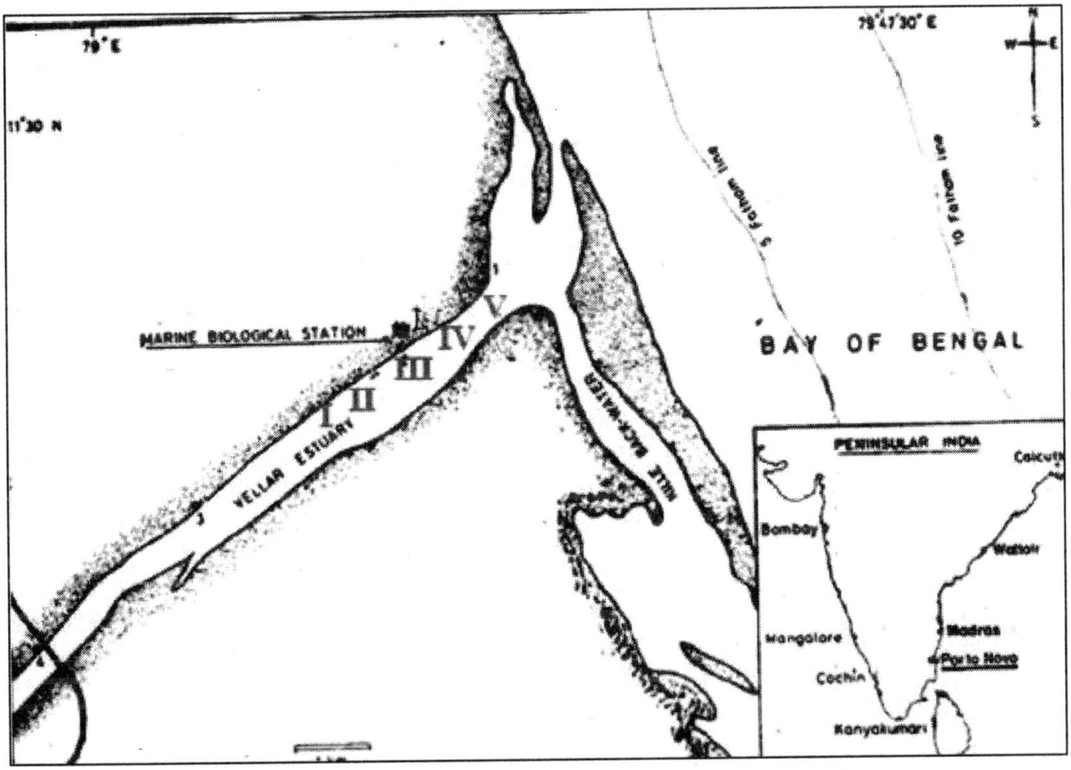

Figure 14.1: Map Showing the Study Area (Transects I–V)

(*i*) Transect I (Lat.11 ° 29' 663" N; Long. 79° 46' 531" E), it lies 10 m east of junction where Killai backwaters joins the Vellar estuary; (*ii*) Transect II (Lat.11 ° 29' 716" N; Long. 79° 46' 586" E), it lies 50 m east of transect I; (*iii*) Transect III (Lat. 11 ° 29' 765" N; Long. 79° 46' 627" E), it lies 50 m east of transect II; (*iv*) Transect IV (Lat. 11 ° 29' 837" N; Long. 79 ° 46' 675" E), it lies 50 m east of transect III and (*v*) Transect V (Lat. 11° 29' 907" N; Long. 79° 46' 699" E), it lies near the mouth region of Vellar estuary.

In each transect, benthic samples were collected in 10 spots across the estuary at fortnightly intervals during high tide. Quantitative sampling was carried out using a long armed Peterson grab. This type of grab is considered to be the most efficient in obtaining good penetrative samples on harder sediments. The grab employed was measured and found to take a sample covering an area of 0.0251m². Two replicate samples were taken at each station and the larger one was taken into account. After collecting samples, they were emptied into a plastic tray. The larger organisms were picked out immediately from the sediment and then sieved through 0.5 mm mesh screen. The procedure adopted for sampling was following the standard method proposed by Mackie (1994).

The organisms retained by the sieve were placed in a labelled container and fixed in 5 per cent formalin. Subsequently, the organisms were stained with Rose Bengal solution (0.1g in 100 ml of distilled water) for enhanced visibility during identification. Once the fixative was added, each sealed sample container was gently upturned and rotated to distribute the formalin evenly throughout the sieved sample. At all stages of the sieving, care was taken to individually remove noticeably fragile animals (*e.g.*, phyllodocids and terebellids).

Once back at the laboratory, the sieved samples were gently but thoroughly washed in freshwater. This removed the formalin and salt, preventing the former from dissolving the shells of delicate molluscs. The samples were then preserved in 75 per cent alcohol for further identification. All the specimens were sorted, enumerated and identified to the advanced level possible with the consultation of available literature. The works of Fauvel (1953), Day (1967), Srikrishnadhas *et al.* (1998) were referred for polychaetes, Barnes (1980) for crustaceans, Lyla *et al.* (1999) for amphipods, Rajagopal *et al.* (1998) for gastropods and Shanmugam *et al.* (1997) for bivalves.

Water quality parameters were measured using standard methods. Temperature was measured using a thermometer with ±0.5°C accuracy; salinity by Hand Refracto meter (Atago co. Ltd, Japan); pH by pH pen (Eutech Instrument, Singapore) and dissolved oxygen was estimated using Winkler's titration method (Strickland and Parsons, 1986). Organic carbon content in sediment samples was analyzed using the standard method of el Wakeel and Riley (1956).

The data were pooled together for seasons and approached to various statistical methods namely, univariate, graphical/distributional and multivariate analyses using the statistical software PRIMER Ver. 6.0 (Clarke and Gorley, 2001).

Results

Temperature

The water temperature recorded in transects is given in Figure 14.1. In transect I, the temperature varied from 27.2°C (October) to 30.4°C (June). In transect II, it fluctuated from 27.2°C (November) to 30.5°C (June). In transect III, it ranged between 27.5°C (November) and 30.6°C (June). In transect IV, the values varied between 27.8°C (November) and 30.4°C (June). At transect V, it fluctuated from 27.8°C (October) to 32.8°C (June). In general, similar seasonal pattern in temperature was noticed in all transects with minimum values during monsoon and maximum values during summer. The temperature which was high in transect V, decreased in transect IV then increased in transect III and decreased through transect II to transect I (Figure 14.2).

Salinity

In transect I, the salinity varied between 8 (November) and 34.1 psu (June). In transect II, it varied from 9.80 (October) to 33.5 psu (June). At transect III, it ranged from 11.2 (October) to 33.5 psu (June). In

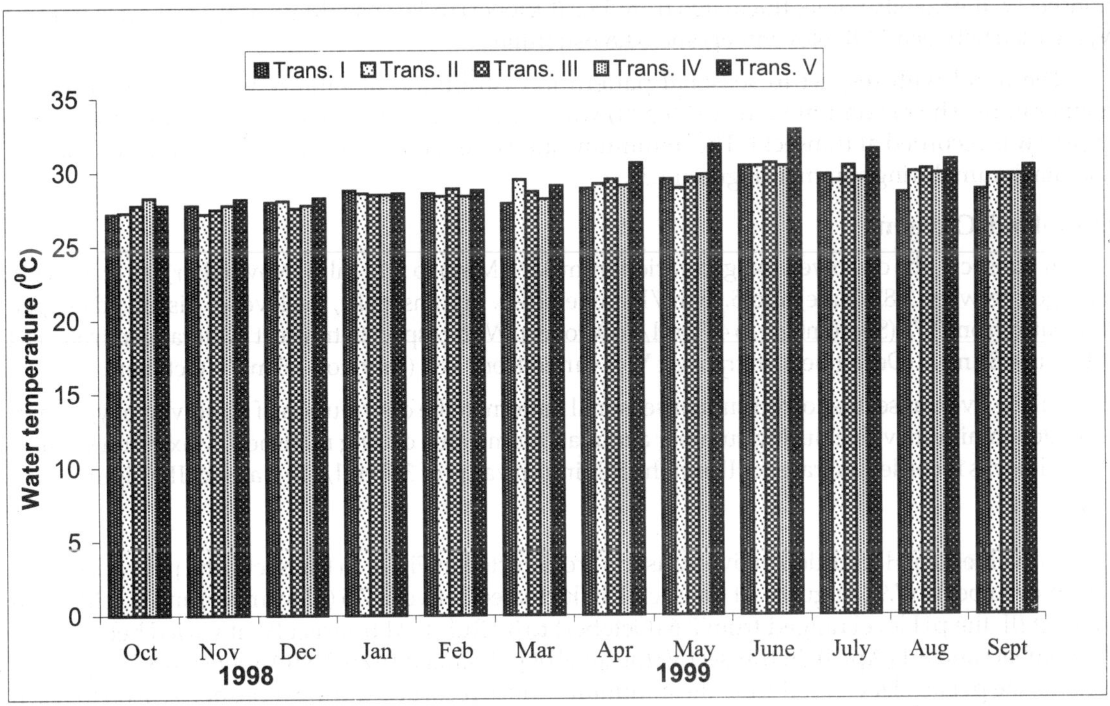

Figure 14.2: Water Temperature Recorded in Transects I–V during the Study Period

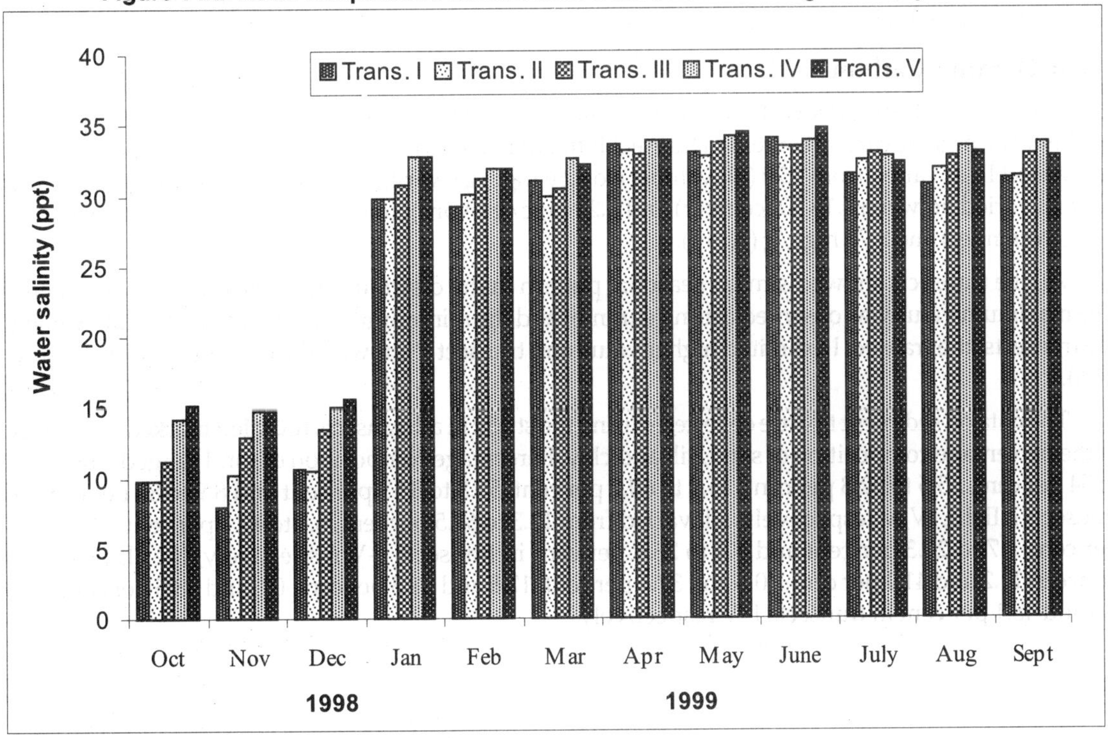

Figure 14.3: Monthly Variations in Water Salinity Observed during the Study Period in Transects I-V

transect IV, the salinity values fluctuated from 14.2 (October) to 34.2 psu (May). With respect to transect V, it varied between 14.8 (November) and 34.8 psu (June).

The trend with respect to seasonal pattern in distribution of salinity was similar to that of temperature. The maximum value (34.8 psu) was recorded in transect V. and the minimum value (8 psu) was recorded at transect I. The minimum values were recorded during the monsoon months and maximum during summer (Figure 14.3).

Dissolved Oxygen

In transect I, the dissolved oxygen varied from 3.89 (May) to 5.12 ml/l (November). At transect II, it ranged between 3.82 (June) and 5.15 ml/l (November). In transect III, the level of dissolved oxygen fluctuated from 3.73 (September) to 4.72 ml/l (October). With respect to transect IV, it ranged from 3.84 (May) to 4.81 ml/l (December). In transect V, it varied from 3.81 (July) to 4.65 ml/l (October).

All the five transects showed similar seasonal pattern in the distribution of dissolved oxygen and exhibited minimum value during summer and maximum value during monsoon. Maximum value of 5.15 ml/l was recorded in transect II and the minimum value of 3.73 ml/l in transect III (Figure 14.4).

pH

The values of pH recorded in five transects are depicted in Fig. 4. In transect I, the pH varied from 7.52 (November) to 7.9 (August). At transect II, it fluctuated between 7.5 (November) and 8.1 (May). In transect III, the pH level ranged from 7.6 (October) to 8.1(July). At transect IV, it varied between 7.6 (November) and 8.1 (April). In transect V, the level of pH ranged from 7.6 (October) to 8.2 (April). In general, the pH level increased from the minimum values recorded during monsoon season to reach the maximum during summer. The maximum value was recorded in transect V and minimum value in transect II (Figure 14.5).

Total Organic Carbon

In transect I, the organic carbon content of the sediment varied from 2.92 (October) to 10.34 mgc/g (May). The organic carbon content in transect II fluctuated from 2.85 (October) to 10.15 mgc/g (May). At transect III, it varied from 2.70 (November) to 9.66 mgc/g (April). In transect IV, the organic carbon content varied between 2.18 (December) and 9.20 mgc/g (April). In transect V, it fluctuated between 1.88 (December) and 8.78 mgc/g (April).

All the transects showed similar seasonal pattern in the distribution of organic carbon content. The maximum value was observed during summer and the minimum value during monsoon. Among the transects, the transect I exhibited higher value and transect V showed the minimum value (Figure 14.6).

The data on sediment texture collected during the study in all transects revealed marked variations in the percentage composition of sand, silt and clay. Percentage composition of sand ranged from 14.3 to 54 per cent, 35.5 to 62.8 per cent, 15.9 to 60.7 per cent, 55.8 to 83.3 per cent and 8.9 to 83 per cent in transects I, II, III, IV, V respectively; silt varied from 13.3 to 60.5 per cent, 6.1 to 42.4 per cent, 11 to 43.5 per cent, 5.7 to 22.3 per cent and 5.4 to 26.4 per cent in transects I–V respectively; the clay ranged between 11.2 and 41.1 per cent, 10.5 and 37.1 per cent, 17.8 and 61.3 per cent, 0.6 and 38.5 per cent and 0.5 and 48.1 per cent in transects I–V respectively.

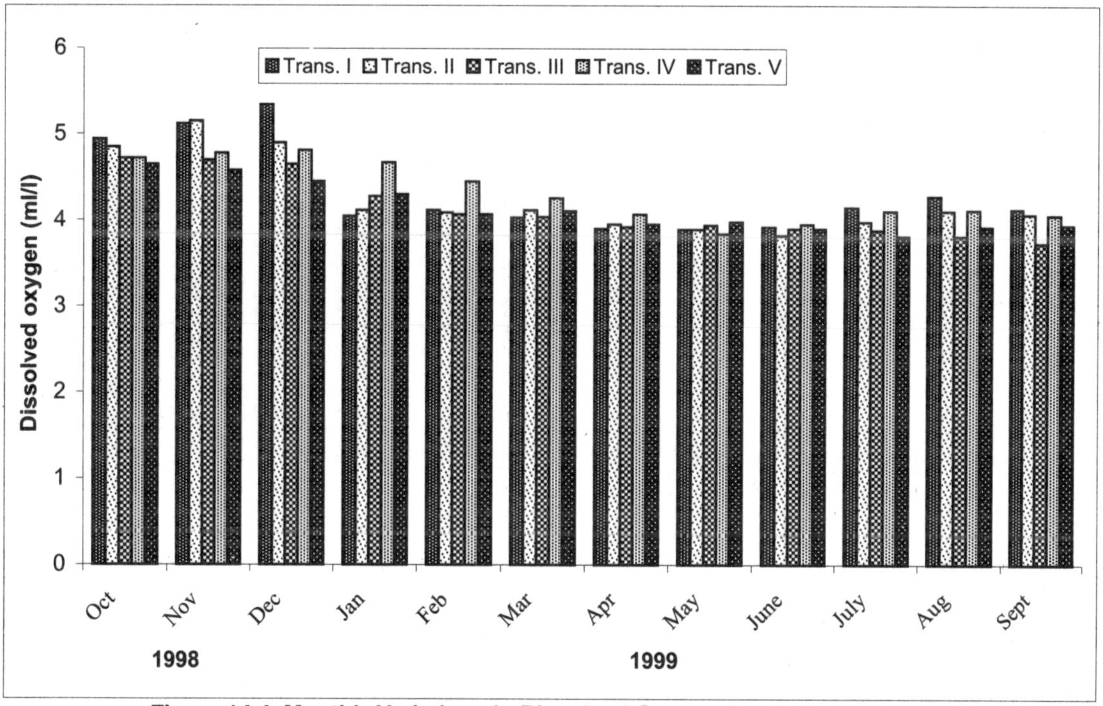

Figure 14.4: Monthly Variations in Dissolved Oxygen Recorded during the Study Period in Transects I–V

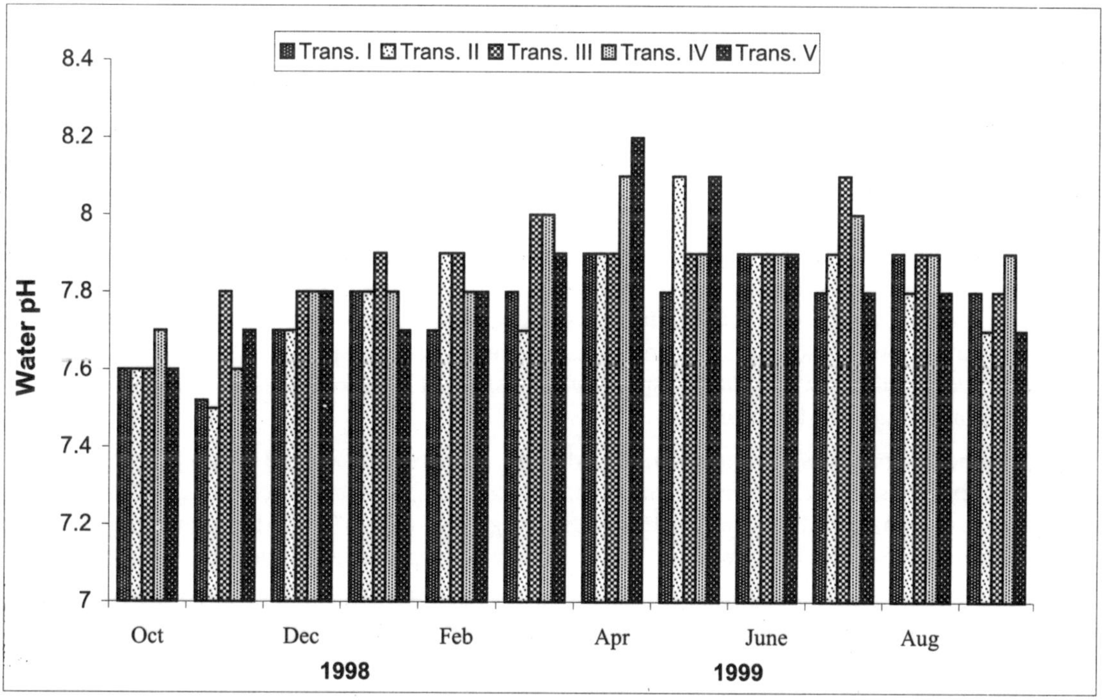

Figure 14.5: Monthly Variations in Water pH Recorded during the Study Period in Transects I–V

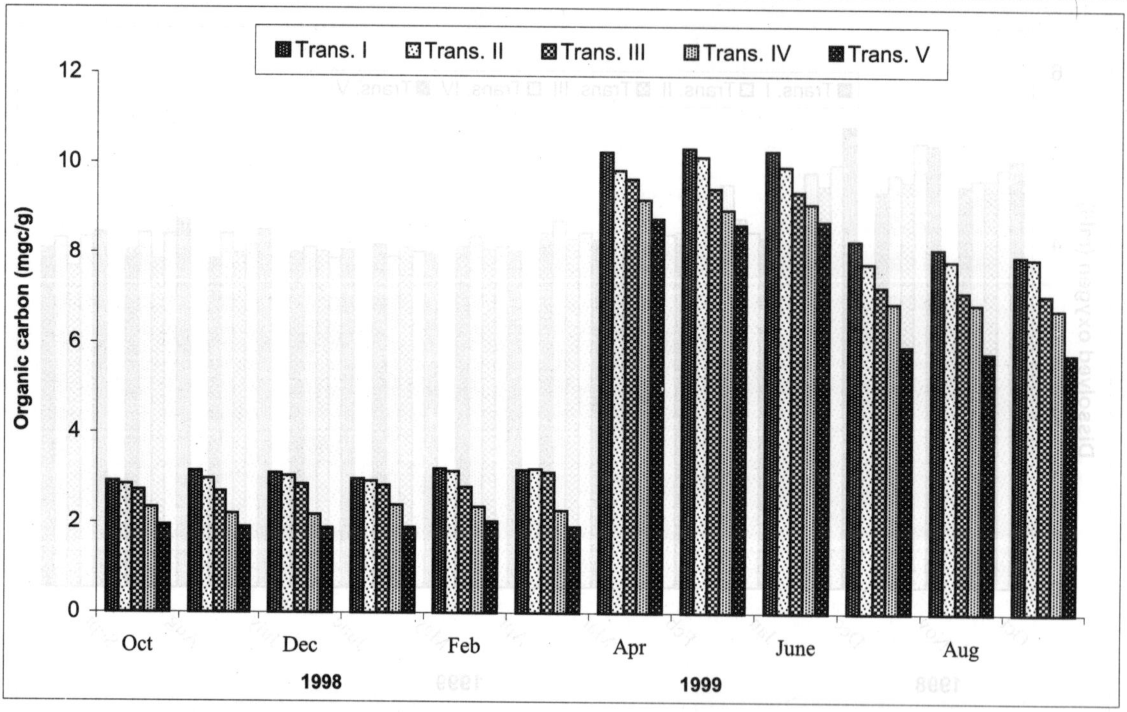

Figure 14.6: Monthly Variations in Total Organic Carbon Recorded during the Study Period in Transects I–V

Species Composition of Macrofauna

As many as 76 species of macrofauna were encountered during the present study from all the 5 transects. Of the 76 species recorded, polychaetes were found to be the largest component in the collection with 37 species. Crustaceans were found to be the next dominant group in the order of abundance with 18 species. The bivalves and gastropods ranked third and fourth respectively with 11 and 8 species. The group "others" came last in the order with 2 species. Among polychaetes, *Nephtys polybranchia, Ancistrosyllis constricta, Prionospio polybranchia, P. pinnata, P.cirrifera, Glycera* sp., *Lumbriconereis simplex, Cossura delta, Ceratonereis costae, Euclymene annandalei, Heteromastus similis, Malacoceros indicus, Diopatra neapolitana, Cirratulus* sp., *Syllis* sp. were found to occur in all the transects round the year. Among crustaceans, the dominant species as *Tanaeus* sp., *Apseudes killaiyensis, Calanus* sp., *Quadrivisio bengalensis, Grandiderella* sp., *Gammaropsis* sp., *Eriopisa chilkensis, Diogenes avarus* were found common throughout the year.

With respect to bivalves, *Meretrix meretrix, M. casta, Anadara granosa* and *Katelysia opima* and the gastropods species such as *Cerithidea cingulata, Turritella* sp., *Umbonium vestiarium* showed their consistency in their distribution in all the transects. The group "others" included some miscellaneous species of brittle stars, foraminiferans and sea anemones which were infrequent in their occurrence.

The population density of benthic organisms varied between 635 and 5,125 organism per square metre. The maximum was during summer in transect I and minimum during pre-monsoon season in transect IV (Figure 14.7).

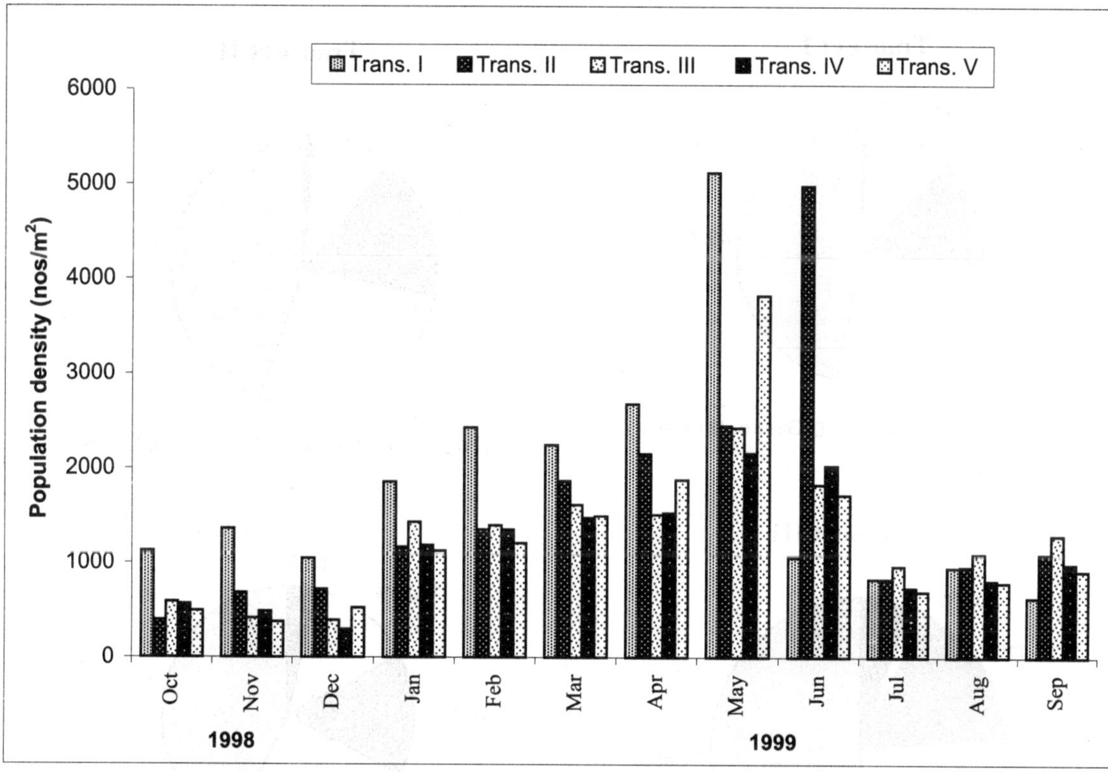

Figure 14.7: Monthly Variations in Population Density of Macrobenthos Recorded in Transects I–V

With regard to percentage composition of macro fauna, in transect I, polychaetes were found to be the dominant group by constituting 46.7 per cent of the total benthic organisms recorded. Crustaceans formed the second dominant group with a percentage occurrence of 27 per cent. Bivalves and gastropods respectively with a percentage contribution of 13.8 per cent and 12 per cent came next. The group "others" came last in the order of dominance with a meagre percentage of 0.4 per cent. As in transect I, at transect II also polychaetes topped with a percentage incidence of 45 per cent of the total benthic organisms enumerated. Crustaceans ranked second with a percentage of 27.1 per cent. Bivalves and gastropods contributed respectively 17.2 per cent and 9.5 per cent to the total benthic organisms collected. The contribution of group "others" was 1.2 per cent. With respect to transect III, polychaetes occupied the top place with a percentage of 43.6 per cent. The crustaceans were found to be the next best with a percentage contribution of 21 per cent. Bivalves and gastropods constituted 17.2 per cent and 13.6 per cent to the total benthic organisms collected. Compared to the other two transects, the contribution of group "others" (4.6 per cent) here was more.

Coming to the transect IV, polychaetes continued to be the dominant group and constituted 42.7 per cent followed by crustaceans, bivalves, gastropods and "others" with a percentage occurrences of 24.7 per cent, 15.4 per cent, 13.2 per cent and 4 per cent respectively.

At transect V also, polychaetes outnumbered with a percentage contribution of 47.5 per cent. As in other transects, crustaceans were found to be the next best with a contribution of 20 per cent. Bivalves and gastropods came next with the percentage occurrences of 15.4 per cent and 14.0 per cent respectively. The contribution of group "others" was 3.2 per cent (Figure 14.8).

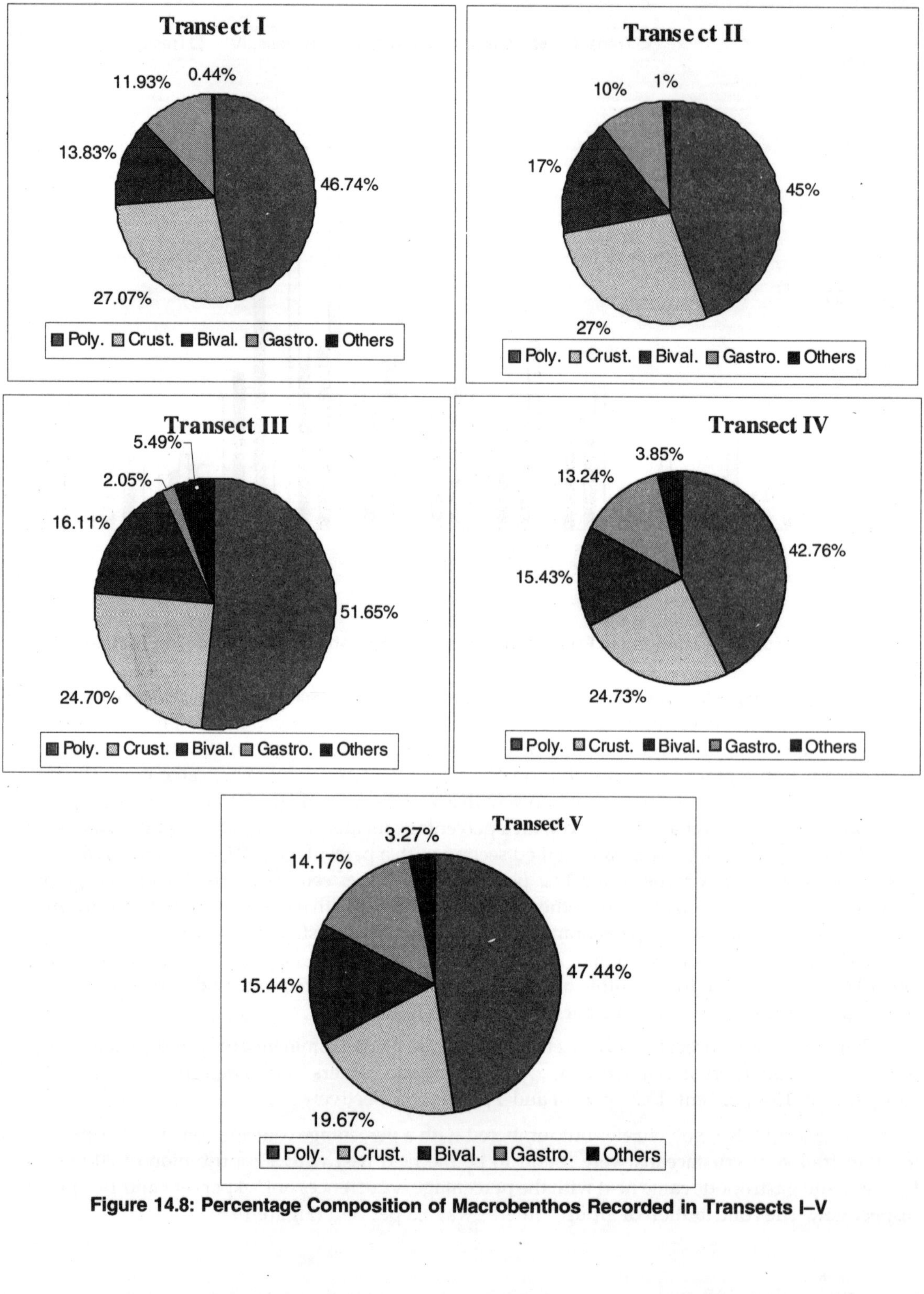

Figure 14.8: Percentage Composition of Macrobenthos Recorded in Transects I–V

Shannon–Wiener index (H′)

During monsoon, the minimum value (2.24) was recorded in transect V and the maximum (2.88) in transect I (Figure 14.9a). During post monsoon, the maximum value (4.33) was observed in transect I and the minimum (3.872) in transect IV (Figure 14.9b). During summer the minimum (4.010) value was encountered in transect V and maximum (4.474) in transect I (Figure 14.9c). In premonsoon season, the minimum value (3.022) was observed in transect V and maximum value in transect I (Figure 14.9d). The diversity which was more in transect I during all the seasons declined through other transects and the minimum was generally recorded in transect V.

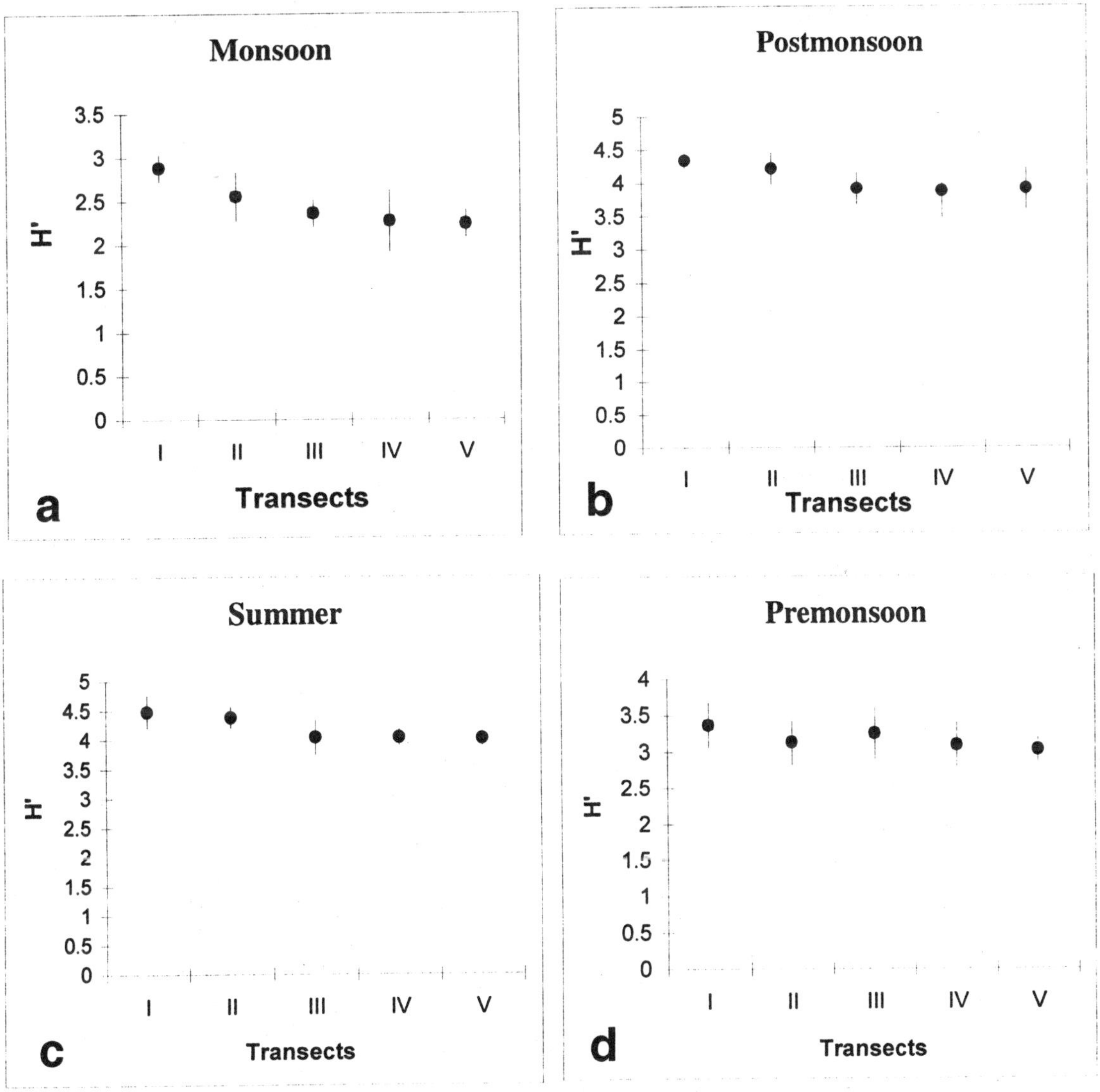

Figure 14.9a-d: Means and 95 per cent Confidence Intervals of Shannon Diversity (H') Recorded during Various Seasons in Transects I–V

Margalef Index (d)

During monsoon, the minimum value (0.787) was recorded in transect V and the maximum value (1.777) in transect I (Figure 14.10a). During postmonsoon the minimum value (2.144) was recorded in transect IV and maximum (2. 827) in transect I (Figure 14.10b). In summer season, the minimum value was observed (2.308) in transect V and maximum (3.118) in transect I (Figure 14.10c). During premonsoon, the minimum value was (1.255) found in transect V and maximum value (1.701) in transect I (Figure 14.10d).

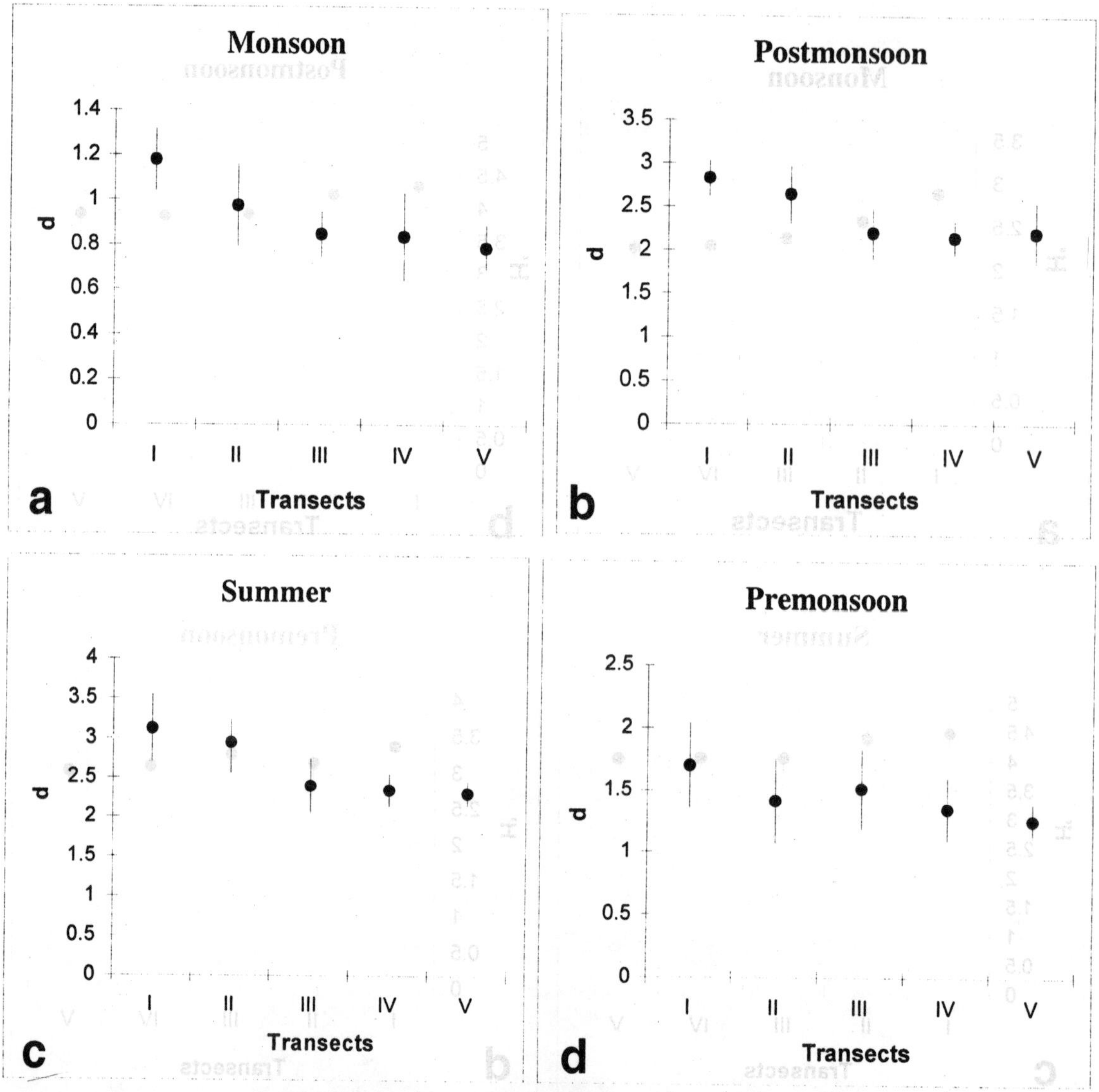

Figure 14.10a-d: Means and 95 per cent Confidence Intervals of Margalef Index (d) Recorded during Various Seasons in Transects I–V

Pielous Evenness Index (d)

As in the other indices, the means and 95 per cent confidence intervals of evenness values were also computed and presented season wise in Figure 14.11. During monsoon, the minimum value was (0.872) recorded in transect IV and the maximum value (0.910) in transect I (Figure 14.11a). During postmonsoon, the minimum value was (0.956) recorded in transect IV and the maximum value (0.968) in transect II (Figure 14.11b). In summer season, the minimum value was (0.966) recorded in transect IV and the maximum (0.970) in transect I (Figure 14.11c). During premonsoon, the minimum value was (0.926) recorded in transect I and the maximum value (0.945) in transect V (Figure 14.11d).

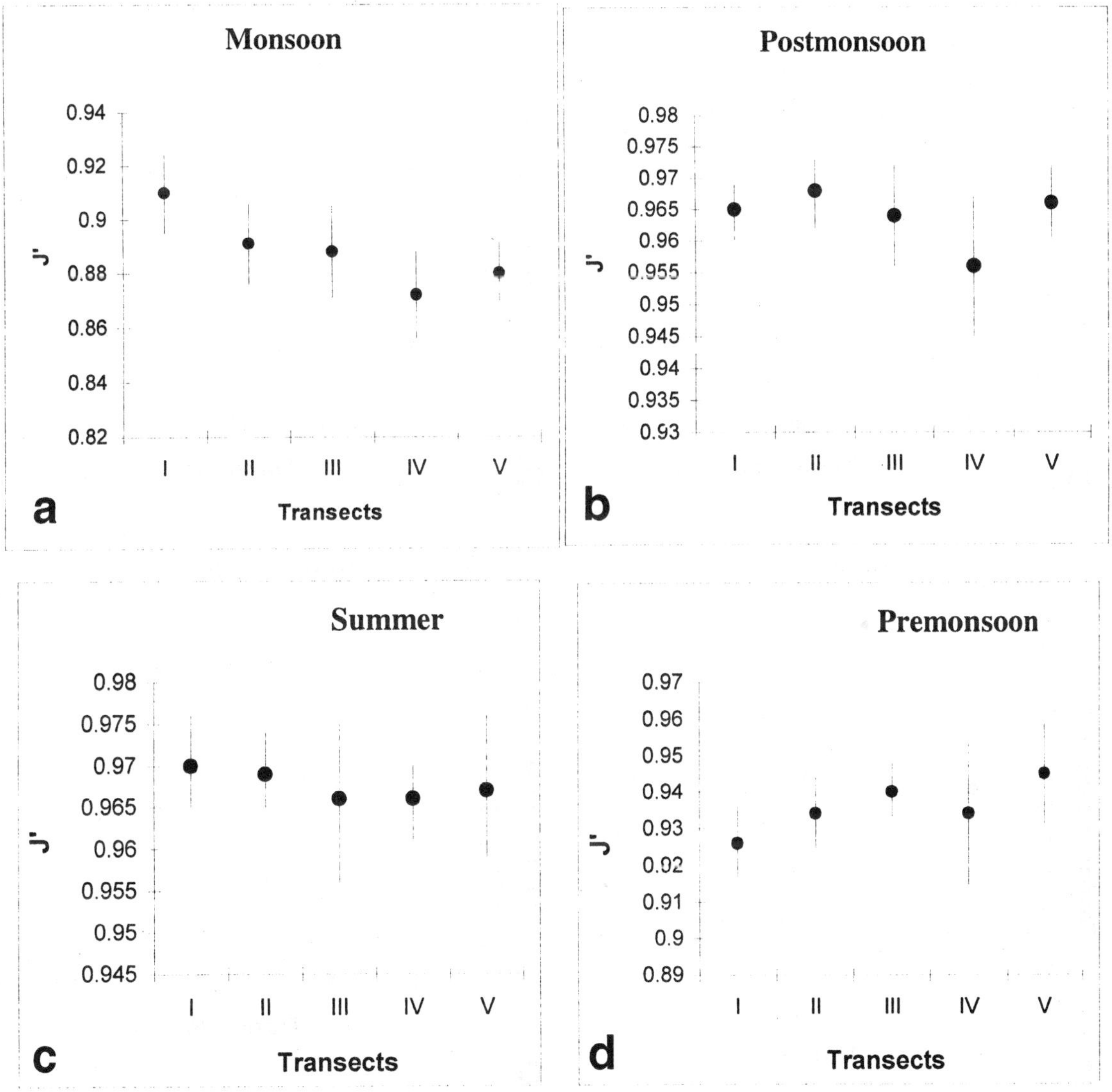

Figure 14.11: Means and 95 per cent Confidence Intervals of Evenness Values (J') Recorded during Various Seasons in Transects I–V

Taxonomic Diversity Index (d)

During monsoon, the values of taxonomic diversity (D) ranged from 64.205 to 75.352. The minimum value was found in transect V and maximum in transect I (Figure 14.12a). In post monsoon season, the delta values were from 80.489 to 81.463 with minimum value in transect III and maximum in transect I (Figure 14.12b). During summer season, the results of taxonomic diversity varied from 81.770 to 82.876. The lowest value was in transect I and the highest in transect III (Figure 14.12c). With

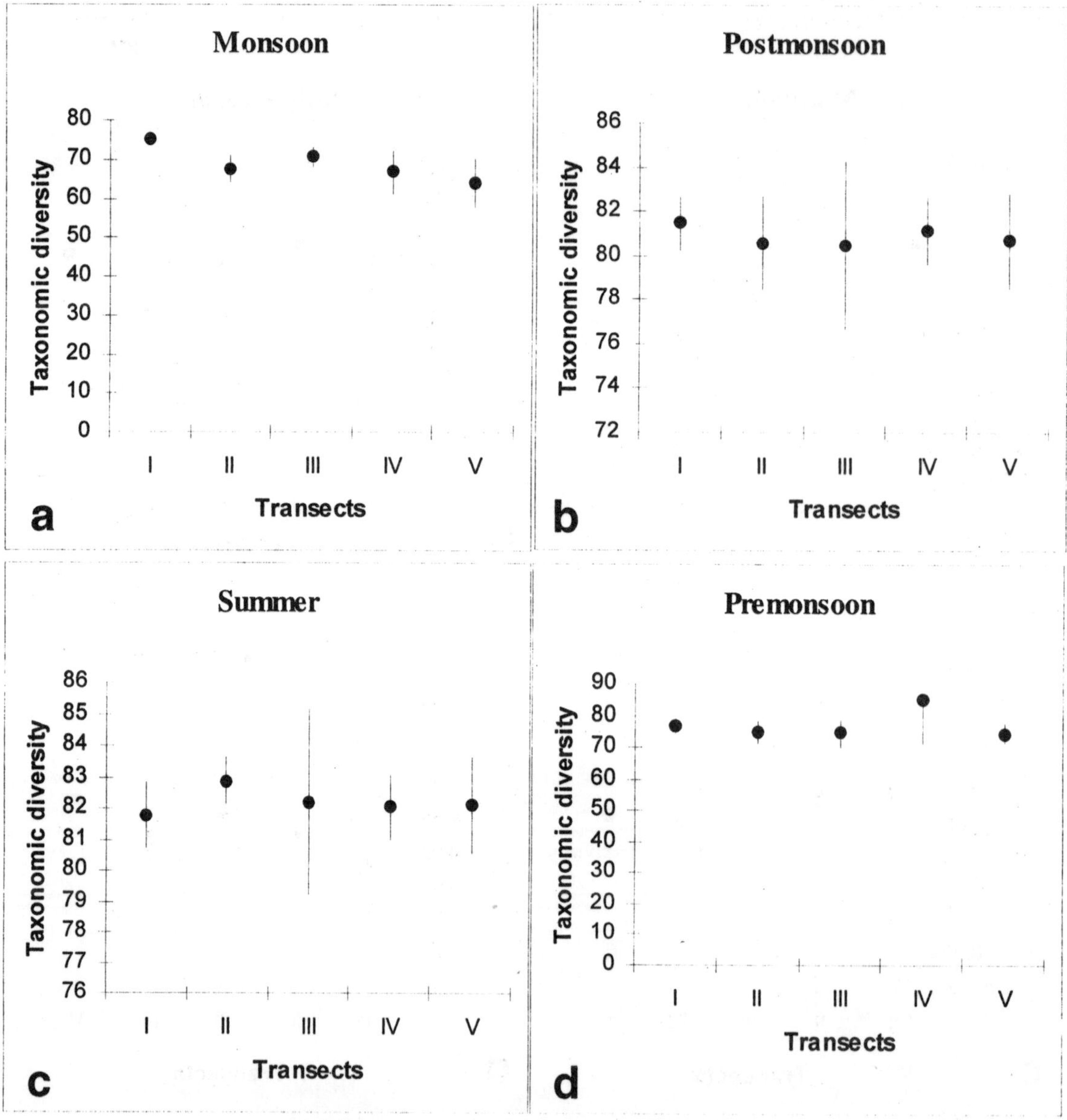

Figure 14.12: Means and 95 per cent Confidence Intervals of Taxonomic Diversity (D) Recorded during Various Seasons in Transects I–V

respect to premonsoon season, the taxonomic diversity values fluctuated between 74.444 and 76.742. The minimum value was noticed in transect V and maximum in transect I (Figure 14.12d).

Phylogenetic Diversity Index (d)

In the present study, the total phylogenetic diversity was computed for all the transects. As in the other univariate indices, the means and 95 per cent confidence intervals were calculated and graphed seasonwise in Figure 14.13.

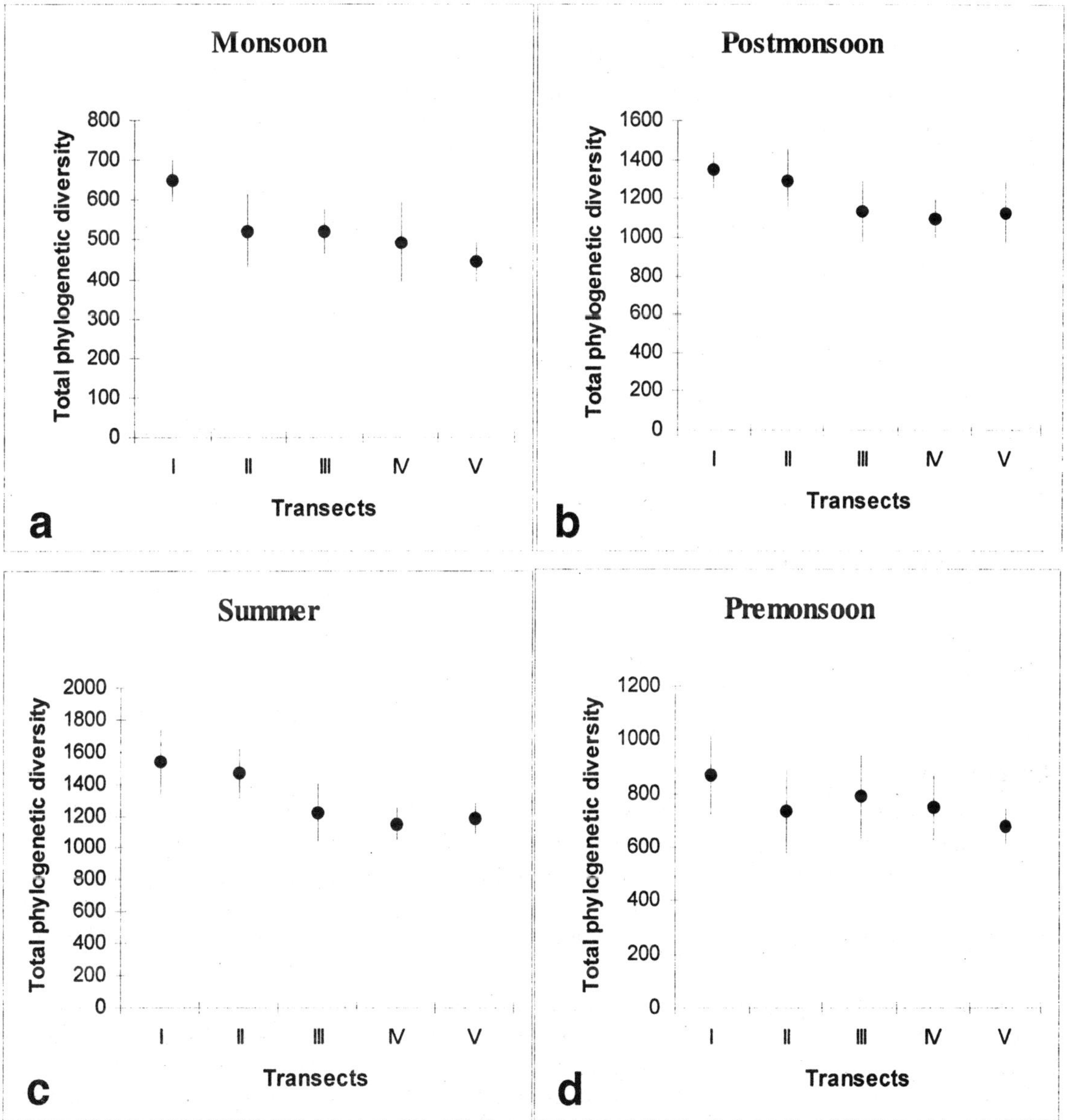

Figure 14.13: Means and 95 per cent Confidence Limits of Total Phylogenetic Diversity Values (S F^+) Recorded during Various Seasons in Transects I–V

During monsoon season, the total phylogenetic diversity ranged from 444 to 648 with minimum value in transect V and maximum in transect I (Figure 14.13a). During post-monsoon, the values fluctuated between 1098 and 1346 with minimum value in transect IV and maximum value in I (Figure 14.13b). In summer months, the total phylogenetic diversity was found to vary from 1150 to 1542. The minimum value was in transect IV and maximum in I (Figure 14.13c). During pre-monsoon period, the values were from 680 to 866 with the lowest value in transect V and the highest in transect I (Figure 14.13d).

The trend emerged in the species diversity was quite true in this index as well and the values were more during all the seasons in transect I and low in transects IV and V.

In the present study, the data pertaining to species abundance and biomass were allowed as inputs to ABC-curve in all transects to see whether they are subjected to any form of disturbances or not. The results are shown graphically.

ABC-curves drawn for the five transects showed that the biomass curve to lie above the abundance curve for the entire length, indicating the dominance of biomass by large species (Figure 14.14 a-e).

In the BIO–ENV method, which was done to measure the agreement between the rank correlations of the biology (Bray–Curtis similarity and environmental matrices Euclidean distance), seven environmental variables such as temperature, salinity, pH, sand, silt, clay and organic carbon were allowed to match the biota of all transects.

In this case, the sand, silt and organic carbon were featured as the major variables explaining the best match (0.68) with faunal distributions, followed by salinity and temperature in influencing the benthic faunal distribution (Table 14.1).

Table 14.1: Harmonic Rank Correlations (rw) between Faunal and Environmental Similarity Matrices

Sl.No.	*No. of Variables*	*Best Variable Combinations*	*Correlation (rw)*
1	3	Sand-silt-organic carbon	0.68
2	4	Silt-clay-salinity-temperature	0.68
3	3	Silt-clay- organic carbon	0.67

Discussion

The surface water temperature follows largely the trend of air temperature. As implied earlier, the temperature in this unstable and unpredictable estuarine environment is largely influenced by the seasonal as well as diurnal changes in air temperature, intensity of solar radiation, evaporation, insolation and freshwater ingression. The temperature is known to influence the chemical characteristics of interstitial waters thereby determining the occurrence and distribution of benthic organisms.

In the present study, the water temperature showed monsoonal minimum and summer maximum as observed earlier by various workers in the southeast coast of India. The low temperature during the wetter months might be due to inclement weather and overcast sky. This is in conformity with the study of Sankaranarayan *et al.* (1978). The maximum salinity value was recorded during summer and the minimum during monsoon period. Higher values during summer could be ascribed to the higher degree of evaporation in the study area and less tidal action with decreased freshwater inflow. The

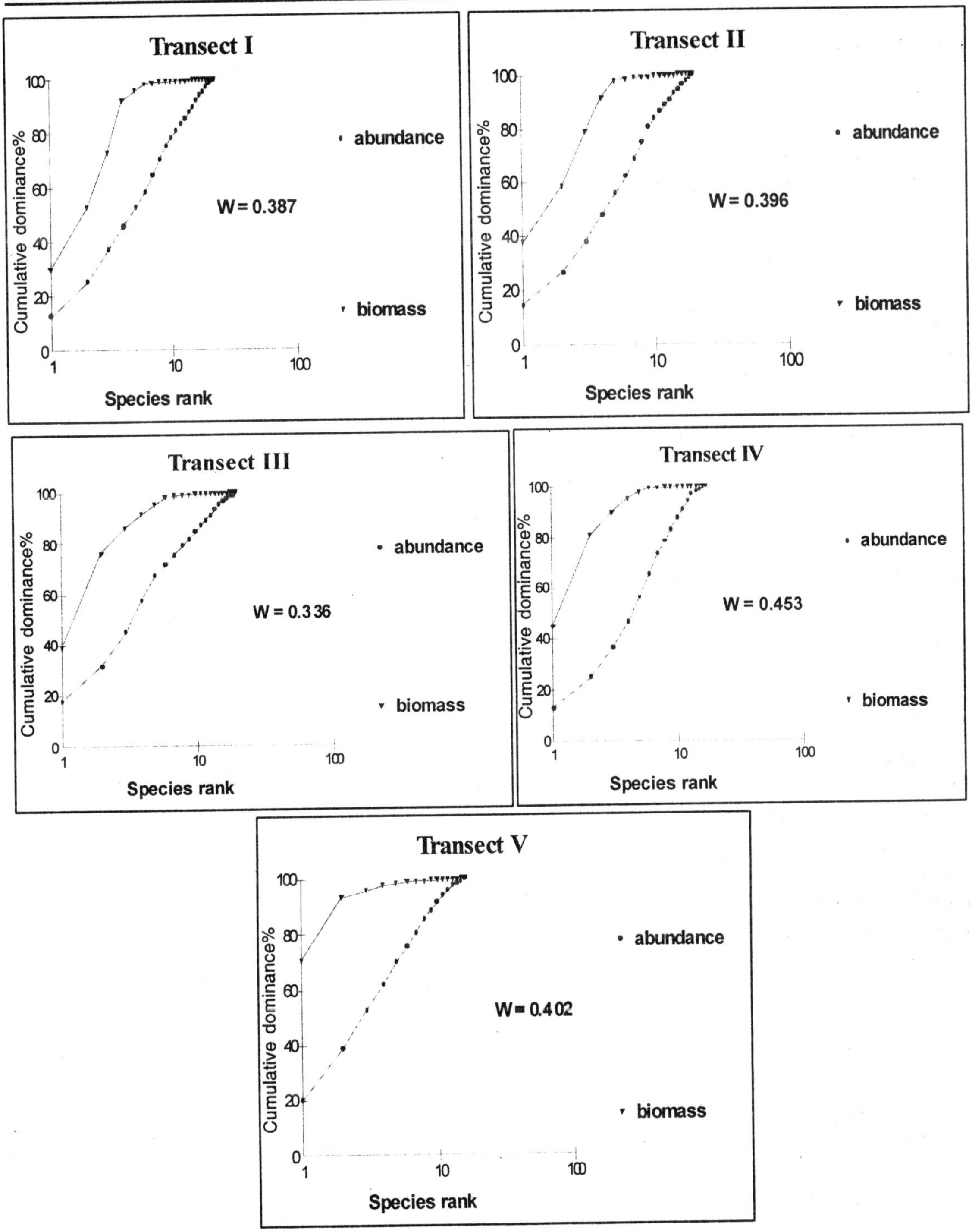

Figure 14.14: ABC-plots (k -dominance curves) for Macrobenthos Recorded in Transects I–V during the Study Period

low values were due to monsoonal downpour and inflow of freshwater from the land. The present findings are in affirmative with that of Thangaraj (1984) and Saravanan (1999) in Vellar estuary; Jegadeesan (1986) in Coleroon estuary; Chandramohan (1990) in the Godavari estuary and Ananthan (1991) in the Ariyankuppam estuary. In the present study, the dissolved oxygen concentration was low during summer and high during monsoon months in all the transects. The low dissolved oxygen concentration observed during summer months could be ascribed to the higher salinity of the water, higher temperature and less inflow of freshwater. Solubility of oxygen in water is a function of temperature and salinity (Carpenter, 1996). The trend noticed in the present study is in conformity with the findings of Mitra *et al.* (1990), Ananthan (1991) and Saravanan (1999).

With respect to the distribution of pH values, it remained alkaline throughout the study period in all the five transects with minimum value during monsoon and maximum in summer season. As observed in the other parameters, the pH also plays its own role in the benthic environment. Most of the natural waters are generally alkaline due to the presence of sufficient quantities of carbonate (Trivedy and Goel, 1984). The influence of neritic waters was also there in the study area. The pH of waters gets drastic change with time due to exposure to biological activity and temperature. Trivedy and Goel (1984) also opined that the pH also changes diurnally and seasonally due to the variations in photosynthetic activity, which increases the pH due to the utilization of carbon dioxide in the photosynthetic process.

The benthic organisms utilize the organic carbon for their metabolic process (Southward and Southward, 1972; Sepers, 1977). It plays an important role in the accumulation and release of different micro pollutants and also reflects more accurately the level of organic pollution. The minimum value (transect V) was found during monsoon and maximum (transect I) in summer season. The higher values recorded during summer may be ascribed to the increased water salinity resulting in rapid flocculation and precipitation of major fraction of terrigenous organic matter (Seiburth and Jensen, 1968). The summer peak of the present study is in harmony with Thangaraj *et al.* (1979) in Vellar estuary.

Sankaranarayanan and Parampunnayil (1979) recorded 7.4 - 38.4 mgc/g in Cochin backwaters. The reason attributed for this was the luxuriant organic productivity of the overlying water mass and nutrients brought by upwelling process.

The faunal composition is influenced by several mutually independent parameters having a limited influence on the number of species (Muss, 1967). Brinkhurst (1987) stated that the sediment particle size and organic content were the principle factors affecting species composition. Similarly, Eleftheriou and Basford (1989) opined that the variation in the currents, type of sediments, and the organic carbon content are important in determining the composition of benthic communities.

Among the benthic faunal groups polychaetes showed dominance in all transects. The preponderance of polychaetes in a benthic sample is reported earlier by many researchers (Alongi, 1990; Frouin, 2000; McCarthy *et al.*, 2000; Pavia, 2001; Lancellotti and Stotz, 2004 and Ellingsen, 2003). Dominance of polychaetes in terms of density and species composition in diverse ecological niches is due to their high degree of adaptability to a wide range of environmental factors. Such a preponderance of polychaetes in the benthic communities of temperate waters were reported earlier by Buchanan and Warwick (1974), Jumars and Fauchald (1978), Fauchald and Jumars (1984) in the Northumberland coast; Thangaraj *et al.* (1979), Fernando *et al.* (1984) and Chandran (1982) in Vellar estuary; Raveenthiranath Nehru (1990) in Coleroon estuary; Sebastin Raja (1990) in Sunnambar estuary and Chakraborthy and Choudhary (1997) in Hooghly estuary, Sagar Island.

Species diversity is a simple and useful measure of a biological system (Redding and Cory, 1985). Sanders (1968) found a high level of agreement between the species diversity and nature of the environment and hence the measure of species diversity regarded as an ecologically powerful tool. In the present study, a marked seasonal variation in the Shannon diversity was noticed with the minimum diversity value (2.240) recorded in transect V during monsoon season and maximum (4.474) in transect I during summer. Shillabeer and Tapp (1989) stated that the estuarine environment is far more dynamic than the fully marine and therefore, there may be a wide range of variations in the diversity of benthos of an estuary. Thus, variations in the species diversity observed in the present study could be deduced to the dynamic nature of the estuarine environment. As in the species diversity, species richness values were also low during monsoon in transect V and high during postmonsoon and summer months in transect I. The reason is that, it is easier to tolerate low salinities at high temperatures than at low temperatures (Sanders, 1968) and as a result more marine forms are able to flourish in tropical estuaries than the temperate waters (Panikker, 1940). Hence, the high species richness noticed during summer and postmonsoon seasons might be ascribed to the entry of marine forms into the estuary and the low values could be attributed to the drastic change in salinity during monsoon season. The trend with respect to richness values of the present study is in harmony with the studies made by Raveenthiranath Nehru (1990) in Coleroon estuary and Sebastin Raja (1990) in Sunnambar estuary. True to this, salinity is well–etched as the major and prime environmental factor controlling the number of species and community structure of estuarine infauna (Carriker, 1967; Kinne, 1971; Boesh, 1977 and Lippson *et al.*, 1979) and this view holds good for the present study.

In the present study the evenness measure (J′) largely followed the trend observed in the species diversity. As in the species diversity, the minimum value was found during monsoon season and maximum during summer in all the five transects. It is generally recognised that when the benthic biota is undisturbed, the representation of species in the benthic community is more or less equal and the evenness would be more. When the disturbance is severe to the biota, the benthic communities become increasingly dominated by one or few species resulting in low evenness. As the environment is found to be pristine in nature, the high evenness values were recorded in the present study.

The taxonomic diversity is defined as the average taxonomic distance between any two organisms chosen at random from the sample and this distance can be visualised simply as the length of the path connecting these two organisms (Warwick and Clarke, 1995).

This index is empirically related to the Shannon (H′) species diversity (Pielou, 1975) but with an added component of taxonomic separation. The taxonomic diversity is seen to be a natural extension of a form of Simpson index, incorporating taxonomic (or phylogentic) information since it mixes taxonomic relatedness with the evenness properties of the abundance distribution. The value of this univariate index appears not to be influenced by sampling effort. This is in stark contrast with those diversity measures (Shannon species diversity), which are strongly influenced by the number of observed species (Warwick and Clarke, 1995). In the present study, the minimum value was observed during monsoon in transect V and maximum during summer in transect II. As it is empirically related to the Shannon species diversity the results of this index is in agreement with the values of species diversity.

In the present study, a comparison of the ABC-curves for species abundances and species biomasses recorded during all the seasons showed the biomass curve to lie above the abundance curve indicating that the transects covered were unstressed. The ABC - curve is helpful in finding out the disturbance to the biota. This method, as originally described by Warwick (1986), involves the plotting of separate k -dominance curves for species abundances and species biomasses on the same graph and making a comparison of the forms of these curves.

Warwick *et al.* (1987) found that this method was applicable to a wide variety of types of perturbation to marine macrobenthic communities, man induced or otherwise, although its applicability to purely toxic pollution (*i.e.*, without organic enrichment) has not been tested. Also, it is possible to work at taxonomic levels higher than that of species (Warwick, 1986).

BIO-ENV yielded the combinations of six environmental entities (sand-silt-clay-organic carbon-salinity-temperature) as best 'defining' the faunal distributions. The associated coefficient of environmental to biotic similarity was 0.68. Similar variables combination was reported earlier by Mackie *et al.* (1995 and 1997).

References

Alongi, B.J., 1990. *Heavy Metals in Soils.* John Wiley and Sons, Inc. New York.

Ananthan, G., 1991. Hydrobiology of Parangipettai and Cuddalore marine environs with special reference to heavy metal pollution. *M. Phil. Thesis*, Annamalai University, India.

Antony, A. and Kuttyamma, V.J., 1983. The influence of salinity on the distribution of polychaetes in the Vembanad estuary, Kerala. *Bull. Dept. Mar. Sci., Univ. Cochin*, 13(122): 121–133.

Barnes, R.D., 1980. *Invertebrate Zoology*. Saunders College, Philadelphia, 1089pp.

Boesch, D.F., 1977. A new look at the distribution of benthos along the estuarine gradient. In: *Ecology of Marine Benthos*, (Ed.) B.C. Coull. University of South Carolina Press, South Carolina, p. 245–266.

Brinkhurst, R.O., 1987. Distribution and abundance of macroscopic benthic infauna from the continental shelf off south western Vancourer Island, British Columbia, Canada. *Can. Tech. Rep. Hydrogr. Ocean Sci.*, 85: 92.

Buchanan, J.B. and Warwick, R.M., 1974. An estimate of benthic macrofauna production in the offshore mud of the Northumberland coast. *J. Mar. Biol. Ass.* U.K., 54: 197–222.

Carikker, M.R., 1967. Ecology of estuarine benthic invertebrate: A perspective. In: *Estuaries*, (Ed.) G.H. Lauff. AAAS Publication, 83: 442–487.

Carpenter, J.H., 1996. New measurements of oxygen solubility in pure and natural water. *Limnol. Oceanogr.*, 11: 264– 277.

Chandramohan, B., 1990. Phytoplankton community structure in a tropical estuarine complex, East coast of India. *Estuar. Coast. and Shelf Sci.*, 24(5): 658– 670.

Chandran, R., 1982. Hydrobiological studies in the Gradient zone of the Vellar estuary. *Ph.D. Thesis*, Annamalai University, India.

Clarke, K. R., and Gorley, R.N., 2006. *PRIMER v6*: User Manual/Tutorial.

Day, J.H., 1967. *A Monograph on the Polychaeta of Southern Africa*. Parts 1 and 2, British Museum (Nat. Hist.), London. 878 pp.

Eleftheriou, A. and Basford, D.J., 1989. The macrobenthic infauna of the offshore Northern North Sea. *J. Mar. Biol. Ass.*, U.K., 69: 123–143.

Ellingsen, K.E., 2002. Soft-sediment benthic biodiversity on the continental shelf in relation to environment variability. *Mar. Eco. Prog. Series*, 232: 15–27.

Fauchald, K. and Jumars, P.A., 1984. The diet of worms: A study of polycheate feeding guides. *Oceanogr. Mar. Biol. A Rev.*, 17: 193–284.

Fauvel, P., 1953. *The Fauna of India Including Pakistan, Ceylon, Burma and Malaya. Annelida: Polychaeta,* Allahabad, 507 pp.

Frouin, P., 2000. Effects of anthropogenic disturbances of tropical soft bottom benthic communities. *Mar. Ecol.Prog. Series,* 194: 39–53.

Gaufen, H.R. and Tarzwell, C.M., 1952. Aquatic invertebrates as indicators of stream pollution. *U.S. Public Health Repts.,* 67: 57–64.

Jayabalan, N., Thangaraj, G.S. and Ramamoorthi, K., 1980. Finfish seed resources of Vellar estuary. *MBAI Proceedings of Symposium on Coastal Aquaculture,* Cochin, India, 12–18 January. Abst. No. 70.

Jegadeesan, P., 1986. Studies on environmental inventory of the marine zone of Coleroon estuary and inshore waters of Pazhayaru, Southeast coast of India. *Ph.D. Thesis,* Annamalai University, India.

Jumars, P.A. and Fauchald, K., 1978. Between community contrasts in successful polychaete feeding strategies. In: *Ecology of Marine Benthos,* (Ed.) B.C. Coull. University of South Carolina Press, Columbia, 41: 1–20.

Kinne, O. I., 1971. Salinity: Animals invertebrates. In: *Marine Ecology I: Environmental Factors Part 2,* (Ed.) O. Kinne. Wiley, London, p. 821–996.

Krumbein, W.C. and Pettijohn, F.J., 1938. Manual of sedimentary petrography. Applenton Century-Crofts, New York, 549pp.

Lancellotti, D.A. and Stotz, W.B., 2004. Effect of shoreline discharges of mine tailings on a marine soft-bottom community in the North Chile. *Mar. Pollut. Bull.,* 48: 303–312.

Lippson, A.J., Haire, M.S., Holland, A.F., Jacobs, F., Jenson, J., Moran-Johnson, R.L., Polgar, T.T. and Richkus, W.A., 1979. *Environmental atlas of the Potomae estuary.* Md: Martin Martietta Corporation Baltimore, 1450 South Rolling Rd., 288 pp.

Lyla, P.S., Velvizhi, S. and Khan, S. Ajmal, 1999. *A Monograph on the Amphipods of Parangipettai Coast.* Annamalai University, India, 78pp.

Mackie, A.S.Y., 1994. Collecting and preserving polychaetes. *Polychaete Res.,* 16: 7–9.

Mackie, A.S.Y., Oliver, P.G. and Rees, E.I.S., 1995. Benthic biodiversity in the southern Irish Sea. Studies in Marine Biodiversity and Systematics From the National Museum of Wales. *BIOMOR Reports,* 1: 263.

McCarthy, S.A., Laws, E.A., Estabrooks, W.A., Baily Brooks, J.H. and Key, E.A., 2000. Intra annual variability in Hawaiian shallow water, soft bottom macrobenthic communities adjacent to a entropic estuary. *Estuar. Coastal Shelf S.,* 50: 245–258.

Mitra, A., Patra, K.C. and Panigrahy, R.C., 1990. Seasonal variations of some hydrographical parameters in tidal creek opening into the Bay of Bengal. *Mahasagar-Bull. Natn. Inst. Oceanogr.,* 23(1): 55–62.

Muss, B.J., 1967. The fauna of Danish estuaries and lagoons: Distribution and ecology of dominating species in the shallow reaches of the mesohaline zone. *Meddr. Danm. Fisk. Harvunders, N.S.,* 5: 1–316.

Panikkar, N.K., 1940. Influence of temperature on osmotic behaviour of some crustacea and its bearing on problems of animal distribution. *Nature,* 146: 366.

Parulekar, A.H., Dhargalkar, V. K. and Singbal, S.Y.S., 1980. Benthic studies in Goa estuaries: Part–III–Annual cycle of macrofaunal distribution, production and trophic relation. *Indian J. Mar. Sci.,* 9: 189– 200.

Pavia, P.C., 2001. Spaital and temporal variation of a near shore benthic community in southern Brazi: implication for the design of monitoring programe. *Estuar. Coastal Shelf S.*, 52: 423–433.

Pielou, E.C., 1975. *Ecological Diversity*. John Wiley and Sons, New York, 165pp.

Rajagopal, S., Khan, S. Ajmal, Srinivasan, M. and Shanmugam, A., 1998. *A Monograph on the Gastropods of Parangipettai Coast*. Annamalai University, India, 38pp.

Raveenthiranaath Nehru, R., 1990. Ecology of macrobenthos in and around Mahendrapalli region of Coleroon estuary, Southeast coast of India. *Ph.D. Thesis*, Annamalai University, India.

Redding, J.M. and Cory, R.L., 1985. Macroscopic benthic fauna of three tidal creeks adjoining the Rhode river, Maryland. *Water Resources Investigation Report, USA*, pp. 39–75.

Reish, D.J., 1980. Use of polychaetous annelids as test organisms for bioassay experiments. In: *Aquatic Invertebrate Bioassays*, (Eds.) A.L. Buikema and J. Cairns. American Society for Testing and Materials Scientific and Technical Paper, 715: 140–154.

Sanders, H.L., 1968. Marine benthic diversity: A comparative study. *American Naturalist*, 102: 243–282.

Sankaranarayanan, V.N. and Panampunnayil, S.V., 1979. Studies on organic carbon, nitrogen and phosphorus in sediments of the Cochin backwater. *Indian J. Mar. Sci.*, 8: 27–30.

Sankaranarayanan, V.N., Rao, D.P. and Antony, M.K., 1978. Studies on some hydrobiological characteristics of the estuarine and inshore waters of Goa during the southwest monsoon, 1972. *Mahasagar-Bull. Natn. Inst. Oceanogr.*, 11(3&4): 125– 136.

Saravanan, S., 1999. Studies on the estuarine hermit crab *Diogenes avarus* Heller (Crustacea: Decapoda: Anomura). *Ph. D. Thesis*, Annamalai University, India.

Sebastin Raja, S., 1990. Studies on the ecology of benthos in Sunnambar estuary, Pondicherry, Southeast coast of India. *Ph.D. Thesis*, Annamalai University, India.

Seiburth, J.M. and Jensen, A., 1968. Studies on algal substances in the sea. I. Gelbstoff (humic material) in terrestrial and marine waters. *J. Exp. Mar. Biol. Ecol.*, 2: 174–189.

Sepers, A.B.J., 1977. The utilisation of dissolved organic compounds in aquatic environments. *Hydrobiologia*, 52: 39–54.

Shanmugam, A., Rajagopal, S. and Nazeer, R.A., 1997. *A Monograph on the Common Bivalves of Parangipettai Coast*. Annamalai University, India.

Shillabeor, N. and Tapp, J.F., 1989. Improvements in the benthic fauna of the Tees estuary after a period of reduced pollution loadings. *Mar. Pollut. Bull.*, 20(3): 119–123.

Southward, A.J. and Southward, E.C., 1972. Distribution and ecology of the hermit crab *Clibanarius erythrops* in the Western Channel. *J. Mar. Biol. Ass. U.K.*, 57: 441–452.

Srikrishnadhas, B., Murugesan, P. and Khan, S. Ajmal, 1998. *A monograph on the Polychaetes of Parangipettai Coast*. Annamalai University, India, 110 pp.

Tenore, K.R., Browne, M.G. and Chesney, E., 1974. Polyspecies aquaculture systems: The detrital trophic level. *J. Mar. Res.*, 32: 425–432.

Thangaraja, G.S., 1984. Ecobiology of the marine zone of the Vellar estuary. *Ph.D. Thesis*, Annamalai University, India.

Thangaraja, G.S., Sivakumar, V., Chandran, R., Santhanam, R., Srikrishnadhas, B. and Ramamoorthi, K., 1979. An environmental inventory of Portonovo coastal zone. In: *Proc. Symp. Environ. Biol.*, Academy of Environmental Biology, India, p. 75–87.

Trivedy, R.K. and Goel, P.K., 1984. Physico-chemical analysis of water. In: *Chemical and Biological Methods for Water Pollution Studies*. Environmental Publications, Karad, India, p. 35–96.

Warwick, R.M., 1986. A new method for detecting pollution effects on marine macrobenthic communities. *Mar. Biol.*, 92: 557–562.

Warwick, R.M, Pearson, T.H. and Ruswahyuni, 1987. Detection of pollution effects on marine macrobenthos: Further evaluation of the species abundance/biomass method. *Mar. Biol.*, 95: 193–200.

Warwick, R.M. and Clarke, K.R., 1993. Comparing the severity of disturbance: A meta-analysis of marine macrobenthic community data. *Mar. Ecol. Prog. Ser.*, 92: 221–231.

Warwick, R.M. and Clarke, K.R., 1995. New 'Biodiversity' measures reveal a decrease in taxonomic distinctness with increasing stress. *Mar. Ecol. Prog. Ser.*, 129: 301–305.

Warwick, R.M., 1993. Environmental impact studies on marine communities: Pragmatical considerations. *Aust. J. Ecol.*, 18: 63–80.

Wolff, W.J., 1983. Estuarine benthos. In: *Ecosystems of the World 26. Estuaries and Enclosed Seas*, (Ed.) Bostwick H. Ketchum. Elsevier Science Publishers B.V., Amsterdam, The Netherlands, p. 151–182.

Chapter 15

Mangrove Biodiversity and its Conservation at Ratnagiri

☆ *A.S. Kulkarni, M.V. Tendulkar, S.M. Nikam and A.S. Injal*

ABSTRACT

Ratnagiri is situated 17° North and 73° East. Bhatye estuary is situated at 73° 15'East and 16° 51'N near Ratnagiri and known for its mangroves, mudflats and clam fauna like *Meretrix meretrix, Katelysia opima* and *Gelonia proxima.* It also happens to be one of the important estuaries as it supports commercial fishery along Ratnagiri coast. 10 species of mangroves and 16 spp. of mangrove associates have been identified in this area. The biodiversity of this area has been studied in detail in period of 2 years and has been documented. The mangrove also supports a rich diversity of benthic organism like nematodes, polychaets, Plankton (Phytoplankton and zooplankton), molluscs, crabs, fishes and avifauna. The impact of anthropogenic activities and other ecological constraints have also been studied. It has been observed that this area also has tremendous potential to develop as an Eco-tourism center and contribute to the economics of this zone. But due to anthropogenic activities it has been estimated that there is decline in mangrove belt and this has lost biodiversity prosperity and is in need to conserve it. With this regards, the present chapter provides an overview of present status of mangrove in this area, its ecological implications and the methods adopted for its conservation.

Introduction

Biodiversity is the basis for human existence. It is of critical importance for meeting the needs of food, health and other needs of the growing human populations. Among the various biodiversity regions, the coastal marine ecosystems are known to be most productive, biologically diverse and exceedingly valuable areas. These unique coastal, tropical forests are now among the most threatened habitats in the world, due to expanding human population and resultant unsustainable economic development. As a result of continuous biotic pressure the mangroves and other marine resources are experiencing habitat loss, changes in species composition, shifts in dominance, loss of biodiversity and threat to survival. (Upadhyay *et al.* 2002).

Ratnagiri is situated at 17° North and 73° east and having an area of about 50,209 sq. miles. Geographically, it is the southern most district of Maharashtra state. The coastline of Ratnagiri district is 250 miles long and marked with several islands, which is a result of drowned topography. The important estuaries along the Ratnagiri coast include Bhatye estuary, Kalbadevi Creek, Jaitapur Creek, Bankot Creek, Sakhartar Creek, Shirgaon Creek etc. Bhatye estuary is situated at 73° 15 east and 16°51 north near Ratnagiri and known for mangroves on the mud flats and the clam fauna like *Meretyrix meretrix, Katelysia opima* and *Gelonia proxima*. It is one of the most important estuarine regions along the Ratnagiri coast and is breeding ground for most of the commercially important fish species. The fishery economic of Ratnagiri largely depends upon Bhatye estuary. Hence this particular area is important from the biodiversity and economics point of view. Mangrove related ecotourism enterprise like mangrove biodiversity, back water safaris, Para Gliding, bird watching, fishing etc., can be developed. The extent of mangrove cover is day by day reducing due to anthropogenic activities. Further literature review showed that no biodiversity estimates was carried out for this area and hence present studies provide the current status of Bhatye estuary.

Matcrials and Methods

Three Zones were selected for sampling, considering the nature of study area. Zone I is a marine zone, Zone II occupies Mangrove Island while Zone III was riverine zone. Six stations were selected within a stretch of 25km, 2 in each Zone. Sampling was done fortnightly covering intermediate phase of the tide to avid tidal effect, if any. Diesel engine boat was used to reach different stations. Mapping of mangroves and mangrove associates was done after selecting the sampling locations. Then each station was estimated for the presence/absence of the flora by visiting the sites frequently. The flora was photographed and also a sample was collected and later on identified in the laboratory. Mangrove species and the associated flora was identified by using taxonomic identification keys mentioned by Rajendran and Sanjeevi, 2004 and Naskar, 2004. Mangrove vegetation was studied using quadrat method as described by Shukla and Chandel (1972).

Molluscs were collected by hand picking when mud flats were exposed at low tides. The animals were preserved and identified by using standard literature. (Macdonald, 1980, Apte, 1992, Rajgopal, *et al.*, 1998 and Ramakrishna and Dey, 2003). Crabs were also collected by hand picking as well as by digging or by pouring dilute formalin inside the burrow. Some crab species were collected inside the mangrove forest by using forceps. Finfishes were collected once in a month by using cast net. The specimens were identified, photographed and preserved in 10 per cent formalin. The specimens were then identified by using standard literature. (Talwar and kicker, 1984) Avifauna was observed by using Olympus binoculars (10 x 50 magnifications). The avifauna was identified by using standard literature for classification and nomenclature of birds (Ali and Ripley, 1995, 2001, Ali, 2002, Pande *et al.*, 2003). Fortnightly census of birds was conducted from March 2004 to February 2005. The censuses were made by visual counts of birds from specific points and also with the help of boat to get closer look of these birds.

For microbiological study rhizosphere and root samples were collected from the mangrove island from the Bhatye estuarine region. The rhizosphere soil samples were diluted in physiological saline and 3 dilutions were plated out on standard late Count Agar in triplicate, in 2 sets each and incubated at room temperature. The inoculated samples were then incubated at 28 ± 2 °C for two day. Further, the distribution of different groups of bacteria was determined on the basis of cultural characteristics and microscopic examination followed by viable colony counting.

Results and Discussion

There is no proper checklist of mangrove species, available from regions of the world. The lists available varied with publications and are rather confusing. (Kathiresan and Rajendran, 2005). The extensive survey conducted during the tenure of the project revealed presence of 10 mangrove species. (Table 15.1) and16 species of mangrove associates. The mangrove species are represented by four families–*Avicenniaceae, Myrsinaceae, Rhizophoraceae* and *Sonneratiaceae.* The mangrove associates are represented by 22 families and the flora along the sandy beach of Bhatye are represented by six families. The study indicated that mangrovephytes studied along the Bhatye estuary present a well established vegetation pattern as it is clearly visible from the checklist as well as from the distribution pattern of the vegetation. The presence of *Avicennia* species indicates the success with which this plant is capable to colonise and establish itself along the estuarine region. *Avicennia officinalis* in some of the locations indicates the success ability of its members to colonise and dominate along with the Rhizophorceae family in this area. While presence of mangroves like *Aegiceras coprniculatum* is an indication of its recent introduction and its existence in physical form throughout its life span, and hence can be considered as an established species of this area. Further growth of species like *Acanthus ilicifolius* is a dominant mangrove associates in this region indicates that the plants that grow in the understorey of tall mangrove trees and can play an important role in their contribution to the mangrove ecosystem by providing shelter to organisms that inhabit the mangroves and also contribute to the organic content of the estuary.

Table 15.1: Mangrove Species Occurring along Bhatye Estuary

Sl.No.	*Botanical Name*	*Family*
1.	*Avicennia alba* Bl.	Avicenniaceae
2.	*Avicennia marina* Stapf and Mold	Avicenniaceae
3.	*Avicennia officinalis* L.	Avicenniaceae
4.	*Aegiceras corniculatum* Blanco.	Myrsinaceae
5.	*Bruigera cylindrica* (L) Blume	Rhizophoraceae
6.	*Bruigera gymnorrhiza* (L) Lamk.	Rhizophoraceae
7.	*Kandelia candel* (L) Druce.	Rhizophoraceae
8.	*Rhizophora mucronata* Lamk.	Rhizophoraceae
9.	*Rhizophora apiculata* Blume.	Rhizophoraceae
10.	*Sonneratia alba* J.Smith.	Sonneratiaceae

Mangrove Associates along Bhatye Estuary

Erythrina indica, Zizypus jujuba, Sisuvium portulacatrum, Vitex Negundo, Ipomea biloba, Derris uliginosa, Derris trifoliate, Pongamia pinnata, Caesalpinia crista, Cordia myxa, Plectranthus volubilis, Acanthus illicifolius, Salvadora persica, Thespesis populnea, Calophyllum inophyllum, Clerodendron inerme.

Molluscs

Katelysia katelysia, Katelysia opima, Meretrix Meretrix, Gelonia proxima, Arca granosa, Crossostrea madrasensis, Dosinia prostrata, Turbo species, Dosinia cretacea, Cardium asiaticum, Solen truncates and *button shells*

Avicennia officinalis

Sonneratia alba

Rhizophora mucronata

Vitex negundo

Clerodendron inerme

Acanthus ilicif

Meretrix meretrix

Gelonia proxima

Arca granosa

Scylla serrata

Sesarma quadrata

Neptunus pelagicus

Boleopthalmus dussumieri

Cephalopholis pachycentron

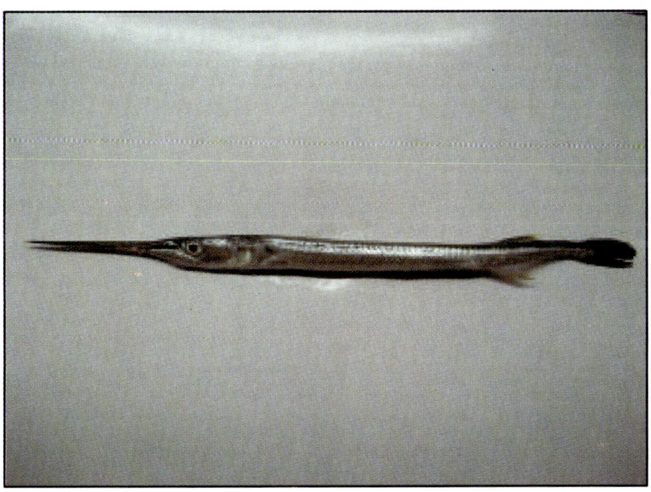

Strongylura strongylura

Halycon smyrnensis
(White throated Kingfisher)

Casmerodius albus
(Large Egret)

Haliastur Indus
(Kite)

Crabs

Scylla serrata, Charybdis callianassa, Fiddler crab, Charbdis orientalis, Gelasimus marionis, Macropthalmus sp., *Matula planipes, Metopograpsus species, Neptunus pelagicus,* and *Sesarma quadrata.*

Fin Fishes

Elasmobranches, Eels, Catfish, Chirocentrus, Sardines, Clupieds, Pomphrets, Mackerel, Seer fish, Tunas, Prawns, Lobsters and Cuttlefish etc.

Avifauna

Haliastur indus (*Kite*), *Casmerodius albus,* (*Large Egret*), *Vanellus indicus* (Red Wattled Lapwing), *Halycon smyrnensis* (Whitethroated Kingfisher), *Egretta gularis* (Black Egret), *Phalacrocorax, Fuscicollis* (Little Cormorant), *Actitis hypoleucos* (Common Sandpiper) and *Mesophoyx intermedia* (Median Egret).

Molluscs are more diverse than the crustaceans, fish and other kinds of organisms. In present studies, molluscs were recorded from station I and II which can be considered as estuarine zone. The bivalves are selectively rich in and around the mangrove environment. The ecosystem provides an ideal niche for the animals due to less water motion, soft substratum and less stress from the predatory organisms, as compared to the other environments. As far as the molluscan fishery is concerned, *Geloina proxima* happens to be one of the important species from fishery point of view. Ratnagiri Taluka stands first in the mangrove clam fishery of the district, though it is having less brackish area, it possesses high potential for the production of mangrove clams.

The Bhatye estuarine region also shows the presence of the various crab species. The study reveals the presence of nine Brachyuran crab species like, *Charybdis callianassa, Charybdis orientalis, Gelasimus marionis, Macropthalmus* spp, *matuta planipes, Metopograpsus* spp, *Neptunus pelagicus, Scylla serrata* and *Sesarma quadrata.* The crab species like, *Gelasimus marionis, Neptunus pelagicus* and *Scylla serrata* are found to be dominant in the mangrove vegetation and the exposed mud flats served as an excellent feeding grounds for these species. It is also observed that the crabs depend directly on mangrove areas for survival, by feeding on the leaves and litter. The crabs have a significant role in detritus formation, nutrient recycling and dynamics of the ecosystem. The digging behaviour, especially in mangrove vegetation by the crabs enhances aeration and facilitates drainage of mangrove soils.

The various fin fishes found along the Bhatye estuarine region are, Elasmobranchs, Eels, Catfishes, Chirocentrus, Sardines, Clupeids, Bombay Ducks, Pomfrets, Mackerels, Seer fishes, Tunas, Prawns, Lobsters and Cuttle fishes, etc. Mangrove habitats usually contain a rich ichthyofauna as mangrove waters provide ideal niche for the juvenile fishes due to less water motion, soft substratum, enormous amount of food, excellent shelter and protection from predatory organisms. The canopy of mangrove provides a cool, stable and humid environment which favors the growth of the young fishes. The Bhatye region shows the presence of almost 22 different species of fin fishes. Thus, it can be said that, the biodiversity of fin fishes in Bhatye contributes a lot to the commercial fisheries especially along the Ratnagiri coast.

The recent studies reveal the presence of 45 species of birds belonging to 9 orders, 18 families and 34 genera. Birds have been considered as useful biological indicators because they are ecologically versatile and live in all kinds of habitats as Herbivores or Carnivores. Thus, many different kinds of birds have been recorded from the Bhatye estuarine area. Among the many, the Egrets are the most abundant in number. Their dominance over the other species of birds is due to their nesting period and location of their nesting sites in the mangroves of Bhatye estuary. As they are the estuarine birds and get their food in this ecosystem, they can be considered as resident and local migrants of this area. The bird species of the family *Scolopacidae* are observed feeding on the mud flats of mangrove island. Most of then are winter migrants. Mud flats are feeding grounds for these visitors, which also indicate rich macro and micro fauna such as crabs, polychaetes and some molluscs on these mud flats.

Birds like Little Cormorant, Western Reef Egrets, Indian Pond Heron, Black Bittern, etc. are mostly seen in the middle zone of the estuary. Their appearance towards the riverine side of the estuary is very rare and shows their presence towards the seaside or towards the mouth of the estuary. Some species such as Kingfishers, Red Watteled Lapwing and some species of Sandpipers are also observed to be present throughout the entire estuary. Thus, the recent studies reveal the presence of rich diversity of avifauna in the Bhatye estuarine region.

Mangroves provide a unique ecological niche to a variety of micro organisms. The fertility of mangrove waters results from the microbial decomposition of organic matter and recycling of the nutrients. The bacteria exist as symbioants with plants and animals, Saprophytes on dead organic matter and as parasites on living organisms, the bacteria perform varied activities in mangrove ecosystems. However the estimated bacterial flora from the mangrove and their associates are *Rhizophora mucronata, Rhizophora apiculata, Avicennia* species, *Sonneratia* species, *Acanthus ilicifolius* and *Garcinia* species.

Major Problems Associated with Ratnagiri Mangroves

- Human settlements - large scale reclamation of forest land and deforestation of mangroves.
- Over-exploitation of aquatic resources - creating problems towards loss of aquatic species diversity.
- Transport facilities - to reach any remote island have lead to exploitation of these forests.
- Over-cutting has caused siltation of river beds.
- Over dumping of domestic wastes along the beds of estuaries.
- Lack of awareness about the importance of mangroves in local community.
- Nesting sites and breeding grounds of different resident and migratory birds are disturbed.
- Lack of conservation or management activities for protection of mangrove forests in the area.

Measures Implemented to Conserve the Mangroves along Ratnagiri Coast

- We are creating awareness among local community and stake-holders about the importance of mangroves and estuarine ecosystem.
- We have been organizing different awareness programmes like drawing competitions, essay competitions, lecture presentation etc. for the school going students.
- Lectures and training programmes are also organized for school teachers regarding the conservation of mangroves.
- On the occasion of World Wet Land Day in collaboration with BNHS, Mumbai, we organize the rally of school and college students, with participation of NSS and NCC cadets along with the Harit Sena members and other social activitists on the main streets of Ratnagiri in order to create a local awareness about the importance of mangroves and its conservation.
- We have developed a mangrove nursery in our college campus.

- ✰ We organize the mangrove plantation campaigns where the nursery saplings are planted at various estuarine regions in association with Cameron International Co. and BNHS, Mumbai.
- ✰ Afforestation of mangrove and associated species in the affected/deforested areas is being practiced with the involvement of local communities and students.

Dedicated research is going on at many research centers across India as well as along the Asian continent. But this research needs greater co-ordination among mangrove researchers. A critique of what has been achieved, the gaps and what needs to be done priority wise for the different sectors of the mangrove ecosystems is lacking and this is more particularly seen along the Konkan coast. In evaluating the fisheries it is necessary to identify the mangrove dependent versus mangrove independent species and their respective roles. The approach in achieving these goals is inter-disciplinary and inter-organizational programmes of basic and applied nature. Further research could also be aimed at the sustainable uses of mangrove areas of Ratnagiri, reforestation in the areas where mangroves are getting degraded.

References

Ali, S., and Ripley, S.D., 1995. *A Pictorial Guide to the Birds of the Indian Subcontinent*. BNHS, Oxford University Press.

Ali, S., and Ripley, S.D., 2001. *Handbook of the birds of India and Pakistan,* 2nd edn. BNHS, Oxford University Press.

Ali, S., 2002. *The Book of Indian Birds,* 13th edn. BNHS, Oxford University Press.

Apte, D., 1992. *The Indian Book of Shells.* BNHS.

Kathiresan K., 2002. Why are mangroves degrading? *Current Science,* 83(10): 1246–1249.

Macdonald Encyclopedia of Shells, 1980. Macdonald and Company, London and Sydney.

Naskar, K., 2004. *Manual of Indian Mangroves.* Daya Publishing House, New Delhi.

Pande, S.S., Tambe, C.M., Francis and Sant, N., 2003. *Birds of Western Ghats, Konkan and Malabar (including birds of Goa).* BNHS, Oxford University Press.

Rajgopal, R., Khan, Ajmal, Srinivasan, M. and Shanmugam, A., 1998. *Gastropods of Parangipettai Coast.* Center of Advanced Study in Marine Biology, Annamalai University, Parangipettai.

Ramakrishna and Dey, A., 2003. *Identification of Schedule Molluscs from India.* Zoological Survey of India, Kolkatta.

Rajendran N., and Sanjeevi, S. Baskara, 2004. *Flowering Plants and Ferns in Mangrove Ecosystems of India.* ENVIS Publications, Parangepettai.

Shukla, R.S. and Chandel, P.S., 1972. *Plant Ecology.* S. Chand and Company, New Delhi.

Talwar, P.K. and Kacker, R.K., 1984. *Commercial Sea Fishes of India.* Zoological Survey of India, Kolkata.

Upadhyay, V.P., Rajiv, R. and Singh, J.S., 2002. Human-mangrove conflicts: The way out. *Current Science,* 83(11): 1328–1335.

Chapter 16

Status of Fishes in Mangroves of Kali Estuary, Karwar, Central West Coast of India

☆ *S.V. Roopa, J.L. Rathod and B. Vasanth Kumar*

ABSTRACT

Mangrove ecosystem harbor variety of flora and fauna, they are breeding, feeding and nursery grounds for many estuarine and marine organisms. In this context a study was undertaken to acess the status of available fin fishes in one of the important mangrove ecosystem of Karnataka. During the present investigation 37 types of fish species representing 20 families were observed. The study was carried out from January 2008 to January 2009 for about 13months, five stations were selected to know the fish availability. During the course of observation highest number of *Etroplus suratensis* (181 No/m^3), *Ambassis gymnocephalus* (164 No/m^3), *Arius arius* (140 No/m^3), followed by *Gerres poieti*(123 No/m^3) and *Ambassis commersoni* (96 No/m^3). During present investigation month of January 2009(179) found to be more productive and August 2008(101) found to be least productive.

Introduction

Covering about 47 per cent of word's mangrove area, containing 85 per cent of word's mangrove species, and occurring in variety of habitats, the mangrove ecosystem plays a vital role in coastal biodiversity of 30 countries bordering the Indian Ocean. This ecosystem supports a rich species diversity of flora and fauna, but it is facing heavy human pressures and natural stresses, leading to a loss in biodiversity. This calls for urgent measures of conservation and management(Kathiresan and Rajendran, 2005). Mangrove forests are among world's most productive ecosystems (Kathiresan and Bingham, 2001). Of late, Uttara Kannada district has become the centre of many developmental activities like sea bird, Kaiga nuclear power plant, and Konkan railway projects, the projects will have substantial impact on the flora and fauna of estuarine and coastal habitat. The mangrove habitat is being converted in to prawn fields which is leading to large scale destruction of mangrove habitat and in turn some fishes have become extinct. Kali estuary (14° 50′ 21″ N and 74° 09′ 05″ E) being one of the major estuarine systems of Uttara Kannada coast is located on the central west coast of India. To record the presently available fin fishes in the riverine as well as estuarine ecosystem an attempt was made. A study was under taken for a period of 13 months from January 2008 to January 2009.

Materials and Methods

A fortnightly sampling of fishes was carried out using drag net of having mesh size 2mm with 3m X 1m size and also with cast net of standard size. Samples were collected all the five study sites during low tide, as the availability was more during low tide (Strickland and Parson, 1975). The study was carried out for a period of 13 months from January 2008 to January 2009. Five study stations were chosen to find out the availability of fishes, three stations (station 1 Kinner, station 2 Kadwad and station 4 Sunkeri) are located present at southern bank of the river. Other stations, station 3 Kanasgeri and station 5 Mavinhole creek are located in the northern bank of river. Three samplings were made in each station, without disturbing the water, a long the low water mark against the wind direction parallel to the shore. The number of fishes obtained in all the three drags was clubbed, represented as number per meter cube (No/m^3) and percentage by number were calculated and presented in the tables. The fishes were preserved in 10 per cent formalin. For identification of fishes Francis Day's Volume-I (1978 and 1980), Talwar and Jhingran (1991) and an illustrated guide on commercial fishes and shell fishes of India 2002 were made use of.

Results

During present observation 37 fish species belonging to 20 families were recorded in the Kali estuarine ecosystem (Table 16.1). Among the fishes observed *Etroplus suratensis* observed in highest number (181No/m^3), *Ambassis gymnocephalus* was second (164No/m^2) prominent fish species. Whereas, *Arius arius* and *Mugil cephalus* ranked third dominant species (140No/m^2). *Gerres poieti* occupied fourth place (123No/m^3), followed by *Ambassis commersoni* (190 No/m^3).

Station 1 Kinner

In this station the order of abundance was 83, 73 and 33 No/m^3. *Etroplus suratensis; A. gymnocephalus* and *G. poieti* respectively. The percentage by numbers also calculated (45.85 per cent, 44.51 per cent and 26.82 per cent respectively) and presented in the Table 16.1. *Sphyraena jello* fishes were observed in least number 1.

Station 2 Kadwad

In this station *A. Gymnocephalus* observed in highest number 38No/m^3 (23.17 per cent), Catfish species were second largest 37 No/m^3 (45.12 per cent) and *Siganus vermiculatus* found to be third dominant species 34 N0/m^3 (100 per cent).

Station 3 Kanasgeri

Arius arius found in highest number 96No/m^3 (68.5 per cent)followed by *A. gymnocephalus* 41 No/m^3 (25 per cent) and *Favonigobius reichei* 30 No/m^3 (53 per cent) respectively second and third dominant species.

Table 16.1: Fishes (No/m³) Percentage by Number in Each Station in Kali Estuary

Sl.No.	Family- fishes	Stn 1		Stn 2		Stn 3		Stn 4		Stn 5		Total
		No	%	No	%	No	%	No	%	No	%	
1.	*Acentrogobius viridipunctatus*	18	20.93	26	30.23	17	19.7	17	19.7	8	9.3	86
2.	*Ambassis commersoni*	32	33.33	8	8.33	15	15.6	31	32.2	10	10.4	96
3.	*Ambassis gymnocephalus*	73	44.51	38	23.17	41	25	8	4.8	4	2.4	164
4.	*Arius arius*	20	14.28	0	0	96	68.5	19	13.5	5	3.5	140
5.	*Belone strongylurus*	18	81.81	4	18.18	0	0	0	0	0	0	22
6.	*catfish*	27	32.92	37	45.12	0	0	9	10.9	9	10.9	82
7.	*Cynoglossus bilineatus*	0	0	5	45.45	0	0	0	0	6	54.5	11
8.	*Cynoglossus semifasciatus*	0	0	9	45	0	0	0	0	11	55	20
9.	*Etroplus suratensis*	83	45.85	16	8.83	16	13	30	16.5	36	19.8	181
10.	*Favonigobius reichei*	13	23.21	0	0	30	53	13	23.2	0	0	56
11.	*Gerres filamentosus*	0	0	9	13.23	12	17.6	12	17.6	35	51.4	68
12.	*Gerres limbatus*	12	33.33	0	0	8	22.2	11	30.5	5	13.8	36
13.	*Gerres lucidus*	10	33.33	7	23.33	3	10	7	23.3	3	10	30
14.	*Gerres macrocanthus*	0	0	0	0	0	0	4	57.1	3	42.8	7
15.	*Gerres poieti*	33	26.82	27	21.95	16	13	28	22.7	19	15.4	123
16.	*Glossogobius giuris*	0	0	28	51.85	7	12.9	5	9.2	14	25.9	54
17.	*Hemiramphus sp.*	0	0	0	0	0	0	13	100	0	0	13
18.	*Leiognathus dussumieri*	21	80.76	5	19.23	0	0	0	0	0	0	26
19.	*Liza tade*	13	43.33	8	26.66	0	0	5	16.6	4	13.3	30
20.	*Liza vaigiensis*	0	0	10	20.4	9	18.3	30	61.2	0	0	49
21.	*Lutjanus argentimaculatus*	0	0	11	37.93	0	0	11	37.9	7	24.1	29
22.	*Lutjanus erythropterus*	7	21.87	11	34.37	0	0	6	18.7	8	25	32
23.	*Lutjanus johni*	8	40	5	25	0	0	7	35	0	0	20
24.	*Monodactylus argenteus*	5	35.71	9	64.28	0	0	0	0	0	0	14
25.	*Mugil cephalus*	22	15.71	24	17.14	28	20	28	20	38	27.1	140
26.	*Scatophagus argus*	7	10.76	20	30.76	9	13.8	29	44.6	0	0	65
27.	*Scomberoides lysan*	0	0	0	0	0	0	0	0	1	100	1
28.	*Secutor ruconius*	8	100	0	0	0	0	0	0	0	0	8
29.	*Siganus canaliculatus*	4	21.05	9	47.36	0	0	0	0	6	31.5	19
30.	*Siganus vermiculatus*	0	0	34	100	0	0	0	0	0	0	34
31.	*Sillago sihama*	6	27.27	2	9.09	0	0	8	36.3	6	27.2	22
32.	*Sphyraena jello*	1	9.09	1	9.09	0	0	6	54.5	3	27.2	11
33.	*Stenogobius Gymnopomus*	8	16	15	30	0	0	11	22	16	32	50
34.	*Terapon jarbua*	0	0	0	0	5	20	13	52	7	28	25
35.	*Tetradon fluviatilis*	0	0	6	54.54	0	0	0	0	5	45.4	11
36.	*Triacanthus biaculeatus*	0	0	0	0	0	0	0	0	9	100	9
37.	*Triacanthus brevirostris*	0	0	10	10		0	0	0	0	0	10

Station 4 Sunkeri

A. commersoni 31 No/m³(32.2 per cent) *E. Suratensis*(30 No/m³ (16.5 per cent) and *Liza vaigiensis* 30 No/m³ (16.5 per cent)and *Gerres macrocanthus* found in least number 4 No/m³ (57.1 per cent).

Station 5 Mavinahole Creek

Here *Mugil cephalus* recorded in highest number 38 No/m³ (27.1 per cent) *E. suratensis* 36 No/m³ (19.8 per cent) and *G. filamentosus* 35 No/m³ (51.4 per cent) whereas, *Scomberoides lysan* found in least number 1 No/m³ (100 per cent) by number.

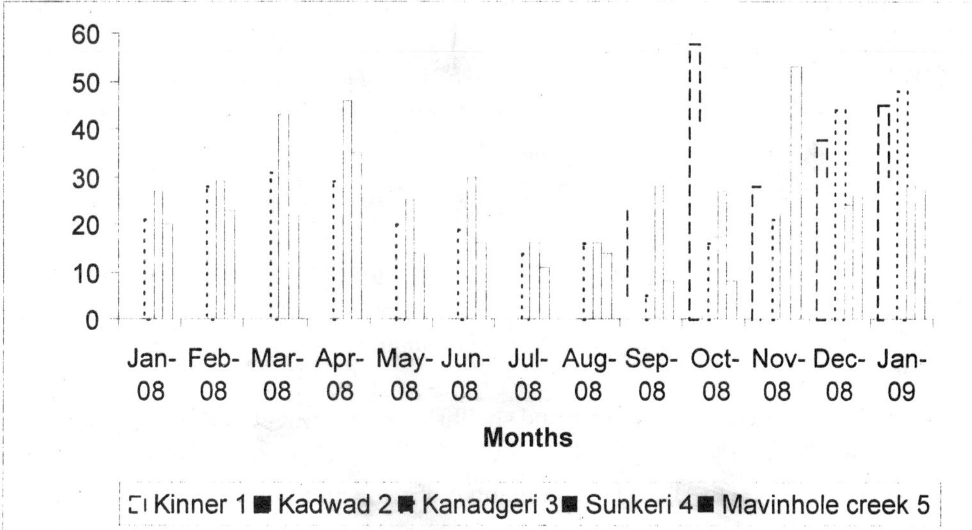

Figure 16.1:Total Fishes Recorded in Each Station during Present Investigation

From the Figure 16.1 it can be stated that January 2009 was found to be more productive because highest fishes were recorded (179), in November 2008 (169) fishes were observed whereas, in December 2008 162 fishes were noticed. Among all the months least number of fishes recorded (101) in the month of August 2008.

From the above findings it can be concluded that Kali estuarine system forms one of the very rich source for majority of fishes.

Discussion

Fish and shellfish seed resources of Kali estuary along with a note on the mariculture potentials in Uttara Kannada was carried out by Nagaraj and Neelakantan (1982). During their observation the most important seeds recorded are of *Penaeus indicus, P. monodon, Metapenaeus dobsoni, M. monoceros, Mugil cephalus, Scatophagus argus, Etroplus suratensis* besides the spats of edible oyster *Crossostrea gryphoides*, the mussel *Perna viridis* and the clam *Meritrix casta*. The earlier workers mainly focused on cultivable fishes of Kali estuary, whereas, present investigation is first its kind from this region, which included all the fishes. Fish and prawn seed resources of mangroves are largely destroyed in Sunderbans, about 40,000 people harvest about 540 million seeds of tiger prawn (*Penaeus monodon*) every year, and in this process, about 10.6 million seeds of other fishes and shrimps are killed, which will have impact on fish diversity and fisheries resources (Choudhuri and Choudhuri, 1994). The total number of mangrove inhabiting faunal species (Kathiresan and Quasim, 2005) in Indian mangroves is 3,111. This includes 55 species of prawns (1.8 per cent of total fauna), 138 species of crabs (4.4 per cent), 308 species of mollusc (9.9 per cent), 745 species of other invertebrates (23.9 per cent), 546 fish species of fish (17.6 per cent), 7 species of fish parasites (0.2 per cent), 13 species of amphibians (0.4 per cent), and 70 species of mammals (2.3 per cent). From this study, it can be concluded that Kali estuary is one of non polluted, which harbors 37 fish species, belongs to 20 families.

References

Choudhuri, A.B. and Choudhuri, A., 1994. *Mangroves of the Sunderbans, India.* IUCN, Glad, Switzerland, 1: 247.

Francis, Day, 1978. *The Fishes of India: Being a Natural History of the Fishes Known to Inhabit the Seas and Freshwaters of India, Burma and Ceylon.* Today and Tomorrow's Book Agency, New Delhi.

Francis, Day, 1980. *Identification of fishes of India.*

Kathiresan, K. and Bingham, B.L., 2001. Biology of mangroves and mangrove ecosystems. *Adv. Mar. Biol.*, 40: 81–151.

Katheresan, K. and Rajendran, N., 2005. Mangrove ecosystems of the Indian Ocean region. *Ind. Jour. Mar. Sci.*, 34(1): 104 –113.

Kathiresan, K. and Quasim, S.Z., 2005. *Biodiversity in Mangroves Ecosystems.* Hindustan Publishers, New Delhi.

Nagaraj, M. and Neelakantan, B., 1982. Fish and shellfish seed resources of Kali estuary along with a note on the mariculture potenties in Uttara Kannada. *Proc. Symp. Coastal Aquaculture*, 1: 383–387.

Strickland, J.D. H. and Parsons, T.R., 1975. A manual of sea water analysis. *Fish. Res., Bd., Canada*, Ottawa, p. 310.

Strickland, J. D. H. and Parson, T. R., 1975. A practical handbook of sea water analysis. *Bull. Fish. Res. Bd. Canada*, 167: 310.

Chapter 17

Marine Organisms as a Source of Drug: An Overview

☆ *J.L. Rathod and P.V. Khajure*

ABSTRACT

Marine organisms comprise approximately a half of the total biodiversity, thus offering a vast source to discover useful therapeutics. In recent years a significant number of novel metabolites with pharmacological properties have been discovered from marine organisms. Although there are only few marine derived products currently available in the market. Several marine products are now in clinical trails. Current research activities, while primarily within the academic laboratories have generated convening evidence that marine natural products have an exceedingly bright future in the discovery of life saving drugs.

Introduction

Life has originated from the oceans that cover over 70 per cent of the earth's surface and contain highly ecological, chemical and biological diversity starting from microorganisms to vertebrates. This diversity has been the source of unique chemical compounds, which hold tremendous pharmaceutical potential. New trends in drug discovery from natural sources emphasize on investigation of the marine ecosystem to explore numerous complex and novel chemical entities. These entities are the sources of new leads for treatment of many diseases such as cancer, AIDS, inflammatory conditions, and a large variety of viral, bacterial and fungal diseases[1]. Because of the highly chemical and physical harsh conditions in marine environment, the organisms produce a variety of molecules with unique structural features and exhibit various types of biological activities. Majority of the marine natural products have been isolated from sponges, coelenterates (sea whips, sea fans and soft corals), tunicates, opisthobranch molluscs (nudibranchs, sea hares, etc.), echinoderms (starfish, sea cucumbers, etc.) and bryozoans (moss animals) and a wide variety of marine microorganisms in their tissues[2].

Sponges, the most primitive multicellular invertebrates, considered as a gold mine during the past 50 years, have fascinated scientists for isolation of promising bioactive compounds for human welfare. Published literature, patents and other scientific records on the genotoxicity and anticancer potentials of marine compounds revealed that few compounds have gone through preclinical evaluations. Interestingly, cytarabine (Cytostar-U) also known as Ara-C, a compound isolated from the Caribbean sponge *Cryptotheca crypta* currently being used with other anticancer drugs in the treatment of acute myelocytic leukaemia (AML) and lymphomas is one of the very few marine anticancer drugs studied in long-term clinical study[3]. Acyclovir, which was synthetically known as Ara-A, was modeled based on sponge-derived spongothymidine or spongouridine. Ara-A is the first sponge-derived antiviral compound in the market. Polyketide Calyculin A (a selective inhibitor of protein phosphatase 1, isolated from sponge *Discodermia calyx*), Manoalide (a potent anti-inflammatory marine natural product and a direct inactivator of venom phospholipase A2), Okadaic acid, a potent inhibitor of protein phosphatases, especially protein phosphatases 1 and 2 respectively isolated from *Luffariella variabilis* and *Halichondria okadai* has reached the market undergoing from basic research to long phases of clinical study[4, 5].

Saclike filter feeder tunicates have been reported to be an importent source in drug discovery. Tetrahydroisoquinolone alkaloid 'Ecteinascidin 743' from *Ecteinascidia turbinate*, cyclic depsipeptides 'Dehydrodidemnin B and Didemnin B' from *Trididemnum solidum*, cyclic peptide 'Vitilevuamide' from *Didemnin cucliferum* and 'Diazonamide' from *Diazona angulata* are a few tunicate compounds in anticancer preclinical or clinical trials. Synthadotin and Soblidotin are two synthetic analogues of Dolastatin isolated from molluscan sp. *Dolabella auricularia* in trials. Alkylamino alcohol 'ES-285' (Spisulosine) isolated from *Mactromeris polynyma* is another compound in preclinical trial; molecular target of this molluscan compound is Rho (GTP-bp). Macrocyclic lactone 'Bryostatin 1' from bryozoan sp. *Bugula neritina*, and soft coral compounds; diterpene glycoside 'Eleutherobin' and Pyrrole alkaloid 'Lamellarin D' anticancer[6] while, anti inflammatory compound 'OAS-100' which is semisynthetic derivative of pseudopterisone A are the hope of new effective therapeutic agents.

Study of marine organisms is a discipline, which endeavours to identify and decipher the troubles regarding not only sustainable exploitation of marine life for human health and welfare but also for marine ecology. Study of marine organisms for their bioactive potential, being an important part of marine ecosystem, has picked up the rhythm in recent years with the growing recognition of their importance in human life. This interdisciplinary study of the life in the oceans ensures an exciting new frontier of scientific discovery and economic opportunity.

Marine biotechnology is the science in which marine organisms are used in full or partially to make or modify products, to improve plants or animals or to develop microorganisms for specific uses. With the help of different molecular and biotechnological techniques, humans have been able to elucidate many biological methods applicable to both aquatic and terrestrial organisms. According to[7], only 10 per cent of over 25,000 plants have been investigated for biological activity. The marine environment may contain over 80 per cent of world's plant and animal species[8]. In recent years, many bioactive compounds have been extracted from various marine animals like tunicates, sponges, soft corals, sea hares, nudibranchs, bryozoans, sea slugs and marine organisms[9,10]. The search for new metabolites from marine organisms has resulted in the isolation of more or less 10,000 metabolites[11], many of which are endowed with pharmacodynamic properties. The deep knowledge about nerve transmission has been learnt using squid and its giant nerve axons and the mesenteries of vision have been unraveled using the eyes of horseshoe crabs, sharks and skates. The surf clam is proving an excellent model for the cell cycle and its regulation while the sea urchin is a model for understanding the molecular basis of cellular reproduction and development. The objective of this review is to highlight some of the recent developments and findings in the area of marine biotechnology with special reference to the biomedical potential of marine natural products.

Occurrence of Marine Natural Products

Natural products have long been used as foods, fragrances, pigments, insecticides, medicines, etc. Due to their easy accessibility, terrestrial plants have served as the major source of medicinally useful products, especially for traditional or folk medicine. According to[12], about 25 per cent of all pharmaceutical sales are drugs derived from plant natural products and an additional 12 per cent are based on microbially produced natural products. The marine environment covers a wide thermal range (from the below freezing temperatures in Antarctic waters to about 350°C in deep hydrothermal vents), pressure range (1-1000 atm), nutrient range (oligotrophic to eutrophic) and it has extensive photic and aphotic zones. This extensive variability has facilitated extensive speciation at all phylogenetic levels, from microorganisms to mammals. Despite the fact that the biodiversity in the marine environment far exceeds that of the terrestrial environment, research into the use of marine natural products as pharmaceutical agents is still in its infancy. This may be due to the lack of ethno-medical history and the difficulties involved in the collection of marine organisms[13]. But with the development of new diving techniques, remote operated machines, etc., it is possible to collect marine samples and during the past decade, over 5000 novel compounds have been isolated from shallow waters to 900-m depths of the sea[14].

Hints from the Physiological Study of Marine Organisms

Life originated in the sea and during evolution, marine organisms have developed into very sophisticated physiological and biochemical systems. During the adaptation to the terrestrial environment, a number of physiological changes have taken place, but in most cases, the basic functions were almost completely retained. The architecture of the shark liver is similar to that of the human liver and the biochemical transformations which take place in a shark's liver, appear to be similar to those that occur in a human liver[15], with slight modifications[16]. The eyes of man and octopus are very similar in structure and function irrespective of the fact that no evolutionary link exists between them[17]. Insulin from fish such as cod exerts the same hormonal activity in mammals as does homologous insulin and insulin from tuna (which has a 40 per cent difference in amino acid residue[18]) that has been used to treat diabetic patients. This suggests that the basic physiological functions of molecules may remain the same regardless of the structural changes, which may possibly occur during evolution[19]. Marine thermococcales have been an important source of high fidelity thermostable DNA polymerases (Pfu, Vent, Pab, etc.)[20] and in addition, the high structural conservation and complementation of DNA replication proteins between euryarchaeal *Pyrococcus* and humans make hyperthermophilic archaea a model of choice to study eukaryotic DNA replication[21]. The knowledge of the physiological and biochemical features of marine organisms might contribute to the identification of natural products of biomedical importance. According to an extract of regenerating fish nerve may induce regeneration of an injured nerve in rabbit.

Role of Marine Microorganisms in Drug Discovery

For more than two decades, there has been an ongoing quest to discover new drugs from the sea. This quest has been manifested in many forms but, as is demonstrated by the other chapters in this book, most efforts have been directed towards chemical studies of marine invertebrates. Although these studies have indeed proven that marine invertebrates are an important source of new biomedical leads, a fact well demonstrated by the number of compounds currently in clinical trials, it has proven notoriously difficult to obtain adequate, reliable, and most importantly, renewable, supplies of these compounds from nature. Compounding the supply problem is the inherent structural complexity of many marine natural products which most often eliminates the possibility of commercially viable syntheses. Because of these problems, and the fact that more than 20 years of extensive research has made it increasingly difficult to isolate novel metabolites from marine invertebrates, a new avenue of study focusing on marine microorganisms has been garnering considerable attention. Although marine microorganisms are not well defined taxonomically, preliminary studies in this field unequivocally indicate that the wealth of microbial diversity in the world's oceans, coupled with its biochemical uniqueness, make this a promising frontier for the discovery of new medicines.

The foundation for chemical studies of marine microorganisms derives from the historical significance of terrestrial microbes as a source of pharmaceuticals. From the discovery of penicillin more than 60 years ago for the treatment of microbial infection, to the more recent anticancer chemotherapeutic agent adriamycin and the important immunosuppressant drug cyclosporine A, microorganisms have yielded over 120 of today's most important medicines. The rate at which new compounds are being discovered from traditional microbial resources, however, has diminished significantly in recent decades as exhaustive studies of soil microorganisms repeatedly yield the same species which in turn produce an unacceptably large number of previously described compounds. In an effort to improve the rates at which new classes of secondary metabolites are discovered, new microbial resources are being sought. These efforts now include studies of marine microorganisms. Although special techniques are required for the isolation and cultivation of marine microorganisms, many of the principles developed for the commercial fermentation industry can be applied making this a controllable resource that does not face the same supply problems associated with the acquisition of compounds from marine invertebrates.

Unlike marine invertebrates, which can be clearly defined, there has been some debate as to what constitutes a marine microorganism. Marine bacteria are most generally defined by their requirement of seawater, or more specifically sodium, for growth [22]. In the case of marine fungi, which in general do not display specific ion requirements, obligate marine species are generally considered to be those that grow and sporulate exclusively in a marine habitat[23]. Although such definitions can prove useful, they tend to select for a subset of the microorganisms that can be isolated from any one environment. This problem is compounded in the case of near-shore or estuarine samples where a large percentage of the resident microbes are adapted to varying degrees of marine exposure. For the purpose of microbial drug discovery, it seems only logical to study all microbes that can be isolated from the marine environment. Once new products are discovered, and then questions about the in situ distributions and metabolic activities of the producing species can be asked.

Most of the new compounds reported from marine microorganisms were obtained from species that can, in principle, be isolated from both land and sea. Although these facultative marine species are clearly a good source of novel metabolites, their ecological roles and degrees of adaptation to the marine environment remain largely unknown. What is clear about these isolates however is that they are yielding unprecedented structures with potent biomedical activities. A finding that suggests that despite, in some cases, apparent taxonomic affiliations with terrestrial species, environmental differences are sufficient for novel compound production.

Microalgae as a Drug Source

Recently, microalgal metabolites are attracting enormous attention, and the topic has been discussed by a number of authors. This author also reviewed the chemical and biochemical aspects of microalgal metabolites[24, 25, and 26]. Therefore, emphasis in this chapter will be placed on some issues concerned with drug screening.

There are a couple of reasons behind the surge of interest in microagal metabolites. First, the traditional drug sources such as terrestrial higher plants and Actinomycetes have been examined very extensively for a length of time, and the screening works are suffering from a high rate of redundancy in both structural type and mechanism of pharmacological action. Here, the microalgal phyla have been recognized to provide chemical and pharmacological novelty and diversity. Another reason for the enthusiasm is the recognition that microalgae are the actual producers of some highly bioactive compounds found in marine invertebrates.

Marine organisms have been looked at as a very promising source of therapeutic drugs. In 1967, the first conference of drugs from the sea was convened at University of Rhode Island [27]. However, more than three decades later, no significant therapeutic drugs have been derived from marine organisms, while several thousand new chemical structures have been discovered from marine organisms[28]. This does not signify, however, that the marine compounds are barren with respect to pharmacological activity. On the contrary, there are strong indications that many of them possess novel bioactivity. The major obstacle to the drug development of marine natural products is the lack of sufficient material for comprehensive pharmacological evaluation. More often than not, compounds are isolated in milligram quantities, and their structures are elucidated and published. This probably means that chances for the compounds to be developed to drugs will be lost for good. No serious enterprise will invest hundreds of millions of dollars for a compound without proper patent protection.

Invertebrates such as sponges, tunicates and soft corals, have been the richest sources of new compounds. With some exceptions, very limited numbers of organisms can be collected for chemical studies. Recollections of the organisms are often impossible or ecologically undesirable, and the procurement of enough material for wide-spectrum assays, not to mention the preclinical or clinical studies, is almost prohibitive. In the case of compounds such as bryostatins[29] or halichondrins[30], the source organisms are relatively abundant, but their low contents (~ppb) limit the advanced clinical studies. For those compounds having complex structures, total synthesis is not practical for a procurement purpose. Thus, without a reproducible resource to supply the compounds, it is impossible to move the compound forward for actual therapeutic use.

The true origins of compounds found in marine invertebrates have been a subject of constant discussion. They may vary case by case, but there are strong hints that dietary or symbiotic algae are one of the participants in the primary production of the metabolites. For example, as early as 1977, the blue-green alga, *Lyngbya majuscula* was recognized as the source of aplysiatoxin found in the sea slug *Aplysia*[31]. Similarly, a series of highly active antitumor compounds, dollastatins, isolated from sea slugs by Pettit's group are considered to be of blue-green algal origin. In fact, a close analogue of dolastatin 10 isolated from *Dollabella auricularia* was found to be a metabolite of a blue-green alga[32]. While the aforementioned compounds are probably introduced into the invertebrates by the food chain, some others may have their origins in the symbiotic organisms. The antitumor, antiviral cyclic depsipeptides, didemnins[33], have been suspected to be produced by photosynthetic prokaryotes, prochlorons, which are symbionts in most *Didemnum* tunicates[34], although no proof for the production has been presented because of the unculturability of the organism. The cytotoxic sponge metabolites, swinholide and analogues, have close relatives such as tolytoxin in blue-green algae [35].

In the eukaryotic algae, various dinoflagellate metabolites are found in shellfish and other invertebrates as toxins. Well-known examples are paralytic shellfish toxins, brevetoxins, ciguatoxins, and dinophysistoxins[36]. From the sponge *Halichondria okadaii*, potent antitumor compounds, halichondrins, have been isolated [37]. The fact that the same sponge also contains okadaic acid, a known dinoflagellate metabolite, suggests that halichondrins with similar polyoxygenated alkyl structures are also of dinoflagellate origin [38].

All these observations led to a high hope to produce the compounds by culturing microalgae.

Metabolites from Marine Cyanobacteria

The fact that cyanobacteria in general and marine forms in particular are one of the richest sources of known and novel bioactive compounds including toxins with wide pharmaceutical applications is unquestionable. Among the five divisions of microalgae, studies of biomedical natural products have been concentrated on only two divisions, *i.e.*, cyanophyta (blue-green algae) and Pyrrophyta (dinoflagellates). Although several metabolites have been isolated from cyanophytes[39,40], most of them are isolated from freshwater species, which are cultured easily in comparison to marine organisms. Lyngbyatoxin-A and debromoaplysiatoxin are two highly inflammatory but structurally different metabolites isolated from toxic strains of *Lyngbya mausculata* collected in Hawaii[41], and anatoxin-a from *Anabaena ciecinalis*. Some of the marine cyanobacteria appear to be potential sources for large-scale production of vitamins of commercial interest such as vitamins of the B complex group and vitamin-E[42]. The carotenoids and phycobiliprotein pigments of cyanobacteria have commercial value as natural food colouring agents, as feed additives, as enhancers of the color of egg yolks, to improve the health and fertility of cattle, as drugs and in the cosmetic industries. Some anti-HIV activity has been observed with the compounds extracted from *Lyngbya lagerhaimanii* and *Phormidium tenue*. More than 50 per cent of the 100 isolates from marine sources are potentially exploitable bioactive substances. The substances tested for were either the ones that killed cancer cells by inducing apoptotic death, or those that affected cell signaling through activation of the members of protein kinase-C family of signaling enzymes[43,44,45]. Cultured *Fusarium chlamydosporum* isolated from the Japanese marine red alga *Carpopeltis affinis* is the source of fusaperazines A and B, two new sulphur containing dioxopiperazine derivatives, and two known compounds which had been originally isolated from the fermentation by the fungus *Tolypocladium* spp.[46]. Chalcomycin-B exhibited activity against a variety of microorganisms and microalgae [47]. Four new epipolysulphanyldioxopiperazines were isolated from a culture of the fungus *Leptosphaeria* spp. originating from the Japanese brown alga *Sargassum tortile*[48]. Absolute stereochemistries were determined by chemical analyses and transformations. Each compound possessed significant cytotoxic activity against the P388 cell-line, while one of the leptosins also exhibited appreciable cytotoxicity against a disease-oriented panel of 39 human cancer cell-lines, and specifically inhibited two protein kinases and topoisomerase-II[49]. Cultures of the marine fungus *Hypoxylon oceanicum*[50] from mangrove wood at Shenzen, China, yielded the macrocyclic polyesters and the linear polyesters[51]. The absolute configurations of the polyesters were deduced from circular dichroism (CD) spectral studies. The compounds exhibited modest activity against the phytopathogenic fungus *Neurospora crassa*. The anti inflammatory and anti-proliferative properties of scytonemin, an extracellular sheath pigment originally isolated from the cyanobacterium *Stigonema* spp. have been reported[52,53,54]. Goniodomin-A, an antifungal polyether macrolide from the dinoflagellate *Goniodoma pseudogoniaulax*[55] has been shown to inhibit angiogenesis by the inhibition of endothelial cell migration and basic fibroblast growth factor (bFGF)-induced tube formation and is active *in vivo*[56]. An immunosuppressive linear peptide microcolin-A, which at nanomolar concentrations suppresses the two way murine mixed lymphocyte reaction, has been isolated from *Lyngbya majusculata*[57]. A unique thiozoline-containing compound, curacin- A, has been purified from the organic extract of a Curacao collection of *L. majusculata*[58]. This compound has been found to be an exceptionally potent antiproliferative agent as it inhibits the polymerization of tubulin, which shows some selectivity for colon, renal and breast cancer-derived cell lines. A series of noval antibiotics agents have been isolated from dianoflagellates, antifungal agents from *Gambierdiscus toxicus*[59] and brevitoxins from *Ptychodiscus brevis*. As they depolarize the excitable membranes and their binding sites on sodium channel the mechanism seems to be different from that of other activators[60]. Okadaic acid, a polyether fatty acid produced by *Prorocentrum* spp., has been a key molecule in studying signal transduction pathways in eukaryotic cells since it is a selective protein phosphatase inhibitor[61].

Metabolites from Seaweeds

Seaweeds are abundant in the intertidal zones and in clear tropical waters. Marine algae have received comparatively less bioassay attention. Presently the seaweed industry consists of two kelps, three *Gelidium* species one *Gracilaria-/Gracilariopsis* species, etc. [62]. In addition, there are a number of seaweeds with economic potential [63]. It will be of great significance if these species could be the major role players in drug development. Alternatively, findings from academic laboratories could result in new cultivation initiatives. Nonetheless, the red alga *Sphaerococcus coronopifolius* was shown to have antibacterial activity; the green alga *Ulva lactuca* was shown to posses an anti-inflammatory compound; and an anti-tumor compound was isolated from *Portieria hornemannii* [64]. *Ulva fasciata* produces a novel sphingosine derivative has been found to have antiviral activity *in vivo*[65]. A cytotoxic metabolite, stypoldione, which inhibits microtubule polymerization and thereby prevents mitotic spindle formation, has been isolated from tropical brown alga, *Stypodium zonale*[66,67]. *P. hornemannii* is found to be a novel source of cytotoxic penta halogenated monoterpene, halomon, which exhibited one of the most extreme examples of differential cytotoxicity in the screening conducted by the National Cancer Institute (NCI), USA. Halomon has been selected for preclinical drug development since this compound shows toxicity to brain, renal and colon tumor cell-lines and preliminary *in vivo* evaluations have been encouraging. An iodinated novel nucleoside has been isolated from *Hypnea valitiae*, which is a potent and specific inhibitor of adenosine kinase. It can be used in the studies of adenosine receptors in a variety of systems, and in studies on nucleotide metabolism and regulation[68].

The green alga *Codium iyengarii* from the Karachi coast of the Arabian Sea has been found as the source of a steroid, iyengadione and two new steroidal glycosides, iyengarosides A and B. Iyengaroside-A displayed moderate activity against a range of bacteria [69]. *Sargassum carpophyllum* from the South China Sea is the source of two new bioactive sterols. These sterols induced morphological abnormality in the plant pathogenic fungus *Pyricularia oryzae*; also exhibited cytotoxic activity against several cultured cancer cell lines[70]. *Sargassum polycystum* collected in the North China Sea yielded a new sterol, stigmast[71]. The fact that there are many algae that can convert simple polyunsaturated fatty acids such as arachidonic acids into complex eicosanoids and related oxylipins has been an exiting development[72]. Derivatives of arachidonic acids are important in maintaining homeostasis in mammalian systems and aberrant production of metabolites of this class occurs in diseases such as psoriasis, asthma, arteriosclerosis, cardiac diseases, ulcers and cancer.

Metabolites from Sponges

Approximately 10,000 sponges have been described in the world and most of them live in marine waters. A range of bioactive metabolites has been found in about 11 sponge genera. Three of these genera (*Haliclona*, *Petrosia* and *Discodemia*) produce powerful anti-cancer, anti-inflammatory agents, but their cultivation has not been studied[73]. The discovery of spongouridine, a potent tumor-inhibiting arabinosyl nucleoside in Caribbean sponge *Cryptotethia crypta*, focused attention on sponges as a source of biomedically important metabolites. The identification of the pharmacophore led to the synthesis of a new class of arabinosyl nucleoside analogues, one of which is arabinosyl cytosine, which is converted into arabinosyl cytosine triphosphate and incorporated into cellular DNA where it inhibits DNA polymerase, is already in clinical use for the treatment of acute mylocytic leukemia and non-Hodgkin's lymphoma. The compound manoalide from a Pacific sponge has spawned more than 300 chemical analogs, with a significant number of these going on to clinical trials as anti-inflammatory agents. An aminoacridine alkaloid, dercitin, has been isolated from the deep-water sponge, *Dercitus* spp. That possesses cytotoxic activities in the low nanomolar concentration range and in animal studies, prolongs the life of mice-bearing ascitic P388 tumours, and is also active against B16 melanoma cells and small cell Lewis lung carcinoma[74]. Halichondrin-B, a polyether macrolide from Japanese sponge *Theonella* spp., has generated much interest as a potential anticancer agent[75]. The theopederins are structurally related to mycalamide-A from marine sponge, *Mycale* spp. collected in New Zealand [76] and onnamide-A from marine sponge, *Theonella* spp. collected in Okinawa[77], which show *in vitro* cytotoxity and *in vivo* antitumour activity in many leukemia and solid tumour model systems[78]. Isoquinolin equinone metabolite cribostatin from the Indian Ocean sponge *Cribrochalina* spp. shows selective activity against all nine human melanoma cells in National Critical Technologies (NCT) panel [79]. Spongstatin, a macrocytic lactone from the Indian Ocean collection of *Spongia* spp., is the most potent substance known against a subset of highly chemoresistant tumour types in the NCT tumour panel [80]. Two new -pyrones (herbarin) along with a new phthalide, herbaric acid, were isolated from two cultured strains of the fungus *Cladosporium herbarum* isolated from the sponges *Aplysina aerophoba* and *Callyspongia aerizusa* collected in the French Mediterranean and in Indonesian waters, respectively [81]. Herbarins displayed activity in the brine shrimp assay. A culture of the fungus *Emericella variecolor* isolated from a sponge collected in the Caribbean Sea off Venezuela yielded varitriol, varioxirane, dihydroterreinand varixanthone, which were characterised by spectroscopic methods and chemical transformations[82]. Varitriol displayed increased potency toward some renal, central nervous system and breast cancer cell-lines in the NCI's 60-cell line panel, while varixanthone displayed antimicrobial activity against a range of bacteria. The antimicrobial glycolipid caminoside-A, isolated from Dominican specimens of *Caminus sphaeroconia*, was found to be a potent inhibitor of the bacterial type-III secretion system[83]. Lembehynes B and C, isolated from an Indonesian species of *Haliclona*, were found to possess neuritogenic activity against neuroblastoma cells[84].

Potent phosphate inhibitors have been isolated from sponges like, okadaic acid from *Halichondria okadai*, motuporin from *Theonella swinhoei* and calyculin-A from *Discodermia calyx*[85,86]. Inhibitors of phospholipase such as manoalide and scalaradial have proved to be useful tools to study the role of this enzyme in the release of arachidonic acid, which is a key molecule, involved in the biochemical processes leading to inflammation [87]. A number of receptor antagonists with potential as biochemical tools or structural leads to the development of therapeutics have been isolated from sponges. Examples include xestobergsterol (isolated from *Xestospongia berguista*), which inhibits immunoglobulin E mediated histamine release from mast cells and is 5000 times more potent than the antiallergic drug disodium cromoglycate [88]. Leucettamine A isolated from *Leucetta microraphis*, is a potent and selective antagonist for the receptor for leukotrine, a non-peptide metabolite of arachidonic acid produced mainly in inflammatory cells [89]. Batzelladine A and B, novel polycyclic guanidine alkaloids from the Carribean sponge *Batzella* spp., exhibit potent inhibition to the binding of HIV glycoprotein, on CD4 receptors of T cells. The series of polymethoxydienes, similar to the alkenes isolated from the cyanophyte *Tolypothrix conglutinata*[90], were isolated from a Philippine specimen of *Myriastra clavosa*, and found to be moderately cytotoxic. *Plakortis nigra*, collected from a depth of 115 m in Palau, was found to contain epiplakinic acid G and H, and the -lactones along with several -carbolines (vide infra). All compounds have been found to inhibit the growth of HCT-116 cells [91]. A peroxylactone originally isolated from a

Plakinastrella species [92] has been synthesized as a racemic mixture [93]. Two new 1,2-dioxolane peroxide acids have been isolated from *Porolithon onkodes*[94]. The moderately cytotoxic thioester irciniamine has been isolated from an *Ircinia* spp. collected in Japan. The previously reported motuporins A–C [95] along with the new congeners, motuporins D–F have been found to inhibit the invasion of breast carcinoma cells into new tissues. These compounds have been isolated from *Xestospongia exigua* collected in Papua New Guinea along with an unresolved mixture of three isomers of motuporins. *Hyrtios erecta* collected from the Egyptian Red Sea has been found to contain salmahyrtisol A and B and sesterstatins, all of which have shown significant cytotoxicity in human cancer cell-lines[96]. A peroxy steroid, from an Okinawan species of the genus *Axinyssa*, has been found to inhibit the growth of several human cancer cell-lines [97]. Three oxygenated sterols have been obtained from a collection of *Polymastia tenax*. The compounds have been found to have significant cytotoxicity to a range of human and murine cancer cell-lines [98].

Metabolites from Cnidarians

The discovery of prostaglandin in corals in the late 1960s contributed greatly to the rapid developments in the field of marine natural products. Palytoxin, which is one of the most potent known toxins, is the product of *Palythoa* species of the family Zoanthidae. It is a useful tool for probing cellular recognition processes since it stimulates arachidonic acid metabolism and down-regulates the response to epidermal growth factor by activating a sodium pump in the signal transduction pathway using sodium as the second messenger. Bioassay-guided fractionation of extracts obtained from soft coral, *Lobophytum crassum*, indicated ceramideas a moderately antibacterial component [99]. New examples of cadinene-skeleton sesquiterpenes, xenitorins A–F, have been isolated from *Xenia puerto-galerae*[100]. The relative stereochemistries of xenitorins A–F are secured by nuclear overhauser enhancement spectroscopy nuclear magnetic resonance (NOESY NMR) experiments. Xentorin A and E exhibited cytotoxicity towards the A and P388 tumour cell-lines. The structure and stereochemistry of alcyopterosin-E, a nitrate ester-containing sesquiterpene isolated from *Alcyonium paessleri*[101], was secured by total synthesis [102]. Lophotoxin from the genus *Lophogorgia* preferentially binds to the nicotinic subunit of acetylcholine receptors and blocks out cholinergic nicotinic pathway in a complex set of interacting neurons. Pseudopetrocin-E, a tricyclic diterpene pentoside from gorgonians of the genus *Pseudopterogorgia*, shows anti-inflammatory and analgesic activities equal in potency to industrial standard indomethicine. A further study of *Subergorgia suberosa* yielded the sesquiterpene suberosols A–D[103]. Relative stereochemistries have been determined by NOESY NMR experiments and all the four metabolites exhibited cytotoxicity towards the P388 murine leukaemia cell-line, while suberosol C and D also exhibited cytotoxicity towards the A 549 and HT-29 tumour cell- lines.

The first chemical study of the soft coral *Lemnalia flava*, collected off Mombasa, Kenya, has yielded lemnaflavoside and three monoacetate derivatives [104]. Clavubicyclone from *Clavularia viridis* exhibited mild cytotoxicity towards MCF-7 and OVCAR-3 tumour cell-lines[105]. Bioassaydirected fractionation of the soft coral *Cespitularia hypotentaculata* yielded diterpene cespitularin A–D, a norditerpene cespitularin E and three further diterpenes, cespitularin F–H, with a novel skeleton. Variable potency and selectivity was observed for the eight compounds towards tumour cell-lines A-549, HT-29 and P388. Two new dolabellane-type diterpenoids as well as the known diterpene clavenone[106] were isolated from *Clavularia* species[107]. An artificial culture of *Erythropodium caribaeorum* has been found to produce a range of diterpenes including the antimitotic agents eleutherobin and aquariolide-A[108]. Saponin was isolated from *Lobophytum* spp. collected from Hainan Island, China. Further[109] investigation of the stony coral *Montipora* spp. from Korea yielded three diacetylene, one of them were the most potent cytotoxin towards a range of tumour cell-lines[110]. *Radianthus macrodactylus*, collected in the Seychelles, yielded three high molecular weight (20 kDa) cytolysins, two low molecular weight cytolysins, RmI (5100 Da) and RmII (6100 Da), and InI, a 7100 Da trypsin inhibitor[111]. The sodium channel toxins Bg II and Bg III, isolated from the sea anemone *Bunodosoma granulifera*[112], have been found to be especially potent towards insect sodium channels[113]. The extracts from *Pseudopterogorgia elizabethae* (contains pseudopterosins) and *Eunicea fusca* (contains fucoside-A) can be used in cosmetic industries[114,115].

Metabolites from Bryozoans

The bioactive compounds are comparatively less in quantity from bryozoans. Most of the extracted products are alkaloids. A sample of *Flustra foliacea* collected in the southern North Sea yielded deformylflustrabromine, which displayed moderate cytotoxicity against the HCT-116 cell-line[116]. The marine bryozoan *Amathia convoluta* collected from the east coast of Tasmania was the source of the tribrominated alkaloids convolutamine-H and convolutindole-A. The compounds displayed potent and selective activity against *Haemonchus contortus*, a parasitic nematode of ruminants[117]. *Watersipora subtorquata* from Tsutsumi Island, Japan, was the source of bryoanthrathiophene. This compound exhibited potent anti-angiogenic activity on bovine aorta endothelial cell (BAEC) proliferation[118]. Asymmetric syntheses of amathamide A and B, alkaloids from the bryozoan *Amathia wilsoni* collected in Tasmania[119], have been accomplished starting from 3-hydroxybenzaldehyde[120]. Bryostatin, a potent anti-cancer compound from *Bugula neritina*[121] shows remarkable selectivity against human leukemia, renal cancer, melanoma and non-small cell lung cancer cell-lines. This compound modulates the signal transduction enzyme protein kinase-C (PKC). The major metabolite convolutamide-A from *Anthia convoluta* exhibits *in vitro* cytotoxicity against L1210 murine leukaemia cells and KB human epidermoid carcinoma cells[122]. *Cribricellina cribreria* has yielded â-carboline alkaloid, which exhibited cytotoxic, antibacterial, antifungal and antiviral activities[123]. Indole alkaloids isolated from *Flustra foliacea* have shown strong antimicrobial activity[124].

Metabolites from Molluscs

More than 2600 scientific studies over the last 20 years testify to the important contribution of toxins extracted from cone snails to medicine and cellular biology. To date, only 100 out of a potential 50,000 toxins have been extracted and analyzed[125]. The *Conus* species have evolved deadly nerve toxins and small, conformationally constrained peptides of 10-30 amino acids. Some of the conotoxins block channels regulating the flow of potassium or sodium across the membranes of nerve or muscle cells; others bind to N-methyl-D-aspartate receptors to allow calcium ions into nerve cells; and some are specific antagonists of acetylcholine receptors responsible for muscle contraction. Thus, conotoxin are valuable probes in physiological and pharmacological studies[126]. Neosurugatoxin isolated from *Babylonia japonica* is useful in characterizing two classes of acetylcholine receptors. Dolastatin, a cytotoxic peptide from *Dolabella auricularia* is an antineoplastic substance[127]. Ulapualide-A, a sponge-derived macrolide isolated from the nudibranch *Hexabranchus sanguineus* exhibits cytotoxic activity against L 1210 murine leukemia cells and antifungal activity, which exceeds that of clinically useful amphotericin-B[128]. Chromodorolide-A isolated from *Chromocloris cavae* exhibits *in vitro* antimicrobial and cytotoxic activities[129]. Onchidal from *Onchidella bieyi* is a useful probe for identifying the active site residues that contribute to binding and hydrolysis of acetyl cholinesterase. A team from the University of Melbourne extracted the conotoxin from a cone-shell snail. It not only inhibits pain as being 10,000 times more powerful than morphine, but also accelerates the recovery of injured nerves[130]. The absolute stereochemistries of membrenones A–C,dihydropyrone-containing polypropionates isolated from the skin of the Mediterranean mollusc *Pleurobranchus membranaceus* [131], have been determined by stereocontrolled syntheses of the enantiomers[132]. The first synthesis of siphonarin-B has confirmed the absolute stereochemistry of the metabolite [133] isolated from the molluscs *Siphonaria zelandica* and *S. atra*[134]. Bursatellanin-P, a 60-kDa protein was purified from the purple ink of the sea hare *Bursatella leachii*[135]. The protein exhibited anti-HIV activity. The first total syntheses of aplyolides B–E, ichthyotoxic macrolides isolated from the skin of sea hare *Aplysia depilans*[136], have been reported confirming the absolute stereochemistry reported for the metabolites[137,138].

Metabolites for Tunicates

Didemnin-B from the Caribbean tunicate *Trididemnum solidum* was the first marine compound to enter human cancer clinical trial as a purified natural product, but was unsuccessful in further trials[139]. Nevertheless, this class of cyclic peptides provides important structural lead for a variety of antiviral, anticancer and immunosuppressant activities [140]. The inhibitor of matrix metalloproteinase (MMP2) from an ascidian of the family Polyclinidae collected off Western Japan was identified as sodium1- (12-hydroxy) octadecanyl sulphate[141]. Two unusual trithiocane derivatives were isolated from the ascidian *Perophora viridis* collected off North Carolina[142]. Relative stereochemistries were deduced from NOESY NMR experiments, while methylthiopropionate (MTPA) derivation of the hydroxyl helped secure the absolute configuration. Both compounds exhibited mild antibacterial activity as well as toxicity towards brinc shrimp. Halocidin was isolated as an antimicrobial peptide (3443 Da) from the hemocytes of the solitary ascidian *Halocynthia aurantium*[143]. Cloning of a peptide precursor from a cDNA library prepared from pharyngeal tissues of the tunicate *Styela clava* identified clavaspirin as an antibacterial peptide[144]. Lepadins D with an unidentified counterion, lepadins E and lepadins F were isolated as antiplasmodial and antitrypanosomal alkaloid constituents of a *Didemnum* spp. ascidian collected from Stanley Reef, the Great Barrier Reef[145]. Coproverdine is a cytotoxic alkaloid isolated by bioassay-directed fractionation of an unidentified ascidian collected at the Three Kings Islands, New Zealand[146]. Ecteinascidin isolated from *Ectenascidia turbinate* shows potent activity *in vivo* against a variety of mouse tumour cells[147]. Cytotoxicity towards a variety of murine and human tumour cell-lines was observed. Rubrolide-M, recently isolated from a Spanish collection of the ascidian *Synoicum blochmanni*[148], was synthesised using palladium catalysed coupling methodology[149]. Eudistomins from *Eudistoma* species exhibit potent antiviral activity *in vitro* and have been synthesized in quantities sufficient for *in vivo* antiviral analysis[150]. Besides eudistomins, a number of potent PKC inhibitors have been isolated from *Eudistoma* spp., which includes staurosporine aglycone, 11-hydroxy staurosporine, trithianes and pentathiepins[151,152,153]. The compound bistratene isolated from *Lissoclinum bistratum* enhances the phosholipid-dependent activity of PKC and may be a useful probe for studying molecular mechanisms of cell growth and differentiation[154] as well as anticancerous drugs. The compound and related congeners were found to exhibit cytotoxicity towards human tumour celllines. Sebastianines A and B isolated as biologically active pyridoacridine metabolites, which show cytotoxic activities towards colon cancer cells, have been extracted from a Brazilian collection of the ascidian *Cystodytes dellechiajei*[155]. A study of the Thai ascidian *Ecteinascidia thurstoni*, using a KCN-pretreatment isolation procedure, identified the known two alkaloids ecteinascidins and the two novel analogues ecteinascidins[156]. The identified ecteinascidins exhibited potent cytotoxicity towards tumour cell-lines and growth inhibition of *Mycobacterium tuberculosis* H37Ra. The sulphated steroid was found to be responsible for sperm activation and attraction in Japanese collections of the ascidians *Ciona intestinalis* and *C. savignyi*[157]. The *in vivo* antitumour activity of the dimeric disulphide alkaloid polycarpine, isolated from the ascidians *Polycarpa clavata*[158] and *P. aurata*[159], and related synthetic analogues has been investigated [160].

Metabolites from Echinoderms

Physiologically active saponins have been studied extensively from sea stars and sea cucumbers[161], but not so useful as drugs because of their tendency to cause cell lysis. Even then, glycosylated ceramides and saponins continue to be the major classes of metabolites identified in echinoderms. A full account of the isolation and characterization of hedathiosulphonic acids A and B, isolated from a deep-sea urchin *Echinocardium cordatum*[162], has been reported[163]. Imbricatine from the sea star *Dermasterias imbricata* is the first benzyltetrahydroisoquinolone alkaloid from a non-plant source and shows in the NCI human cell-line screen. A study of the starfish *Diplopteraster multipes* indicated a range of sterol sulphates[164,165]. Lysastroside-A a new steroidal glycoside was isolated from the starfish *Lysastrosoma anthosticta* collected in the Sea of Japan. Ten new saponins, certonardosides A–J were isolated from the starfish *Certonardoa semiregularis* collected off the Coast of Komun Island, Korea. The absolute configurations of the side chains were secured by the 1H NMR analysis of MTPA esters. All compounds were evaluated for a range of antiviral properties towards HIV, herpes simplex (HSV), Coxsachie (CoxB), encephalomyocarditis virus (EMCV) and vesicular stomatitis virus (VSV), but only mild potency was observed for certonardosides-I and certonardosides-J. Linckosides A and B, neuritogenic steroidal glycosides, were reported from an Okinawan collection of the starfish *Linckia laevigata*[166]. In the search for antagonists of the chemokine receptor subtype-5 (CCR5) as possible anti-HIV agents, bioassay-guided fractionation of an Andaman and Nicobar Island, India, collection of the sea cucumber *Telenata ananas* afforded two triterpene glycosides[167]. Both compounds exhibited inhibitory activity in a CCR5, while no activity was observed towards the related chemokine receptor CXCR2. A new route for the synthesis of a ceramide sex pheromone isolated from the female Hair Crab, *Erimacrus isenbeckii*[168,169], was reported [170], while squaric acid ester-based methodology was used in a new synthesis of echinochrome-A, a polyhydroxylated napthoquinone pigment commonly isolated from sea urchin spines [171].

Metabolites from Fish, Sea Snakes and Marine Mammals

Metabolites extracted from fish, sea snakes and aquatic mammals are scanty. Various fish species are used to extract fish oil, rich in omega-3 fatty acids, which are used in the preparation of various kinds of drugs for the remedies of human beings, such as arthritis and many others. Through out the world about 500 species of fish are considered toxic. The most spectacular substance of pharmacological importance extracted from fish is tetradotoxin (TTX), the puffer or fugu poison. Other toxins isolated include ciguatoxin from electric rays, which is served as a potent antidote for pesticide poisoning [172]. TTX isolated from puffer fish and many other marine organisms has become a useful tool for researchers studying the voltage-gated sodium channel, and tetradotoxins also plays an important role in many biological experiments. A new class of water-soluble broad-spectrum antibiotics, squalamines has been isolated from the stomach extracts of dogfish shark, *Squalus acanthias*[173].

The sea snakes belong to the family Hydrophiidae. An anticancerous drug, namely "Fu-anntai", which has antiblastic effects on cervical carcinoma, stomach cancer, rhinocarcinoma and leukaemia cells, has been extracted from them in China [174]. A group of scientists in Australia have extracted a novel drug from rat snake[175].

Conclusions

As described in this article, the rich diversity of Marine biota with its unique physiological adaptations to the harsh marine environment provides a fruitful source for the discovery of life saving drugs. With the implementation of scuba diving tools and the development of the sophisticated instruments for the isolation and elucidation of structure of natural products from the marine organisms, a new and exciting vista is open for the exploration of precious drugs. However it must be acknowledged that supply problems still hamper the development of many promising metabolites of marine origin and have stimulated research on alternative methods for the marine metabolite production, Isolation and cultivation of suspected microbial producers of bioactive natural products could provide much needed answer to the supply problem by using molecular biological approach. It is also under investigation to transfer a bacterial gene cluster responsible for the biosynthesis of desired natural products to a vector suitable for large scale fermentation. In summary words ocean could play any important role in supplying life saving drugs in future. Although substantial progress has been made in identifying novel drugs from the marine sources, great endeavors are still needed to explore these molecules for clinical applications.

References

1. Jain, R. and Tiwari, A., 2007. *Curr. Sci.*, 93: 444–445.
2. Kijjoa, A. and Sawangwong, P., 2004. *Mar. Drugs,* 2: 73–82.
3. Schwartsmann, G., Da Rocha, A.B., Mattei, J. and Lopes, R., 2003. *Expert Opin. Investig. Drugs,* 12: 1367–1383.
4. Wakimoto, T., Matsunaga, S., Takai, A. and Fusetani, N., 2002. *Chem. Biol.*, 9: 309–319.
5. Lombardo, D. and Dennis, E.A., 1985. *J. Biol. Chem.*, 260: 7234–7240.
6. Arif, J.M., Al-Hazzani, A.A., Kunhi, M. and Al-Khodairy, F., 2004. *J. Biomed. Biotechnol.*, 2: 93–98.
7. Harvey, A., 2000. Strategies for discovering drugs from previously unexplored natural products. *Drug Discov Today*, 5: 294–300.
8. McCarthy, P.J. and Pomponi, S.A., 2004. A search for new Pharmaceutical Drugs from marineorganisms. *Marine Biomed. Res.*, p. 1–2.
9. Donia, M. and Hamann, M.T., 2003. Marine natural products and their potential applications as antiinfective agents. *The Lancet.*, 3: 338–348.
10. Haefner, B., 2003. Drugs from the deep. *Drug Discov. Today*, 8: 536–544.
11. Fuesetani, N., 2000. In: *Drugs from the Sea*, (Ed.) M. Fuesetani. Karger, Basel, 1: 1–5.
12. Joffe, S. and Thomas, R., 1989. Phytochemicals: A renewable global resource. *Biotech News Information*, 1: 697–700.
13. Faulkner, D., 1992. Biomedical uses for natural marine chemicals. *J. Oceanus*, 35: 29–35.
14. McCarthy, P.J. and Pomponi, S.A., 2004. A search for new pharmaceutical drugs from marine organisms. *Marine Biomed. Res.*, p. 1–2.
15. Wolf, S.G., 1978. In: *Drugs from the Sea,* (Eds.) P.N.Kaul and C.S. Siderman. The University of Oklahoma Press, Norman, pp. 7–15.
16. Halvey, S., 1990. In: *Microbiology: Applications in Food Biotechnology*, (Eds.) B.H. Nga and Y.K. Lu. Elsevier Applied Science Press, New York, pp. 123–134.
17. Salisbury, F., 1971. Doubts about the modern synthetic theory of evolution. *Am. Biol. Teach.*, 33: 335–336.
18. Grant, P.T. and Mackie, A.M., 1977. Drugs from the sea: Facts and fantasy. *Nature*, 267: 786–788.

19. Halvey, S., 1990. In: *Microbiology: Applications in Food Biotechnology*, (Eds.) B.H. Nga and Y.K. Lu, Elsevier Applied Science Press, New York, pp. 123–134.

20. Hamilton, S.C., Farchaus, J.W. and Davis, M.C., 2001. DNA polymerases as engines for biotechnology. *Biotechniques*, 31: 370–376, 378–380, 382–393.

21. Hunneke, G., Raffin, J.P., Ferrari, E., Jonsson, Z.O., Deitrich, J. and Hubscher, U., 2000. The PCNA from *Thermococcus fumicolans* functionally interacts with DNA polymerase delta. *Biochem. Biophys. Res. Commun.*, 276: 600–606.

22. Macleod, R.A., 1965. The question of the existence of specific marine bacteria. *Bacteriol. Rev.*, 29: 9–23.

23. Kohlmeyer, J. and Kohlmeyer, E., 1979. *Marine Mycology: The Higher Fungi*. Academic Press, London.

24. Shimizu, Y., 1993. Microalgal metabolites. *Chem. Rev.*, 93: 1685–1698.

25. Shimizu, Y., 1993. Dinoflagellates as sources of bioactive molecules. In: *Marine Biotechnology, Vol. 1: Pharmaceutical and Bioactive Natural Products*, (Eds.) D.H. Attaway and O.R. Zaborsky, Plenum, New York, pp. 391–410.

26. Shimizu, Y., 1996. Microalgal metabolites: A new perspective. *Ann. Rev. Microbiol.*, 50: 431–465.

27. Freudenthal, H.D. (Ed.), 1968. *Drugs from the Sea: Transactions of the Drugs from the Sea Symposium University of Rhode Island 1967*. Marine Technology Society, Washington, D.C.

28. Faulkner, D.J., 2000. Marine natural products. *Nat. Prod. Rep.*, 17: 7–55, and previous reports in this series.

29. Philip, P.A., Rea, D., Thavasu, P., Carmichael, J., Stuart, N.S.A., Rakett, H., Talbot, D.C., Ganesa, T., Pettit, G.R., Balkwill, F., Harris, A.L., 1993. Phase I study of bryostatin 1, assessment of interleukin-6 and tumor necrosis factor-alpha induction *in vivo*. *J. Natl. Cancer Inst.*, 85: 1812–1818.

30. Hirata, Y. and Uemura, D., 1986. Halichondrins: Antitumor polyether macrolides from a marine sponge. *Pure Appl. Chem.*, 58: 701–710.

31. Mynderse, J., Moore, R., Kashiwagi, M. and Norton, T., 1997. Antileukemia activity in the Oscillatoriaceae: Isolation of debromoaplysiatoxin from *Lyngbya*. *Science*, 196: 538–540.

32. Harrigan, G.G., Luesch, H., Yoshida, W.Y., Moore, R.E., Nagle, D.G., Paul, V.J., Mooberry, S.L., Corbett, T.H. and Valeriote, F.A., 1998. Symplostatin 1: A dolastatin 10 analogue from marine cyanobacterium *Symploca hydnoides*. *J. Nat. Prod.*, 61: 1075–1077.

33. Rinehart, K.L., Gloer, J.B., Hughes, R.G., Renis, H.E., McGovern, J.P., Swynernberg, E.B., Stringfellow, D.A., Kuentzel, S.L. and Li, L.H., 1981. Didemnins: Antiviral and antitumor depsipeptides from a Caribbean tunicate. *Science*, 212: 933–935.

34. Lewin, R.A. and Cheng, L. (Eds.), 1989. *Prochloron: A Microbial Enigma*. Chapman and Hall, New York.

35. Carmeli, S., Moore, R.E. and Patterson, G.M.L., 1990. Tolytoxin and new scytophycins from three species of *Scytonema*. *J. Nat. Prod.*, 53: 1533–1542.

36. Hall, S. and Strichartz, G. (Ed.), 1990. *Marine Toxins*. American Chemical Society, Washington, D.C.

37. Tachibana, K., Scheuer, P., Tsukitani, Y., Kikuchi, H., Van Engen, D., Clardy, J., Gopichand, Y. and Schmitz, F.J., 1998. Okadaic acid: A cytotoxic polyether from two marine sponges of the genus *Halichondria*. *J. Am. Chem. Soc.*, 103: 2469–2471.

38. Murakami, Y., Oshima, Y. and Yasumoro, T., 1982. Identification of okadaic acid as a toxic component of a marine dinoflagellate *Prorocentrum lima*. *Nippon Suisan Gakkaishi*, 48: 69–72.

39. Moore, R.E., Patterson, M.L. and Carmichael, W.W., 1988. In: *Biomedical Importance of Marine Organisms*, (Ed.) D.G. Fautin, California Academy of Sciences, San Francisco, pp. 143–150.

40. Beltron, E.C. and Nielan, B.A., 2000. Geographical segregation of Neurotoxin-producing Cyanobacterium *Anabaena circinalis*. *App and Environ Microbiol.*, 66: 4468–4474.

41. Cardillina, J.H. II, Marner, F.J. and Moore, R.E., 1979. Seaweed dermatitis: Structure of lyngbyatoxin A. *Science*, 204: 193–195.

42. Plavsic, M., Terzic, S., Ahel, M. and van den Berg, C.M.G., 2004. Folic acid in coastal waters of the Adriatic Sea. *Mar. Freshw. Res.*, 53: 1245–1252.

43. Fujiki, H. and Sugimura, T., 1987. New classes of tumor promoters: Telocin, aplysiatoxin and palytoxin. *Adv. Cancer Res.*, 59: 223–264.

44. Wender, P.A., Koehler, K.E., Sharkey, N.A., Dell'Aquilla, H. L. and Blumberg, P.M., 1986. Analysis of phorbol ester pharmocophor on protein kinase C as a guide to the retional design of new classes of anlogas. *Proc. Natl. Acad. Sci.*, USA, 83: 4214–4218.

45. Abstract Mass–CT97–0156, 2004. Marine cyanobacteria as a source for bacterioactive (apoptosis modifying) compounds with potential as cell biology reagents and drugs. *Short Popular Version*, p. 1–2.

46. Asolkar, R.N., Maskey, R.P., Helmke, E. and Laatsch, H., 2002. Marine bacteria. XVI. Chalcomycin B, a new macrolide antibiotic from the marine isolate *Streptomyces* sp. B7064. *J. Antibiot* (*Tokyo*), 55: 893–898.

47. Lin, Y., Li, H., Jiang, G., Zhou, S., Vrijmoed, L.L.P. and Jones, E.B.G., 2002. A novel γ-lactone, eutypoid-A and other metabolites from marine fungus *Eutypa* sp. (#424) from the South China Sea. *Indian J. Chem, Sect. B: Org. Chem. Incl. Med. Chem.*, 41B: 1542–1547.

48. Yamada, T., Iwamoto, C., Yamagaki, N., Yamanouchi, T., Minoura, K., Yamori, T., Uehara, Y., Andoh, T., Umemura, K. and Numata, A., 2002. Leptosins M–N1, cytotoxic metabolites from a *Leptosphaeria* species separated from a marine alga. Structure determination and biological activities. *Tetrahedron*, 58: 479–487.

49. Afiyatullov, S.S., Kalinovsky, A.I., Kuznetsova, T.A., Isakov, V.V., Pivkin, M.V., Dmitrenok, P.S. and Elyakov, G.B., 2002. New diterpene glycosides of the fungus *Acremonium striatisporum* isolated from a sea cucumber. *J. Nat. Prod.*, 65: 641–644.

50. Abbanat, D., Leighton, M., Maiese, W., Jones, E.B.G., Pearce, C.J. and Greenstein, M.J., 1998. Cell wall-active antifungal compounds produced by the marine fungus *Hypoxylon oceanicum* LL–15G256. I. Taxonomy and fermentation. *J. Antibiot.* (*Tokyo*), 51: 296–302.

51. Schlingmann, G., Milne, L. and Carter, G.T., 2002. Isolation and identification of antifungal polyesters from the marine fungus *Hypoxylon oceanicum* LL–15G256. *Tetrahedron*, 58: 6825– 6835.

52. Proteau, P.J., Gerwick, W.H., Garcia-Pichel, F. and Castenholtz, R., 1993. The structure of scytonemin, an ultraviolet sunscreen pigment from the sheaths of cyanobacteria. *Experientia*, 49: 825–829.

53. Stevenson, C.S., Capper, E.A., Roshak, A.K., Marquez, B., Grace, K., Gerwick, W.H., Jacobs, R.S. and Marshall, L.A., 2002. Scytomenin–a marine natural product inhibitor of kinases key in hyperproliferative inflammatory diseases. *Inflammation Res.*, 51: 112–118.

54. Stevenson, C.S., Capper, E.A., Roshak, A.K., Marquez, B., Eichman, C., Jackson, J.R., Mattern, M., Gerwick, W.H., Jacobs, R.S. and Marshall, L.A., 2002. The identification and characterization of the marine natural product scytonemin as a novel antiproliferative pharmacophore. *J. Pharmacol. Exp. Ther.*, 303: 858–866.

55. Murakami, M., Makabe, K., Yamaguchi, S., Konosu, S. and Walchi, R., 1988. A novel polyether macrolide from the dinoflagellate *Goniodoma pseudogoniaulax*. *Tetrahedron Lett.*, 29: 1149–1152.

56. Abe, M., Inoue, D., Matsunaga, K., Ohizumi, Y., Ueda, H., Asano, T., Murakami, M. and Sato, Y., 2002. Goniodomin A: An antifungal polyether macrolide, exhibits antiangiogenic activities via inhibition of actin reorganization in endothelial cells. *J. Cell Physiol.*, 190: 109–116.

57. Koehn, F.E., Longley, R.E. and Reed, J.K., 1992. Microcolin A and B, new immunosuppressive peptides from the blue green alga *Lyngbya majuscula*. *J. Nat. Prod.*, 55: 613–619.

58. Gerwick, W.H., Proteau, P.J., Nagh, D.G., Hamel, E., Blobhin, A. and Slate, D.L., 1994. Structure of cruacin A: A novel antimitotic, antiproliferative and brine shrimp toxic natural product from the marine cyanobacterium *Lyngbya majusula*. *J. Org. Chem.*, 59: 1243–1245.

59. Nagai, H., Murata, M., Torigoe, K., Satake, M. and Yasumoto, T., 1992. Gambieric acids, new potent antifungal substance with unprecedented polyether structures from a marine dinoflagellate *Gambierdiscus toxicus*. *J. Org. Chem. Commun.*, 57: 5448–5453.

60. Shimizu, Y., 1993. In: *Marine Biotechnology*, (Eds.) D. Attaway and O. Zeborsky. Plenum Press, New York, 1: 391–410.

61. Cohen, P., Holmes, C. and Tsukitani, Y., 1990. Okadaic acid: A new probe for the study of cellular regulation. *Trends Biochem. Sci.*, 15: 98–102.

62. Anderson, R.J., Bolton, J.J., Molloy, F.J. and Rothmann, K.W.G., 2003. Commercial seaweeds in southern Africa. In: *Procedings of the 17th International Seaweed Symposium*, (Eds.) R.O. Chapman, R.J. Anderson, V.J. Vreeland and I.R. Davison. Oxford University Press, Oxford, p. 1–512.

63. Critchley, A.T., R.D. Gillespie and K.W.G. Rotman. 1998. In: *Seaweed Resources of the World*, (Eds.) M. Critchley and A.T. Ohno. Japan International Cooperation Agency, Japan, p. 413– 425.

64. Faulkner, D.J., 2002. Marine natural products. *Nat. Prod. Rep.*, 19: 1–48.

65. Garg, H.S., Sharma, T., Bhakuni, D.S., Pramanik, B.N. and Bose, A.K., 1992. An antiviral sphingosine derivative from green alga *Ulva fasciata*. *Tetrahedron Lett.*, 33: 1641–1644.

66. Gerwick, W.H. and Fenical, W., 1981. Ichthyotoxic and cytotoxic metabolites of the tropical brown alga *Stypopodium zonale*. *J. Org. Chem.*, 46: 21–27.

67. Jacobs, R.S., Culver, P., Langdon, R., O'Brien, T. and White, S., 1985. Some pharmacological observations on marine natural products. *Tetrahedron.*, 41: 981–984.

68. Ireland, C., Copp, B., Foster, M., McDonald, L., Radisky, D. and Swersey, J., 1993. In: *Marine Biology*, (Eds.) D. Attaway and O. Zeborsky. Plenum Press, New York, 1: 1–43.

69. Ali, M.S., Saleem, M., Yamdagni, R. and Ali, M.A., 2002. Steroid and antibacterial steroidal glycosides from marine green alga *Codium iyengarii* Borgesen. *Nat. Prod. Lett.*, 16: 407–413.

70. Tang, H.-F., Yi, Y.-H., Yao, X.-S., Xu, Q.-Z., Zhang, S.-Y. and Lin, H.-W. Bioactive steroids from the brown alga *Sargassum carpophyllum*. *J. Asian Nat. Prod. Res.*, 4: 95–105.

71. Xu, S.-H., Ding, L.-S., Wang, M.-K., Peng, S.-L. and Liao, X., 2002. Studies on the chemical constituents of the algae *Sargassum polycystum*. *Youji Huaxue* (*Chinese J. Org. Chem.*), 22: 138–140.

72. Gerwick, W.H. and Bernart, M.W., 1993. In: *Marine Biotechnology*, (Eds.) D. Attaway and O. Zaborsky. Plenum Press, New York, 1: 101–152.

73. Blunt, J.W., Copp, B.R., Munro, M.H.G., Northcote, P.T. and Prinsep, M.R., 2004. Marine natural products. *Nat. Prod. Rep.*, 21: 1–49.

74. Burres, N.S., Sazech, S., Gunavardana, G.P. and Clement, J.J., 1989. Antitumor activity and nucleic acid binding properties of dercitin. *Cancer Res.*, 49: 5267–5274.

75. Fuestani, N., Sugawara, T. and Matsunago, S., 1992. Potent antitumor metabolites from a marine sponge. *J. Org. Chem.*, 57: 3828–3832.

76. Perry, N.G., Blunt, J.W. and Munro, H.H.G., 1988. Mycalamide A, and antiviral compound from a New Zealand sponge of the genus Mycale. *J. Am. Chem. Soc.*, 110: 4850–4851.

77. Sakemi, S., Ichiba, T., Kohmoto, S. and Saucy, G., 1988. Isolation and structure elucidation of onnamide A, a new bioactive metabolite of a marine sponge. *Theonella* sp. *J. Am. Chem. Soc.*, 110: 4851–4853.

78. Burres, N.S. and Clement, J.J., 1989. Antitumor activity and mechanism of action of the novel marine natural products mycalamide-A and -B and Onnamide. *Cancer Res.*, 49: 2935–2940.

79. Pettit, G.R., Collins, J.C., Herald, D.L., Doubek, D.L., Boyd, M.R., Schmidt, J.M., Hooper, D.L. and Tackett, L.P., 1992. Isolation and structure of cribostatins1 and 2 from blue marine sponge, *Cribrochalina* sp. *Can. J. Chem.*, 70: 1170–1175.

80. Petitt, G.R., Cichacz, Z.A., Gao, F., Herald, C.L. and Boyd, M.R., 1993. Isolation and structure of the remarkable human cancer cell growth inhibitors spongistatins 2 and 3 from an Eastern India Ocean *Spongia* sp. *J. Chem. Soc.* (*Lond. Chem. Commun.*), 1: 1166–1168.

81. Judulco, R., Brauers, G., Edrata, R.A., Ebel, R., Sudarsono, V. Wray and Proksh, P., 2002. New metabolites from sponge-derived fungi *Curvularia lunata* and *Cladosporium herbarum*. *J. Nat. Prod.*, 65: 730–733.

82. Malmstrom, J., Christophersen, C., Barrero, A.F., Oltra, J.E., Justica, J. and Rosales, A., 2002. Bioactive metabolites from a marine-derived strain of the fungus *Emericella variecolor*. *J. Nat. Prod.*, 65: 364–367.

83. Linington, R.G., Robertson, M., Gauthier, A., Finlay, B.B., van Soest, R., Anderson, R.J., 2002. An antimicrobial glycolipid isolated from the marine sponge *Caminus sphaeroconia*. *Org. Lett.*, 4: 4089–4092.

84. Aoki, K., Takahashi, M., Hashimoto, M., Okuno, T., Kurata K. and Suzuki, M., 2002. Structure-activity relationship of neuritogenic spongean acetylene alcohols, lembehynes. *Biosci. Biotechnol. Biochem.*, 66: 1915–1921.

85. De Silva, S.D., Williams, D.E., Andersen, R.J., Klix, H., Holmes, C.F.B. and Allen, T.M. A potent protein phosphatase inhibitor isolated from the Papau New Guinea sponge *Theonella swinhoie* Gray. *Tetrahedron Lett.*, 33: 1561–1564.

86. Kato, Y., Fusetani, N., Matsunaga, S. and Hashimoto, K. Calculin A, a novel antitumor metabolite from marine sponge *Discodermia calyx*. *J. Am. Chem. Soc.*, 108: 2780–2781.

87. Potts, B.C.M., Faulkner, D.J. and Jacobs, R.S., 1992. Phospholipase A2 inhibitors from marine organisms. *J. Nat. Prod.*, 55: 1701–1717.

88. Shoji, N., Umeyama, A., Shin, K., Takedo, K. and Ashihara, S., 1992. Two unique pentacyclic steroids with Cis C/D ring junction from *Xestospongia bergquistia* Fromont, powerful inhibitors of histamine release. *J. Org. Chem.*, 57: 2996–2997.

89. Chan, G.W., Mong, S., Hemling, M.E., Freyer, A.J., Offen, P.H., DeBrosse, C.W., Sarau, H.M. and Westley, J.W., 1993. New leukotrine B4 receptor antagonist: Leucettamine A and related imidazole alkaloids from the marine sponge *Leucetta microraphis*. *J. Nat. Prod.*, 56: 116–121.

90. Mynderse, J.S. and Moore, R.E., 1979. Isotactic polymethoxy-1-alkenes from the blue-green alga *Tolypothrix conglutinata* var. chlorata. *Phytochemistry*, 18: 1181–183

91. Sandler, J.S., Colin, P.L., Hooper, J.N.A. and Faulkner, D.J., 2002. Cytotoxic β-carbolines and cyclic peroxides from the Palauan sponge *Plakortis nigra*. *J. Nat. Prod.*, 65: 1258–1261.

92. Qureshi, A., Salvá, J., Harper, M.K. and Faulkner, D.J., 1998. New cyclic peroxides from the Philippine sponge *Plakinastrella* sp. *J. Nat. Prod.*, 61: 1539–1542.

93. Jung, M., Ham, J. and Song, J., 2002. First total synthesis of natural 6-epiplakortolide E. *Org. Lett.*, 4: 2763–2765.

94. Kuramoto, M., Fujita, T. and Ono, N., 2002. Ircinamine: A novel cytotoxic alkaloid from *Ircinia* sp. *Chem. Lett.*, p. 464–465.

95. Williams, D.E., Lassota, P. and Andersen, R.J., 1998. Motuporamines A-C, cytotoxic alkaloids isolated from the marine sponge *Xestospongia exigua* (Kirkpatrick). *J. Org. Chem.*, 63: 4838–4841.

96. Yousaf, M., El Sayed, K.A., Rao, K.V., Lim., C.W., Hu, J.-F., Kelly, M., Franzblau, S.G., Zhang, F., Peraud, O., Hill, R.T. and Hamann, M.T., 2002. 12,34 Oxamanzamines, novel biocatalytic and natural products from manzamine producing indo-pacific sponges. *Tetrahedron*, 58: 7397–7402.

97. Iwashima, M., Terada, I., Iguchi, K. and Yamori, T., 2002. New biologically active marine sesquiterpenoid and steroid from the Okinawan sponge of the genus *Axinyssa*. *Chem. Pharm. Bull.*, 50: 1286–1289.

98. Santafé, G., Paz, V., Rodríguez, J. and Jiménez, C., 2002. Novel cytotoxic oxygenated C29 sterols from the Colombian marine sponge *Polymastia tenax*. *J. Nat. Prod.*, 65: 1161–1164.

99. Vanisree, M. and Subbaraju, G.V., 2002. Alcyonacean Metabolites VIII: Antibacterial metabolites from *Labophytum crassum* of the Indian Ocean. *Asian J. Chem.*, 14: 957–960.

100. Duh, C.-Y., Chien, S.-C., Song, P.-Y., Wang, S.-K., El-Gamal, A.A.H. and Dai, C.F., 2002. New cadinene sesquiterpenoids from the Formosan soft coral *Xenia puerto-galerae*. *J. Nat. Prod.*, 65: 1853–1856.

101. Palermo, J.A., Rodríguez Brasco, M.F., Spagnuolo, C. and Seldes, A.M., 2000. Illudalane sesquiterpenoids from the soft coral *Alcyonium paessleri*: The first natural nitrate esters. *J. Org. Chem.*, 65: 4482–4486.

102. Witulski, B., Zimmermann, A. and Gowans, N.D., 2002. First total synthesis of the marine illudalane sesquiterpenoid alcyopterosin E. *Chem Commun.*, p. 2984–2985.

103. Wang, G.-H., Ahmed, A.F., Kuo, Y.-H. and Sheu, J.-H., 2002. Two new subergane-based sesquiterpenes from a Taiwanese gorgonian coral *Subergorgia suberosa*. *J. Nat. Prod.*, 65: 1033–1036.

104. Rudi, A., Levi, S., Benayahu, Y. and Kashman, Y., 2002. Lemnaflavoside: A new diterpene glycoside from the soft coral *Lemnalia flava*. *J. Nat. Prod.*, 65: 1672–1674.

105. Iwashima, M., Terada, I., Okamoto, K. and Iguchi, K., 2002. Tricycloclavulone and clavubicyclone, novel prostanoid-related marine oxylipins, isolated from the Okinawan soft coral *Clavularia viridis*. *J. Org. Chem.*, 67: 2977–2981.

106. Mori, K., Iguchi, K., Yamada, N., Yamada, Y. and Inouye, Y., 1988. Bioactive marine diterpenoids from Japanese soft coral of *Clavularia* spp. *Chem. Pharm. Bull.*, 36: 2840–2852.

107. Iguchi, K., Sawai, H., Nishimura, H., Fujita, M. and Yamori, T., 2002. New dolabellane-type diterpenoids from the Okinawan soft coral of the genus *Clavularia*. *Bull. Chem. Soc., Jpn.*, 75: 131–136.

108. Taglialatela-Scafati, O., Deo-Jangra, U., Campbell, M., Roberge, M. and Andersen, R.J., 2002. Diterpenoids from cultured *Erythropodium caribaeorum*. *Org. Lett.*, 4: 4085–4088.

109. He, X.-X., Su, J.-Y., Zeng, L.-M., Yang, X.-P. and Liang, Y.-J., 2002. Studies on the secondary metabolite of the soft coral *Lobophytum* sp. *Huaxue Xuebao (ACTA Chim Sinica)*, 60: 334–337.

110. Alam, N., Hong, J., Lee, C.-O., Choi, J.S., Im, K.S. and Jung, J.H., 2002. Additional cytotoxic diacetylenes from the stony coral *Montipora* sp. *Chem. Pharm. Bull.*, 50: 661–662.

111. Monastyrnaya, M.M., Zykova, T.A., Apalikova, O.V., Shwets, T.V. and Kozlovskaya, E.P., 2002. Biologically active polypeptides from the tropical sea anemone *Radianthus macrodactylus*. *Toxicon.*, 40: 1197–1217.

112. Loret, E.P., del Valle, R.M.S., Mansuelle, P., Sampieri, F. and Rochat, H., 1994. Positively charged amino acid residues located similarly in sea anemone and scorpion toxins. *J. Biol. Chem.*, 269: 16785–16788.

113. Bosmans, F., Aneiros, A. and Tytgat, J., 2002. The sea anemone *Bunodosoma granulifera* contains surprisingly efficacious and potent insect-selective toxins. *FEBS Lett.*, 532: 131–134.

114. Roussis, V., Wu, Z., Fenical, W., Stobel, S.A., Van Duyne, D.G. and Clardy, J., 1990. New anti-inflammatory pseudopterosins from the marine octocoral *Pseudopterogorgia elisabethae*. *J. Org. Chem.*, 55: 4916–4922.

115. Lilies, G., 1996. Gambling on marine biotechnology. *Bioscience*, 46: 250–253.

116. Lysek, N., Rachor, E. and Lindel, T., 2002. Isolation and Structure Elucidation of Deformylflustrabromine from the North Sea Bryozoan *Flustra foliacea*. *Z. Naturforsch., C: Biosci.*, 57: 1056–1061.

117. Narkowicz, C.K., Blackman, A.J., Lacey, E., Gill, J.H. and Heiland, K., 2002. Convolutindole A and convolutamine H, new nematocidal brominated alkaloids from the marine bryozoan *Amathia convoluta*. *J. Nat. Prod.*, 65: 938–941.

118. Jeong, S.-J., Higuchi, R., Miyamoto, T., Ono, M., Kuwano, M. and Mawatari, S.F., 2002. Bryoanthrathiophene: A new antiangiogenic constituent from the bryozoan *Watersipora subtorquata* (d'Orbigny, 1852). *J. Nat. Prod.*, 65: 1344–1345.

119. Blackman, A.J. and Matthews, D.J., 1985. Amathamide alkaloids from the marine bryozoan *Amathia wilsoni* Kirkpatrick. *Heterocycles*, 23: 2829–2833.

120. Ramirez Osuna, M., Aguirre, G., Somanathan, R. and Molins, E., 2002. Asymmetric synthesis of amathamides A and B: novel alkaloids isolated from *Amathia wilsoni*. *Tetrahedron-Asymmet.*, 13: 2261–2266.

121. Pettit, G.R., 1991. In: *Progress in the Chemistry of Organic Natural Products*, (Eds.) W. Herz, G.W. Kinby, W. Steglich and C. Tamm. Springer Verlag, Berlin, 57: 153–195.

122. Zhang, H., Shigemori, H., Ichibashi, M., Kosaka, T., Pettit, G.R., Kamano, Y. and Kobayashi, J., 1994. Convolutamides A-F, novel α-lactam alkaloids from the marine bryozoan *Amathia convoluta*. *Tetrahedron.*, 50: 10201–10206.

123. Princep, M.R., Blunt, J.W. and Munro, M.H.G., 1991. New cytotoxic β-carboline alkaloids from the marine bryzoans *Cribricellina cribraria*. *J. Nat. Prod.*, 54: 1068–1076.

124. Holst, P.B., Anthoni, U., Christophersen, C. and Neilson, P.N., 1994. Marine alkaloids, two alkaloids, flustramine E and debromoflustramine B, from the marine bryozoan *Flustra foliacea*. *J. Nat. Prod.*, 57: 997–1000.

125. Pickrell, J., 2003. "Wonder Drug" snails face threats, Expert warn. *National Geographic News*, p. 1–2.

126. Myers, P.A., Cruz, L.Z., Rivier, J.E. and Olivera, B.M., 1993. Conus peptides as chemical probes for receptors and ion channels. *Chem. Rev.*, 93: 1923–1936.

127. Pettit, G.R., Singh, S.B., Hogan, F., Lloyd-williams, P., Herald, C.L., Burbett, D.D. and Clewlow, P.J., 1989. The absolute configuration and synthesis of natural (-)-dolostatin10. *J. Am. Chem. Soc.*, 70: 5463–5465.

128. Rorsener, J.A. and Scheuer, P.J., 1986. Ulapualids A and B: Extraordinary antitumor macrolides from nudibranch egg masses. *J. Am. Chem. Soc.*, 108: 846–847.

129. Morris, S.A., De Silva, E.D. and Anderson, R.J., 1990. Chromodorane diterpenes from the tropical dorid nudibranch *Chromocloris cavae*. *Can. J. Chem.*, 69: 768–771.

130. Holmes, I., 2002. Snail toxin could ease chronic pain. *Nature Science Update*, March 29.

131. Ciavatta, M.L., Trivellone, E., Villani, G. and Cimino, G., 1993. Membrenones: new polypropionates from the skin of the Mediterranean mollusc *Pleurobranchus membranaceus*. *Tetrahedron Lett.*, 34: 6791–6794.

132. Sampson, R.A. and Perkins, M.V., 2002. Total synthesis of (-)-(6S, 7S, 8S, 9R, 10S, 2'S) membrenone-A and (-)-(6S, 7S, 8S, 9R, 10S)-Membrenone-B and structural assignment of membrenone-C. *Org. Lett.*, 4: 1655–1658.

133. Paterson, I., Chen, D.Y.-K. and Franklin, A.S., 2002. Total synthesis of siphonarin B and dihydrosiphonarin B.*Org. Lett.*, 4: 391–394.

134. Hochlowski, J.E., Coll, J.C., Faulkner, D.J., Biskupiak, J.E., Ireland, C.M., Zheng, Q.-T., He, C.-H. and Clardy, J., 1984. Novel metabolites of four *Siphonaria* species. *J. Am. Chem. Soc.*, 106: 6748–6750.

135. Rajaganapathi, J., Kathiresan, K. and Singh, T.P., 2002. Purification of anti-HIV protein from purple fluid of the sea Hare *Bursatella leachii* de Blainville. *Mar. Biotechnol.*, 4: 447–453.

136. Spinella, A., Zubía, E., Martinez, E., Ortea, J. and Cimino, G., 1997. Structure and stereochemistry of aplyolides A-E, lactonized dihydroxy fatty acids from the skin of the marine mollusk *Aplysia depilans*. *J. Org. Chem.*, 62: 5471–5475.

137. Spinella, A., Caruso, T. and Coluccini, C., 2002. First total synthesis of natural aplyolides B and D, ichthyotoxic macrolides isolated from the skin of the marine mollusc *Aplysia depilans*. *Tetrahedron Lett.*, 43: 1681–1683.

138. Caruso, T. and Spinella, A., 2002. First total synthesis of natural aplyolides C and E, ichthyotoxic macrolides isolated from the skin of the marine mollusc *Aplysia depilans*. *Tetrahedron Asymmet.*, 13: 2071–2073.

139. Davidson, B.S., 1993. Ascidians: producers of amino acid derived metabolites. *Chem. Rev.*, 93: 1771–1791.

140. Sakai, R., Stroch, J.C., Sullins, D.W. and Rinehart, K.L., 1995. Seven new didemnins from the marine tunicate *Trididemnin solidum*. *J. Am. Chem. Soc.*, 117: 3734–3748.

141. Fujita, M., Nakao, Y., Matsunaga, S., Nishikawa, T. and Fusetani, N., 2002. Sodium 1-(12-hydroxy) octadecanyl sulfate, an MMP2 inhibitor, isolated from a tunicate of the family Polyclinidae. *J. Nat. Prod.*, 65: 1936–1938.

142. Rezanka, T. and Dembitsky, V.M., 2002. Eight-membered cyclic 1,2,3-trithiocane derivatives from *Perophora viridis*, an Atlantic tunicate. *Eur. J. Org. Chem.*, p. 2400–2404.

143. Jang, W.S., Kim, K.N., Lee, Y.S., Nam, M.H. and Lee, I.H., 2002. Halocidin: A new antimicrobial peptide from hemocytes of the solitary tunicate *Halocynthia aurantium*. *FEBS Lett.*, 521: 81–86.

144. Lee, I.-H., Zhao, C., Nguyen, T., Menzel, L., Waring, A.J., Sherman, M.A. and Lehrer, R.I., 2001. Clavaspirin: An antibacterial and haemolytic peptide from *Styela clava*. *J. Peptide Res.*, 58: 445–456.

145. Wright, A.D., Goclik, E., König, G.M. and Kaminsky, R., 2002. Lepadins D–F: Antiplasmodial and antitrypanosomal decahydroquinoline derivatives from the tropical marine tunicate *Didemnum* sp. *J. Med. Chem.*, 45: 3067–3072.

146. Urban, S., Blunt, J.W. and Munro, M.H.G., 2002. Coproverdine: A novel, cytotoxic marine alkaloid from a New Zealand ascidian. *J. Nat. Prod.*, 65: 1371–1373.

147. Sakai, R., Rinehart, K.L., Guan, Y. and Wang, A.H.J., 1992. Seven new didemnins from the marine tunicate *Tridemnin solidum*. *Proc. Natl. Acad. Sci., USA*, 89: 11456–11460.

148. Ortega, M.J., Zubía, E., Ocaña, J.M., Naranjo, S. and Salvá, J., 2000. New rubrolides from the ascidian *Synoicum blochmanni.Tetrahedron*, 56: 3963–3967.

149. Bellina, F., Anselmi, C. and Rossi, R., 2002. Total synthesis of rubrolide M and some of its unnatural congeners. *Tetrahedron Lett.*, 43: 2023–2027.

150. Rinehart, K.L., Shield, L.S. and Cohen-Parsonsm, M., 1993. In: *Marine Biotechnology*, (Eds.) D. Attaway and O. Zaborsky. Plenum Press, New York, 2: 309–342.

151. Kinnel, R. and Scheuer, P., 1992. 11 hydroxy-staurosporine: A highlycytotoxic, powerful protein kinase C inhibitor from a tunicate. *J. Org. Chem.*, 57: 6327–6329.

152. Carte, B.K., Chan, G., Freyer, A., Hemling, M.E., Hofmann, G.A., Mattern, M.R., Compagone, R.S. and Faulkner, D.J., 1994. Pentatheipins and trithianes from two *Lissoclinum* species and a *Eudistoma* sp.: inhibitors of protein kinase C. *Tetrahedron*, 50: 12785–12792.

153. Horton, P.A., Longly, R.E., McConnel,O.J. and Ballas, L.M., 1994. Staurosporine aglycone (K252–c) and acryriaflavin A from the marine ascidian *Eudistoma* sp. *Experientia*, 50: 843–845.

154. Foster, M.P., Mayne, C.L., Dunkel, R., Pugmire, R.J., Grant, D.M., Kornprobst, J., Verbist, J., Biard, J. and Ireland, C.M., 1992. Revised structure of bistramide A (bistrane A): application for a program for the analysis of 2D INADEQUATE spectra. *J. Am. Chem. Soc.*, 114: 1110–1111.

155. Torres, Y.R., Bugni, T.S., Berlinck, R.G.S., Ireland, C.M., Magalhães, A., Ferreira, A.G. and da Rocha, R.M., 2002. Sebastianines A and B, novel biologically active pyridoacridine alkaloids from the Brazilian ascidian *Cystodytes dellechiajei*. *J. Org. Chem.*, 67: 5429–5432.

156. Suwanborirux, K., Charupant, K., Amnuoypol, S., Pummangura, S., Kubo, A. and Saito, N., 2002. Ecteinascidins 770 and 786 from the Thai tunicate *Ecteinascidia thurstoni*. *J. Nat. Prod.*, 65: 935–937.

157. Yoshida, M., Murata, M., Inaba, K. and Morisawa, M., 2002. A chemoattractant for ascidian spermatozoa is a sulfated steroid. *Proc. Natl. Acad. Sci., U.S.A.*, 99: 14831–14836.

158. Kang, H. and Fenical, W., 1996. Polycarpine dihydrochloride: a cytotoxic dimeric disulfide alkaloid from the Indian Ocean ascidian *Polycarpa clavata*. *Tetrahedron Lett.*, 37: 2369–2372.

159. Abas, S.A., Hossain, M.B., van der Helm, D., Schmitz, F.J., Laney, M., Cabuslay, R. and Schatzman, R.C., 1996. Alkaloids from the Tunicate *Polycarpa aurata* from Chuuk Atoll. *J. Org. Chem.*, 61: 2709–2712.

160. Popov, A.M., Novikov, V.L., Radchenko, O.S. and Elyakov, G.B., 2002. The cytotoxic and antitumor activities of the imidazole alkaloid polycarpin from the ascidian *Polycarpa aurata* and its synthetic analogues. *Dokl Biochem Biophys.*, 385: 213–218.

161. Dubois, M.A., Higuchi, R., Komori, T. and Sasaki, T., 1988. Structure of two new oligoglycoside sulfates, pectinoside E and F, and biological activities of 6 new pectinosides. *Liegbig's Annalen der Chemis*, p. 845–850.

162. Takada, N., Watanabe, M., Suenaga, K., Yamada, K., Kita, M. and Uemura, D., 2001. Isolation and structures of hedathiosulfonic acids A and B, novel thiosulfonic acids from the deep-sea urchin *Echinocardium cordatum*. *Tetrahedron Lett.*, 42: 6557–6560.

163. Kita, M., Watanabe, M., Takada, N., Suenaga, K., Yamada, K. and Uemura, D., 2002. Hedathiosulfonic acids A and B, novel thiosulfonic acids from the deep-sea urchin *Echinocardium cordatum*. *Tetrahedron*, 58: 6405–6412.

164. Levina, E.V., Andriyashchenko, P.V., Kalinovsky, A.I., Dmitrenok, P.S. and Stonik, V.A., 2002. Steroid Compounds from the Far Eastern Starfish Diplopteraster multiples. *Russ. J. Bioorg. Chem.*, 28: 189–193.

165. Levina, E.V., Andriyashchenko, P.V., Kalinovsky, A.I., Dmitrenok, P.S., Stonik, V.A. and Prokof'eva, N.G., 2002. Steroid compounds from the starfish *Lysastrosoma anthosticta* collected in the Sea of Japan. *Russ. Chem. Bull.*, 51: 535–539.

166. Qi, J., Ojika, M. and Sakagami, Y., 2002. Linckosides A and B, two new neuritogenic steroid glycosides from the Okinawan starfish *Linckia laevigata*. *Bioorg. Med. Chem.*, 10: 1961–1964.

167. Hegde, V.R., Chan, T.-M., Pu, H., Gullo, V.P., Patel, M.G., Das, P., Wagner, N., Parameswaran, P. S. and Naik, C.G., 2002. Two selective novel triterpene glycosides from sea cucumber, *Telenota Ananas*: inhibitors of chemokine receptor-5. *Bioorg. Med. Chem. Lett.*, 12: 3203–3205.

168. Asai, N., Fusetani, N., Matsunaga, S. and Sasaki, J., 2000. Sex pheromones of the hair crab *Erimacrus isenbeckii*. Part 1: Isolation and structures of novel ceramides. *Tetrahedron*, 56: 9895–9899.

169. Asai, N., Fusetani, N. and Matsunaga, S., 2001. Sex Pheromones of the Hair Crab *Erimacrus isenbeckii*. II. Synthesis of Ceramides. *J. Nat. Prod.*, 64: 1210–1215.

170. Masuda, Y., Yoshida, M. and Mori, K., 2002. Pheromone, synthesis. Part 217. Synthesis of (2S, 2'R, 3S, 4R)-2-(2'-hydroxy-21'-methyldocosanoylamino)-1,3,4-pentadecanetriol, the ceramide sex pheromone of the female hair crab, *Erimacrus isenbeckii*. *Biosci. Biotechnol. Biochem.*, 66: 1531–1537.

171. Peña-Cabrera, E. and Liebeskind, L.S., 2002. Squaric acid ester-based total synthesis of echinochrome A. *J. Org. Chem.*, 67: 1689–1691.

172. Oliviera, J.S., Pires Junior, O.R., Morales, R.A.V., Bloch Junior, C., Schwartz, C.A. and Freitas, J.S., 2003. Toxicity of puffer fish-two species (*Lagocephalus laevigatus*, linaeus 1766 and *Sphoeroides spengleri*, Bloch 1785) from the southeren Brazilian coast. *J. Venom. Anim. Toxins Incl. Trop.*, 9: 76–82.

173. Moore, K.S., Wehrli, S., Roder, H., Rogers, M., Forrest, J.N., McCrimmon, D. and Zasloff, M., 1993. Squalamine: An aminosterol antibiotic from the shark. *Proc. Natl. Acad. Sci., U.S.A.*, 90: 1354–1358.

174. Sci-Edu, 2000. New cancer drug extracted from marine organism. *People's Daily*, 1–4. (www.fpeng.peopledaily.com.cn/200012/05/eng).

175. Anonymous, 2003. Venom hunt finds 'harmless' snakes: A potential danger. *Science Daily*, 1–2. (www.sciencedaily.com/release/2003/12).

Chapter 18

Haematological Responses of *Anabas testudineus* (Bloch) exposed to Sublethal Concentrations of Plant Extract of *Calotropis gigantea* (L.) R.Br.

☆ *K. Sree Latha*

ABSTRACT

Blood parameters are highly sensitive and have gained momentum in recent years in view of their importance in assessing the condition of fish and their responses to environmental change. Biopesticides are heading the list of pollutants posing a great potential risk.The present study deals with sublethal effects of plant extract of *Calotropis gigantea* (L.) R.Br. on haematological responses of freshwater fish, *Anabas testudineus*(*Bloch*).Decreased values($P<0.001$)were observed in haemoglobin content, oxygen combining capacity, Total red blood corpuscles count (RBC), haematocrit (Ht) and mean corpuscular haemoglobin concentration (MCHC) and increased values (reverse trend) ($P<0.001$) were observed in Mean corpuscular haemoglobin (MCH) and mean corpuscular volume (MCV) and total white blood cell count.

Introduction

Toxic pollutants rendered almost lifeless by a lethal combination of agricultural fertilizers, pesticides, industrial pollutants, sewage runoff. The process kills the food chain from the bottom up, rendering the area virtually lifeless. The interaction among various chemical pollutants in the aquatic system may be synergistic, antagonistic or additive, and may cause acute toxic effect on different species. Biopesticides of any type have been approved for use by producers in the developing world where, chemical use is largely or totally uncontrolled. They can be insecticidal or antimicrobial, some are herbicidal or fertilizers. Naturally occurring insecticide chemicals could also be potential pollutants. The bioactive chemical substances are found in plants species that show deleterious effects on fishes and other non-target organisms leading to death. Earlier studies revealed mostly about mortality patterns of fishes exposed to chemical compounds and very less data is available on the biopesticidal effect on fishes. The hazards of biopesticides usage have awakened the modern man to realize the risk involved in constant use of pesticides. Numerous biochemical indices of stress have been proposed to assess the health of non-target organisms exposed to toxic chemicals in aquatic ecosystems (Niimi, 1990). The physical properties of fish blood are very sensitive to environmental changes (Huges and Nemcsok, 1988) and are commonly used (Wedemeyer and Yastuke, 1977). The use of haematological methods as indicators of sub lethal stress can supply valuable information concerning the physiological reaction of fish in changing environment, the reason for this is the close association between the circulatory system of the fish and the external environment (Cassilas and smith, 1977), which induces changes in erythropoietic activity. Hence, an attempt is made to investigate the effects of plant extract of *C. gigantea* (L.)R.Br., the aerial parts of the plant and the extracts are used as fertilizer, and also insecticidal, good ovicidal and larvicidal in nature, effective fish poison, arrow poison, infanticide, homicidal,slows anti implantation activity (Gurdeep Chatwal, 1981; Agarwal, 1992;Azariah *et al.*, 1988; Savita and Anandhi, 2006; Patole and Mahajan, 2006).The present study was undertaken to trace the effects of sub-lethal concentrations of plant extract of *Calotropis gigantean* (L.) R.Br. on blood of *Anabas testudineus* (*Bloch*), freshwater fish, commonly edible and widely cultivated in India. The parameters studied includes haemoglobin content (Hb), oxygen combining capacity, total red blood corpuscles count (RBC), total white blood cell count, haematocrait (Ht), mean corpuscular haemoglobin (MCH), mean corpuscular volume (MCV), mean corpuscular haemoglobin concentration (MCHC).

Materials and Methods

The live specimens of Air-breathing fish, *Anabas testudineus* (*Bloch*) (11± 2.2 grams weight, 8±2 cms. length) were collected from local fish farm and acclimatized for 15 days under laboratory conditions with fresh goat liver as their diet. Bioassay to determine the 96 hour LC_{50} was conducted in plexi-glass aquarium employing the technique as described by APHA (1998). An aqueous plant extract was prepared collecting the stem and leaves by treating 5 grams powder with 100ml of water, heating at 50-60 ° C on water bath over a period of 6-8hrs, filtered through Whatman filter paper No. 1 and concentrated on water bath by slow evaporation (Mahajan and Patole, 2003; Tiwari and Singh, 2003). The 96 hour LC_{50} for plant extract of was 10ml/5lit. The experimental fishes were exposed to three different sub lethal concentrations *viz.*, 1/5th,1/10th and 1/20th of LC_{50} values of plant extract *i.e.*, TU1 (0.5 ml/5litre); TU2 (1 ml/5litre); TU3 (2 ml/5litre) for 96hrs exposure was carried out. The physico-chemical characteristics of water obtained from head tank used in aquarium was analyzed during experimentation period as per standard method of APHA (1998). Three fishes randomly picked were introduced in each of the three test chambers containing the above mentioned concentrations, a fourth test chamber also containing three fishes, served as control. The blood parameters haemoglobin content (Hb), oxygen combining capacity(OCC), total red blood corpuscles count (RBC's), haematocrit (Ht) and mean corpuscular haemoglobin concentration (MCHC), mean corpuscular haemoglobin (MCH) and mean corpuscular volume (MCV) and total white blood cell count (WBC's) were studied by collecting blood samples directly from ductus cuvies, situated beneath the operculum (Smith *et al.*, 1952) using 2 ml sterile disposable syringe rinsed with anticoagulant (EDTA). The blood was mixed well in a vial containing anticoagulant (EDTA) (Blaxhall and Daisley, 1973). Haemoglobin content of blood was determined by Sahli's method.

Oxygen combining capacity of blood was calculated by multiplying the haemoglobin content with the oxygen combining power of 1.25 ml of oxygen per gram of Haemoglobin (Decie and Lewis, 1963). The total red blood corpuscles count and total number of leukocytes was done using Haemocytometer and Neubauer chamber. Haematocrit was determined using haematocrit tube,centrifuging sample for 15 minutes at 3500 rpm.The reading was made from the graduation of the tube and expressed as the volume of the erythrocytes per 100cm^3 (Snieszko, 1960). Mean corpuscular volume of red cells in cubic micrometers was determined (Dacie and Lewis, 1963).

$$\text{MCV} = \frac{\text{Packed cell volume per liter of blood}}{\text{Erythrocytes in millions per mm}^3}$$

Mean corpuscular haemoglobin (MCH) in micro grams was determined (Decie and Lewis, 1963).

$$\text{MCH} = \frac{\text{Heamoglobin in grams per litre of blood}}{\text{Erythrocytes in millions per mm}^3}$$

Mean corpuscular haemoglobin concentration (MCHC) (*Decie and Lewis, 1963*) was determined.

$$\text{MCHC} = \frac{\text{Haemoglobin in g per 100 ml}}{\text{Haemoglobin in g per 100 ml}} \times 100$$

Results and Discussion

The blood smear of plant extract treated fish showed a number of morphological changes in the erythrocytes. The blood exhibited, more pronounced departures in the shape of the cells from the elliptical forms (of the controls) and vacuolation of the cytoplasm in TU1. And in TU2 and TU3 showed drastic changes in the shape of the nucleus, characterized by lysis of cell membrane compared to the control with the increase in concentration of plant extract. The haemoglobin content was lower in the plant extract exposed fish with respect to controls was observed but the decrease was significant in TU1 ($P<0.01$), TU2 ($P<0.05$) and $P<0.001$in TU3.The percentage of decrease varies from -6.1 to -7.97 per cent (Figure 18.1).The RBC counts were lower in treated fish compared to controls, but the level of decrease was gradual on all the three plant extract treated groups, TU1 and TU2 with $P< 0.01$ was observed.Decrease was highly significant in TU3 ($P<0.001$). The percentage change observed in plant extract treated fish was -10.78 to -19.6 (Figure 18.1).

The oxygen combining capacity was lower in the treated groups of fish, compared to controls, $P<0.01$ in TU1 for plant extract; Decrease was highly significant in TU2,TU3 ($P<0.001$) The percentage of decrease varies from -6.1 to -8 (Figure 18.1).The decrease in haematocrit value was significant ($P<0.05$) in TU1,for TU2 and TU3 it is $P<0.001$. The percentage of decrease varies from -1.3 to -3.0 (Figure 18.1). MCHC shows a decreasing trend ($P<0.05$) in all treatment groups of plant extract. The percentage of decrease in MCHC varies from -4.77 to -5.6 (Figure 18.1).

Gradual increase in MCV was observed in the three sub lethal concentrations of plant extract treated fish, values were significant ($P<0.05$) in TU1, TU2 TU3. The percentage of increase varies from 9.7 to 20.68 (Figure 18.2). MCH follow the similar trend as observed in MCV. Changes were significant ($P< 0.05$) in all treatment groups of plant extract. The percentage of increase varies from 1.99 to 13.20 (Figure 18.2).

Haematological changes of *Anabas testudineus* exposed to sublethal concentrations of Plant extract of *Calotropis gigantea* for 96 hrs.

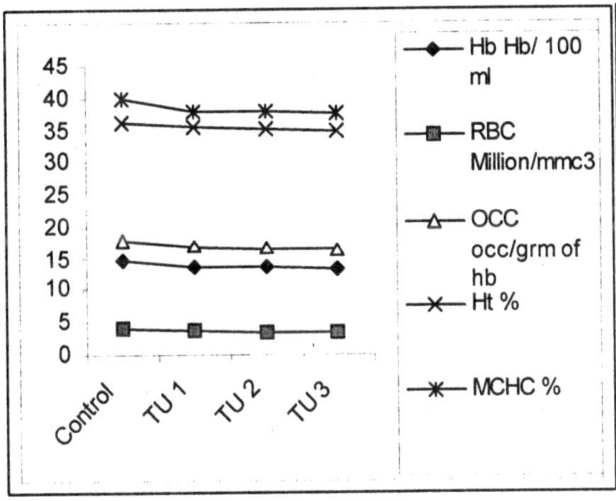

Figure 18.1: RBC's, Hb, OCC, MCHC and Ht

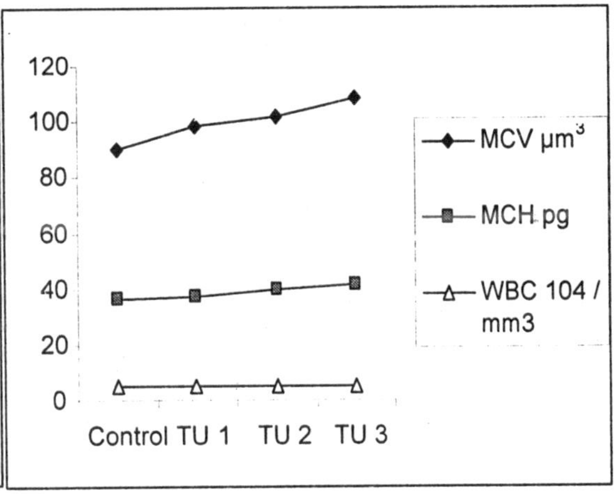

Figure 18.2: MCV, MCH and WBC's

Gradual increase in total WBC level was observed in fishes exposed to the three sublethal concentrations, when compared to the control. But the increase was significant (P<0.01) in TU1, TU2, TU3 of plant extract treated fish. The percentage of increase in WBC's varies from 4.16 to 10.4 (Figure 18.2). Haemolysis of red blood cells provides simple and rapid way of studying the effect of pollutants on biological membranes (Harington *et al.*, 1971) numerous investigations have considered membrane model a measure of pollutant's cytotoxicity (Allison *et al.*, 1966). The red blood cell membrane haemolysis has proved to be a simple and rapid way of attempting to find the possible correlation between toxicity and haemolytic activity (Macnab and Harington, 1967). In the present study a clear trend was observed linking plant extract concentration with membrane damage. The impact was the most severe on fishes exposed to the highest of the three sub-lethal concentrations of plant extract of *Calotropis gigantea*. The present experiments revealed that Haemoglobin content, RBC count,OCC,Ht and MCHC were significantly lesser in the plant extract exposed fishes. The most pronounced depression in RBC count, Oxygen combining capacity of blood is direct function of its Hb content. Similar observation in rats exposed to monocrotophos and its analogues (Siddiqui *et al.*, 1991) and in *Anabas testudineus* exposed to monocrotophos (Santha Kumar, 1998), *Anabas testudineus* exposed to biopesticide prepared from *Calotropis gigantea* (Bharathi, 2005) was found. Decrease in Hb content, RBC count, OCC, Ht and MCHC was also observed in fishes exposed to other pesticides such as fenproparthrin (Ahamad Figar *et al.*, 1995); paraquant (Ibrahim *et al.*, 1995); malathion (Ikhatair–ud- Din *et al.*, 1996) praquant (Ibrahim *et al.*, 1995); Methyl parathion (Nath Rabindra and Banerjee, 1996). Haematological studies disclose possible reaction of blood and blood forming organs to the pesticide treatment (Siddiqui *et al.*, 1987; Srinivasan and Radha Krishnamurthy, 1983). Janardhan and Sisodia (1990) reported decrease in the Hb concentration, erythrocyte counts in rats exposed to monocrotophos. The reduction in RBC number may be due to microcytic or normocytic anaemia (Tuschiya, 1973). In most vertebrates, including fish erythropoietic activity is regulated by erythropoietin produced in the kidney (Gordon *et al.*,1967). Hypoxia constitutes the fundamental stimulus for erythropoiesis with the kidney as the probable sensing organ for low blood oxygen tensions (Jacobsen and Krantz, 1968). In the present studies, *Anabas testudineus* showed a decrease in RBC counts suggesting a decrease in erythropoietic activity. A structurally intact and normally functioning kidney is essential for erythropoietin revealed progressive dystrophic changes in the kidney tubules of *Anabas testudineus* exposed to the three sublethal concentrations of monocrotophos (Kumar, 1998). Kidney damage usually causes decrease in erythropoietin levels which in turn decrease RBC production and Hb synthesis even under pesticide induced stress condition. The significant decrease in Ht value in fish is correlated with a decrease in RBC, which might be due to its effect on blood forming organs (Srinivasan and Radhakrshnamurthy, 1983). The slight increase and decrease observed in MCHC and MCH values cannot be attributed to cells shrinking or swelling but rather to a proportional decrease in Red blood cells and haemoglobin concentration. The significant increase in MCH and MCV is result of swelling may be due to beta adrenergic stimulation caused by exposed stress condition experienced by the test organisms to plant extract of *Calotropis gigantea* (Butler *et al.*, 1978 and Santha Kumar,1998). These findings in *Anabas testudineus* exposed to plant extract of *Calotropis gigantea* are in partial agreement with the results of other researchers (Dalela *et al.*, 1981; Mishra and Srivastava, 1983; El-Doismaty 1987; Gill *et al.*, 1991a, b; Dutta *et al.*, 1992c, Santha Kumar,1998;Josphene paulina,2003). The leukocytes play a crucial role in defending the animal against the invading toxins. They contribute to the immune response by destroying harmful bacteria and inactivating harmful foreign materials in the tissues and blood. The results of our experiment reveals that exposure to plant extract of *Calotropis gigantea* caused increase in WBC count.The increase in WBC count may be attributed to the response of the fish to plant extract of *Calotropis gigantea*, where plant extract may act as an antigen. The significant increase in the total leukocyte count corroborates with the earlier work on fishes (Ahamad Figar *et al.*, 1995; Ibrahim *et al.*, 1995; Ikhatair-Ud-Din *et al.*, 1996; Nath Rabindra and Banerjee, 1996). WBC is inextricably involved in the regulation of immunological function and prolonged exposure of *Anabas testudineus* to plant extract may inflict immunological deficiency.

Acknowledgements

Author expresses deep sense of gratitude to Dr. Kadem Ramudu, Department of Zoology, Tagore Arts College, Lawspet, Puducherry,and Prof. R.V. Gopala Rao, HOD, Department of Zoology, Dr. S.R.K. Govt. Arts College, Yanam for the inspiring advice and constant encouragement throughout the work.

References

Agarwal, O.P., 1992. *Chemistry of Organic Natural Products*. Goel Publishing House, Meerut, India, 1: 74.

Ahamad, Figar, Ali Syed, S. and Abdul, Shakoori, 1995. Sub lethal effects of Danitol a synthetic pyrethrinoid on Chinese grass carp, *Folia. Biol.*, 43(3–4): 151–159.

Allision, A.C., Harigtan, J.S. and Birbeck, 1966. *Journal of Exp. Mep.*, 24: 1141–1153.

APHA, 1998. *Standard Methods for Examination of Water and Wastewater* 20th edn. American Public Health Association, New York.

Azariah, J., Azariah, H., Mallikesan, Sumathi, M.V. and Sunderraj, C., 1988. Toxicology of plant *Calotropis gigantean*. *Toxicon.*, 26(1): 15.

Bharathi, M., 2005. The impact of biopesticide prepared from *Calotropis gigantea* on fish *Anabas testudineus* (Bloch). *M.Phil. Thesis*, Pondicherry University, Puducherry, India.

Blaxhall, P.C. and Daisley, 1973. Routine heamatological methods for use with fish blood. *J. Fish Biol. R.*, p. 771–781.

Butler, P.J., Taylor, E.W., Capra, M.F. and Davidson, 1978. The effect of hypoxia on the level of circulating catacholamines in *Scylivrhinus canicula. J. Comp. Physiol.*, 127: 325–330.

Cassilas, E. and Smith, L.S., 1977. Effect of stress on blood coagulation and haematology in rainbow trout. *J. Fish Biol.*, 10: 481–491.

Chatwal, Gurdeep, 1981. *Organic Chemistry of Natural Products.* Himalaya Publishing House, Mumbai, India, 1: 111.

Dalela, R.C., Bhatnagar, M.C. and Verma, S.R., 1981. Effects of *in vivo* haematological alteration in *Mystus vittatus* following subacute to pesticide and their combinations. *Envir. Pollu.*, 21: 3–8.

Decie, J.V. and Lewis, S.M., 1963. *Practical Haematology.* J&A Ltd., Churchill, London.

Dutta, H.M., Doger, J.V.V., Singh, N.K. and Roy, 1992c. Malathion induced changes in serum protein and haematological parameters of *Heteroneustes fossilis. Bull. Envir. Contam. Toxicol.*, 49: 91–97.

El-Doimaty, N.A., 1987. Stress response of juvenile *Clarias* larva elicited by copper. *Comp. Biochem. Physiol.*, 88C: 259–262.

Gill, T.S., Pande, J. and Tewari, 1991a. Effects of endosulfan and phosphomidon poisoning on the peripheral blood *of Barbus conchonius. J. Env. Sci. Health*, 26(2): 249–255.

Gill, T.S., Pande, J. and Tewari, 1991b. Haemo pathological changes associated with aldicarb poisoning in fish. *Bull. Envir. Contam. Toxicol.*, 47(4): 628–633.

Gordon, A.S., Cooper, G.W. and Zanjani, E.D., 1967. The kidney and erythropoiesis. *Sem. Haemat.*, 4: 337–343.

Harington, J.S., Miller, K. and Mac Nab, G., 1971. *Environ Res.*, 4: 95–117.

Huges, G.M. and Nemcsok, J., 1988. Effect of low pH alone and combined copper sulphate on blood parameters of rain bowtrout. *Env. Pollu.*, 55: 89–95.

Ibrahim, I.G., Mohamad, A.M. and Zaki, Z.T., 1995. Changes induced by paraquat herbicide in muscle protein and blood of two fishes from Egypt. *Al-Azar Bull. Sci.*, 6(1): 637–647.

Ikhatair-Ud-Din, K. Hafeez and Mohmad Abdul, 1996. Effect of malathion on blood of *Cyprinion. Pak. J. Zool.*, 28(1): 45–49.

Jacobsen, L.D. and Krantz, S.B., 1968. Summary on erythropoietin. *Am. Ny. Ac. Sci.*, 149: 578–583.

Janardhan, A. and Sisodia, P., 1990. Monocrotophos: Short term toxicity in rats. *Bull. Environ. Contaminat. Toxicol.*, 44: 230–239.

Josphene Paulina, C., 2003. The toxic effect of latex of *Calotropis gigantea* and its recovery with additive nutrients on *Anabas testudineus* (*Bloch*). *M.Phil. Thesis*, Pondicherry University, Puducherry, India.

Macnab and Harrington, 1967. *Nature* (London), 214: 522–523.

Mahajan, R.T. and Patole, S.S., 2003. Effect of plant extracts on oxygen consumption rate in fish: *Nemacheilus evezardi. J. Freshwater Biol.*, 15(1–4): 109–113.

Mishra, J. and Srivastava, A.K., 1983. Malathion induced haematological changes in *Clarias batracus* exposed to malathion. *Env. Poll. Ser. Ecol. Biol.*, 22: 149–158.

Nath, Rabindra and Banerjee, 1996. Effect of methyl parathion and cypermethrin on *Heteropneustes fossilis*. *Env. Ecol.*, 14(1): 163–165.

Niimi, A.J., 1990. Review of biological methods and other indicators to assess fish health in aquatic ecosystems containing toxic chemicals. *J. Great Lakes Res.*, 16: 529–541.

Patole, S.S. and Mahajan, R.T., 2006. Comparative Icthyotoxic activity of some Indigenous plants in fish: *Lepidocephalichthys*. *J. Aqua. Biol.*, 21(1): 168–172.

Santhakumar, M., 1998. Studies on toxicological effects of monocrotophos in an air breathing fish *Anabas testudineus(Bloch)* under Sublethal exposure. *M.Phil. Thesis*, Pondicherry University, Puducherry, India.

Savita and Anandhi, P., 2006. *Agrobios*, 5 (6): 19.

Siddiqui, M.K.J., Rahaman, M.F., Mustafa, M. and Balerao, U.T., 1991. Comparative study of blood changes and brain acetylcholinesterase inhibition by monocrotophos in rats. *Ecotoxicol. and Env. Safety*, 21: 283–289.

Smith, G.C., Lewis, W.M. and Kaplan, 1952. A comparative morphologic and physiologic study of fish blood. *Progre. Fish. Cul.*, 14: 169–172.

Snieszko, S.F., 1960. Microhaematocrit as a tool in fishery research and management. Spec. Scient. Rep. U.S. Fish. Wildlf. Serv. No: 341.

Srinivasan, K. and Radha Krishnamurthy, R., 1983. Effect of HCH isomers on some haematological parameters. *J. Food Sci. Techl.*, 20: 322.

Tiwari, S. and Singh, A., 2003. Control of common predatory fish, *Channa punctatus*, through *Nerium indicum* leaf extracts. *Chemosphere*, 53(8): 865–875.

Tuschiya, K., 1973. Lead. In: *Handbook on Toxicology of Metals*, (Eds.) L. Friberg, G.A. Nordberg and V.B. Vonk. Elsvier North Holland Biomedical Press, Amsterdam.

Wedemeyer, G.A. and Yastuke, W.T., 1977. Clinical methods for assessment of the effects of environmental stress in fish health. *U.S. Fish and Wildlife Service*, Washington DC, 89.

Chapter 19

Adaptive Changes in Respiratory Movements of Air-breathing Fish, *Anabas testudineus* (Bloch) Exposed to Latex and Plant extract of *Calotropis gigantea* (L.) R.Br. and Recovery of Toxicity with Additive Nutrients

☆ *K. Sree Latha*

ABSTRACT

Behaviour is the recordable and observable activity of the living organisms. In the present study adaptive changes in respiratory movements of air-breathing fish, *Anabas testudineus* (Bloch) exposed to sublethal concentrations of latex and plant extract of *Calotropis gigantea* (L.) R.Br. and recovery of toxicity with additive nutrients were studied, the parameters-surfacing behaviour and opercular movements were estimated, a significant increase ($P>0.001$) in surfacing behaviour and opercular movements with decreasing ($P>0.001$) trend was observed in all the sublethal concentrations of latex and plant extract compared to control. And also, among the three groups of treated fish,the per cent of suppression or elevation of the activity observed was partially recovered in the fish treated with latex and supplements,these results indicated adaptability of fish exposed to three test concentrations of latex by slow recovery in the levels of its energetics.

Introduction

Biopesticides of any type in the developing world, where chemical use is largely or totally uncontrolled they can be insecticidal or antimicrobial, herbicidal are proved as potential pollutants, if not used properly. The large scale discharge of pollutants from the place of actual use, find their way to aquatic system through spray drift, washing from the atmospheric precipitation, erosion and surface run off water, where they ravage the biotic life including fishes and may alter the physical, chemical and biological nature of water. Naturally occurring insecticides such as Pyrethrum, Nicotine, Rotenone, Hellebore, Ryania, and Sabadilla are proved as potential pollutants (Savita and Anandhi, 2006; Patole and Mahajan, 2006). Presence of these toxic chemicals (pesticides, fungicides and fertilizers) in freshwater media, may cause death or sublethal effects including respiratory metabolism on the non-target organisms like fish etc. (Bhattacharya, 1985; Desai and Joshi, 1985; Sastry and Malik, 1979; Bergeri *et al.*, 1984). The bioactive chemical substances are found in plants species that show deleterious effects on fishes are Saponins, which lower the surface tension preventing uptake of oxygen leading to death of fish.Glycosides, like alkaloids, they possess physiological activity (Cardiac glycosides) are poisonous compounds (Agarwal, 1992; Gurdeep Chatwal, 1981). The species of Calotropis are laticiferous shrubs or small trees which are poisonous (Azariah *et al.*, 1988; The Wealth of India, 1992). Alkaloids of *Calotropis* species has pronounced physiological activity, although, many possess curative properties, they are powerful poisons, protective substances against animal or insect attack (Agarwal, 1992), and also effective fish poison, insecticidal, good ovicidal and larvicidal in nature (The Wealth of India, 1992; Khare, 2004). *Calotropis gigantea* (L.) R.Br alkaloids are namely uscharin, voruscharin (Robert Raffauf, 1970) and Terpenoids are optically and biologically active namely, insecticidal, antihelmintic or antiseptic in action. These compounds induce histomorphological changes in gills, liver, RBC's, stomach, kidneys and brain (Agarwal, 1992; Patole and Mahajan, 2006). Most of the air-breathing fishes use a dual system of gas exchange through the gills/skin system in water and accessory organs from air. Air breathing becomes a more important alternative for them, when the environment is less favourable. *Anabas testudineus* (Bloch) was chosen for the study for its easy availability, economically and ecologically important and its tolerance to wide range of environmental factors.

Materials and Methods

The live specimens of air-breathing fish, *Anabas testudineus* (Bloch) (11± 2.2 grams weight, 8±2 cms length) were collected from local fish farm and acclimatized for 15 days under laboratory conditions with fresh goat liver as their diet. Feed alone with water as control and supplements (Vitamin C, Vitamin E, Glucose, Fructose and Egg albumin plus glycine) added was 100mg/5litre volume along with feed, in combination with three sublethal concentrations of latex of *C. gigantea* (L.) R.Br. was also done separately. Bioassay to determine the 96 hour LC_{50} were conducted (in plexi-glass aquarium) employing the technique as described by APHA (1998). Latex of *C. gigantea* (L.) R.Br. is collected from cut end of the stem tip and used for the analytical work (Rooj *et al.*, 1984). And aqueous plant extract with water was prepared collecting the stem and leaves by treating 5 grams powder with 100ml of water, heating at 50-60 °C on water bath over a period of 6-8hrs, filtered through whatman filter paper and concentrated on water bath by slow evaporation.(Mahajan and Patole, 2003; Tiwari and Singh, 2003). The 96 hour LC_{50} for latex was found 0.1ml/5lit. and for plant extract of was 10ml/5lit. The experimental fishes were exposed to three different sub lethal concentrations *viz.*, $1/5^{th}$,$1/10^{th}$ and $1/20^{th}$ of LC_{50} values TU1 (0.005 ml/5litre); TU2 (0.01 ml/5litre) and TU3 (0.02 ml/5litre) of latex, for plant extract TU1 (0.5 ml/5litre); TU2 (1 ml/5litre); TU3 (2 ml/5litre) and three different sub lethal concentrations *viz.*, $1/5^{th}$,$1/10^{th}$ and $1/20^{th}$ of LC_{50} values TU1 (0.005 ml/51itre); TU2 (0.01 ml/5litre) and TU3 (0.02 ml/5litre) of latex plus supplements for 96hrs exposure was carried out. The physico-chemical characteristics of water obtained from head tank used in aquarium was analyzed during experimentation period as per standard method of APHA (1998). Three fishes, randomly picked, were introduced in each of the nine test chambers containing the above mentioned concentrations, a tenth test chamber also containing three fishes, served as control. The depth of the water was kept constant at 25 cms in all aquariums. All the fishes accepted the feed, ate readily and showed no fright responses. The number of times each test individuals surfaced was observed for a known period (5-10 min) at 9.30 AM., 11.30 AM and 3.30 PM. The number of opercular movements was observed for 1 minute using a stopwatch. The observation was done again for 1 minute and mean values were taken to get the behavioral action of the particular fish. Significance between control and experimental values was determined by Students 't'test.

Results and Discussion

No mortality were observed in the fishes exposed to three different sub lethal concentrations *viz.*, 0.005 ml, 0.01 ml, 0.02 ml of latex, for plant extract 0.5 ml, 1 ml, 2 ml and three different sub lethal concentrations 0.005 ml, 0.01 ml and 0.02 ml of latex plus supplements for 96hrs exposure. Figure 19.1 shows the surfacing frequency of control and latex, plant extract,latex plus supplements exposed *Anabas testudineus*. Generally,the surfacing frequency showed progressive increase with increasing concentrations of latex and plant extract in different days. The increase was significant ($P<0.001$) in TU2and TU3 compared to control.

Opercular movements with number of beats per minute showed progressive decrease with increasing concentration of latex and plant extract (Figure 19.2). The decrease was significant ($P<0.001$) in TU2 and TU3 compared to control. In *Anabas testudineus,* the branchial chambers have been modified for air-breathing. *Anabas* consumes more oxygen from air (54 per cent) by surfacing then from water (46 per cent) of its total volume of oxygen during bimodal respiration.In the present study, the drop in opercular movements and corresponding increase in frequency of surfacing of fish clearly indicates the fish adaptively shifts towards aerial respiration (by obtaining atmospheric oxygen by surfacing).

Behavioural changes of *Anabas testudineus* exposed to sublethal concentrations of Latex, Latex+Supplements and Plant extract of *Calotropis gigantea* for 96 hrs

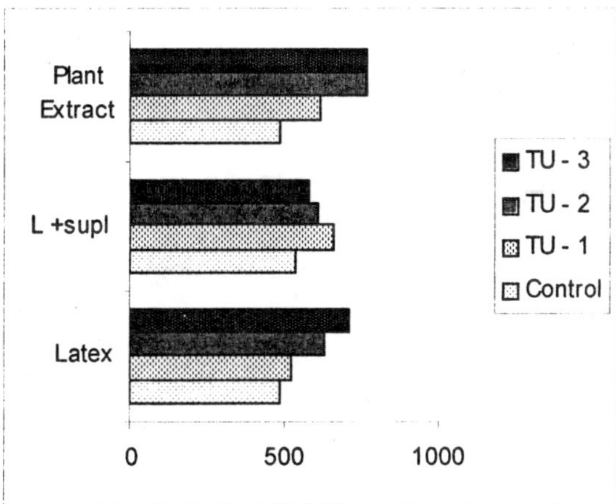

Figure 19.1: Number of Surfacing/Day

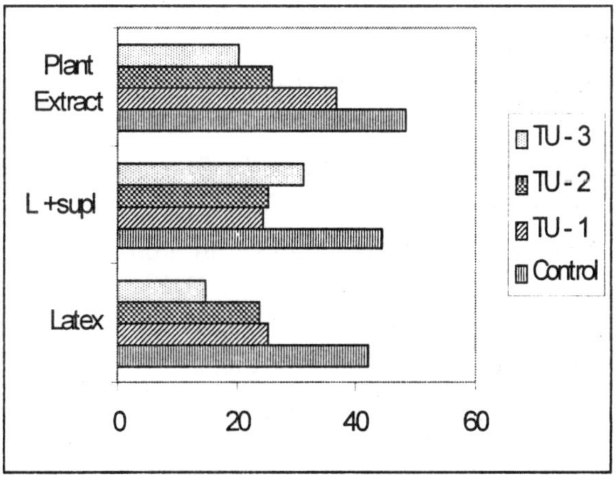

Figure 19.2: Opercular Movements/min

Changes in ventilation rate and surfacing frequencies are the symptoms noticed in the fish after exposed to latex and plant extract, and these activities help the fish to avoid the contact with poison and fight against stress (Roy and Munshi, 1987). Specialised chomoreceptor cells present in the lining of suprabranchial chamber help in detecting the noxious substance present in water (Hughes and Munshi, 1973). Bimodal species are noted for their resistance to environmental stress and aquatic hypoxia (Dehadrai and Tripathi, 1976). Air-breathing render fish more resistance to toxicants by reducing the gill ventilation and there by reducing contact with toxicant at major site of uptake of noxious substances should be added to the list of other possible advantage of bimodal respiration in air-breathing fishes. The present studies revealed that, exposure of air-breathing fish to latex in the aquatic environment cause drastic changes in respiratory movements with respect to oxygen uptake. Hence by surfacing,the gills ventilatory movements are important observable responses depending on the availability of oxygen which in turn was influenced by the latex and plant extract contamination and induced damage in tissues which reflects on behavioral pattern, also due to shift in respiratory metabolism as a result of stress induced by latex and plant extract which ultimately lead to deterioration of the general health of the fish. Surfacing frequency in latex plus supplements treated fish showed significant decrease in values ($P<0.001$), and opercular movements showed significant increase in values ($P<0.001$). Similar observations were found in *Channa orientalis* with cadmium (Borane *et al.*, 2008) and in *Tilapia mossambica* with lindane toxicity (Koundinya, 1978; Narsimha, 1983). Addition of glucose and fructose will enhance energy and thereby fish protected from stress condition. The addition of Albumin, Glycine, Vitamin C and E (Halver, 1972; Seemuller, 1992; Kalp *et al.*, 1994; Battacherjee *et al.*, 2003) makes the tissue resistant and also disease resistant including anti-oxidant property preventing tissue damage and providing tensile strength, in the present study certain amount of recovery could be seen in latex plus supplements treated fish, enabling to withstand stress (compensatory step to overcome) in toxic environment and the nutrients, that provide strength, rapid tissue repair in trauma or diseased condition and support to fish.

Acknowledgements

Author expresses deep sense of gratitude to Dr. Kadem Ramudu, Department of Zoology, Tagore Arts College, Lawspet, Puducherry, and Prof. R.V. Gopala Rao, HOD, Department of Zoology, Dr. S.R.K. Govt. Arts College, Yanam for the inspiring advice and constant encouragement throughout the work.

References

Agarwal, O.P., 1992. *Chemistry of Organic Natural Products.* Goel Publishing House, Meerut, India, 1: 74.

APHA, 1998. *Standard Methods for Examination of Water and Wastewater*, 20th edn. American Public Health Association, New York.

Azariah, J., Azariah, H., Mallikesan Sumathi, M.V. and Sunderraj, C., 1988. Toxicology of plant *Calotropis gigantea. Toxicon.*, 26(1): 15.

Battacherjee, C.R., Dey, S. and Goswami, P., 2003. Protective role of ascorbic acid against lead toxicity in blood of mice. *Bull. Env. Contam. Toxicol.*, 70: 1189–1196.

Bengeri, K.V., Shivraj, K.M. and Patil, H.S., 1984. Toxicity of dimethyl parathion to *Labeo rohita* and oxygen uptake rate. *Envi. Ecol.*, 20: 1–4.

Bhattachary, 1985. Toxicity of carbofuran and phehthoate on *Channa punctatus* and *Anabas testudineus. J. Env. Biol.*, 6: 129–137.

Borane, V.R., Patel, R.D. and Zambare, S.P., 2008. Ascorbic effect on cadmium induced alterations in behavior of *Channa orietalis*. *J. Aqua. Biol.*, 23(2): 155–158.

Chatwal, Gurdeep, 1981. *Organic Chemistry of Natural Products*. Himalaya Publishing House, Mumbai, India, 1: 111.

Dehadrai, P.V. and Tripathi, S.D., 1976. Environment and ecology of freshwater air breathing teleosts. In: *Respiration of Amphibious Vertebrates*, (Ed.) G.H. Huges. Academic Press, London, p. 39–72.

Desai, A.K. and Joshi, U.M., 1985. Histological observation of *Tilapia mossambica* after exposure to monocrotophos. *Toxicol. Letter*, 21: 325–331.

Halver, J.E., 1972. Role of ascorbic acid in fish diseases and tissue repair. *Fishery Bull. World Life Service*, US, 64: 1.

Hughes, G.M. and Morgan, M., 1973. The structure of fish gills in relation to their respiratory functions. *Biol. Rev.*, 48: 419–475.

Kalp, D., Weiser, H. and Rambeck, W.A., 1994. The influence on vitamin C on cadmium retention in pigs. *Rev. Med. Vet.*, 145: 291–297.

Khare, C.P., 2004. *Encyclopedia of Indian Medicinal Plants*. Springer Publications, India, p. 120–121.

Koundinya, P.R., 1978. Study on physiological responses on *Tilapia mossambica* to pesticide impact. *Ph.D. Thesis*, S.V. University, Tirupathi, India.

Mahajan, R.T. and Patole, S.S., 2003. Effect of plant extracts on oxygen consumption rate in fish: *Nemacheilus evezardi*. *J. Freshwater Biol.*, 15(1–4): 109–113.

Narsimha, Murthy, 1983. Studies on toxic potentiality of lindane on *Tilapia mossambica*. *Ph.D. Thesis*, S.V. University, Tirupathi, India.

Patole, S.S. and Mahajan, R.T., 2006. Comparative Icthyotoxic activity of some indigenous plants in fish: *Lepidocephalichthys*. *J. Aqua. Biol.*, 21(1): 168–172.

Rooj, N.C., Ojha, J. and Datta Munshi, 1984. Effect of certain biocidal plant sap on gills of certain Hillstream fishes of Ranchi (India). *Biol. of India*, 6: 320–322.

Roy, P.K. and Munshi, J.S.D., 1987. Toxicity of malathion on *Cirrhinus mrigala*. *Biol. Bull. India*, 9(1): 50–56.

Sastry, K.V. and Malik, P.V., 1979. Studies on effect of demacron on digestive system of *Channa punctatus*. *Env. Contam. Toxicol.*, 8: 397–398.

Savita and Anandhi, P., 2006. *Agrobios*, 5(6): 19.

Seemuller, K.R., 1992. Cadmium influenced dietary components in rats and pigs. *Ph.D. Thesis*, Ludwig Maxmilans University, Germany.

The Wealth of India, 1992. *Raw Materials*. Publications and Information Directorate, CSIR, India, p. 78–84.

Tiwari, S. and Singh, A., 2003. Control of common predatory fish, *Channa punctatus*, through *Nerium indicum* leaf extracts. *Chemosphere*, 53(8): 865–875.

Chapter 20

Role of Purna Fish Co-operative Society to Rejuvinate Yeldari Reservoir Fishery in Parbhani District of Maharashtra

☆ *S.D. Niture and S.P. Chavan*

ABSTRACT

There are 28 fish co-operative societies working on different reservoirs in District Parbhani. There was no uniformity in the working pattern among these societies. There was no any ideal working system found in these co-operative societies except Purna Fish Co-operative Society At. Sawali, Taluka Jintur, District Parbhani working on Yeldari Reservoir in Parbhani District. Purna Fish Co-operative Society Sawali was established in year 2003, with some modern trends of management like selection of capital investing company for the stocking of fingerlings of Indian major carps and prawn juveniles, use of high speed motor boats for patrolling to prevent fish poaching, establishment of special vigilance squad, use of gill-nets of mesh size more than 8 cm., Ban on the use of cast nets and mosquito nets etc.The results of applications of improved management techniques and socio-economic upliftment of fisher community is discussed in this chapter.

Introduction

Yeldari reservoir is a large sized reservoir of 6272 hectare area (Sugunan,1995). The reservoir has catchments area of 7330 sq. km. It was constructed in year 1968 on river Purna at Yeldari camp near village Yeldari in Parbhani district of Maharashtra. The reservoir lies in between latitude 19° 15′N and longitude 76° 45′ E.

Reservoir fishery is the main source of income for the fisher community living in Parbhani district of Maharashtra state. The State fisheries department has given the fishing rights of the reservoirs in this region on lease basis to the private parties or to the fish co-operative societies (Sakhare, 2003). Purna Matsyavyvasai Sahakari Sanstha Maryadit, Sawali (Purna Fish Co-operative Society) of Yeldari reservoir emerged as an active fish co-operative society with implementation of some modern trends of management. The society has made an agreement with capital investing private companies only to stock the fish seed and prawn juveniles in to the Yeldari reservoir and to purchase the daily harvested catch from the reservoir. For the year 2006-07 the agreement of the society was with a company named Intellect Agri Product Company, Andheri East, Mumbai; and for the year 2007-08 the agreement was with another company named Aqua Fisheries and Food Products, Andheri West, Mumbai.To harvest the fish and prawn from the reservoir the society has issued fishing license to the member fishermen of the society and non-member fishermen from 34 villages around the Yeldari reservoir. The society has charged Rs. 05/month as fishing license fees to all fishermen involved in the fishing in Yeldari reservoir. The fisher community involved in fishing in Yeldari reservoir was largely benefited for their assured daily wages and overall income as compared to earlier condition.

Materials and Methods

The study was conducted from June 2006 to May 2008. The information about the working pattern of Purna Fish Co-operative Society was collected from the Secretary of the Society, the fishermen involved in the fishing activity and the administrative staff from the office of the society at Yeldari reservoir. The information about the nets and gears used in fish and prawn harvesting from the reservoir; methods of fishing, fish collection, preservation and marketing of the catch, past and present socio-economic status of the fisher community was collected from the fisher communities living in 34 villages around Yeldari reservoir; The data was collected through personal interviews of the fishermen and from the direct observations of the activity. The data about fish and prawn catch, fish and prawn price fluctuation at various levels of marketing, fisher community tribes and casts and their population from the Yeldari reservoir fishery was represented in the tabular form.

Results and Discussion

Fish and Prawn Species

29 fish species found in Yeldari reservoir belonging to 20 genera falling in 4 orders (Sakhare, 2003). The commonly found fish species include *Labeo rohita, Catla catla, Wallago attu, Mastacembelus armatus, Mystus blaker* and *Channa* sp. The juveniles of prawn species *Macrobrachium rosenbergii* were firstly stocked in to the Yeldari reservoir by Purna Fish Co-operative Society in year 2006.

Purna Fish Co-operative Society, Sawali of Yeldari Reservoir

Purna fish co-operative society is a registered co-operative society, Reg. No. PBN/JTR/AGR/O/1937, dated 17/03/2003, with the name Purna Matsyavyvasai Sahakari Sanstha Maryadit Sawali Taluka Jintur, District Parbhani. Total members of the society were 52. Total number of fishermen involved in the fishing process other than members of the society were 964 belongs to 34 villages present around Yeldari reservoir (Table 20.3). Purna Fish Co-operative Society has taken Yeldari reservoir on lease from July 2005 to June 2010 for fishing from the Maharashtra State fisheries department.

Modern Trends in Working Pattern Adopted by Purna Fish Co-operative Society

1. Purna Fish Co-operative Society had called the tenders from different traders, private companies, Private agents for the fish seed stocking in to the reservoir and to purchase the harvested catch up to the date 15/12/2005. From the received 6 tenders, the co-operative

Table 20.1: The Fish Catch at Yeldari Reservoir from June, 2006 to May, 2008

Sl.No.	Months and Year	Yeldari Fish Collection Centre — Fish Catch/Day of Month in kg		Bamni Fish Collection Centre — Fish Catch/Day of Month in kg		Khadki Fish Collection Centre — Fish Catch/Day of Month in kg		Average Fish Catch/Day of Month in kg	Total Fish Catch of Month in kg
		Minimum	Maximum	Minimum	Maximum	Minimum	Maximum		
1.	Jun-06	–	–	–	–	–	–		
2.	Jul-06	30	155	–	–	–	–	92.5	2867.5
3.	Aug-06	28	944	–	–	–	–	486	15066
4.	Sep-06	179	1612	–	–	–	–	895.5	26895
5.	Oct-06	21	446	–	–	–	–	233.5	7238.5
6.	Nov-06	120	483	–	–	–	–	302.5	9045
7.	Dec-06	374	674	–	–	–	–	524	16244
8.	Jan-07	338	713	–	–	–	–	525.5	16290.5
9.	Feb-07	233	612	158	342	–	–	672.5	18830
10.	Mar-07	235	674	141	441	91	332	954	29667
11.	Apr-07	45	467	67	342	95	280	648	19440
12.	May-07	60	281	20	343	104	348	578	17918
13.	Jun-07	117	598	51	254	242	369	815.5	24465
14.	Jul-07	233	928	166	732	135	755	1474.5	45709.5
15.	Aug-07	233	1013	73	606	109	788	1411	43741
16.	Sep-07	334	785	130	495	313	902	1479.5	44385
17.	Oct-07	386	1038	182	602	362	776	1673	51863
18.	Nov-07	381	1176	94	326	337	904	1609	48270
19.	Dec-07	439	1014	38	196	316	814	1408.5	43663.5
20.	Jan-08	329	1055	27	134	231	560	1168	36208
21.	Feb-08	714	1434	36	234	316	742	1788	50295
22.	Mar-08	208	1081	34	253	193	1107	1438	44578
23.	Apr-08	304	930	40	212	250	870	1303	39090
24.	May-08	210	880	70	250	210	850	1235	38285
	Total	5551	18993	1327	5762	3304	10397	22667	690054.5

Table 20.2:The Prawn Catch at Yeldari Reservoir From Jun, 2006 to May, 2008

Sl.No.	Months and Year	Yeldari Prawn Collection Centre Frawn Catch/Day of Month in kg		Bamni Prawn Collection Centre Prawn Catch/Day of Month in kg		Khadki Prawn Collection Centre Prawn Catch/Day of Month in kg		Average Prawn Catch/Day of Month in kg	Total Prawn Catch of Month in kg
		Minimum	Maximum	Minimum	Maximum	Minimum	Maximum		
1.	Jun-06	–	–	–	–	–	–	–	–
2.	Jul-06	1.75	5.25	–	–	–	–	3.5	52.5
3.	Aug-06	2	11	–	–	–	–	6.5	100.75
4.	Sep-06	0.25	3	–	–	–	–	1.625	24.375
5.	Oct-06	0.25	2	–	–	–	–	1.125	17.4375
6.	Nov-06	0.25	4	–	–	–	–	2.125	31.875
7.	Dec-06	1.25	6	–	–	–	–	3.625	56.1875
8.	Jan-07	0.25	4	–	–	–	–	2.125	32.9375
9.	Feb-07	0.25	1.25	0.25	1			1.375	19.25
10.	Mar-07	0.25	5	0.25	1	0.25	0.75	3.75	58.125
11.	Apr-07	0.25	21	0.75	10	0.25	30.5	31.375	470.625
12.	May-07	3	26	4	233	19	123	204	3162
13.	Jun-07	69	386	15	73	49.75	101	346.875	5203.125
14.	Jul-07	87	199	34	104	44	139	303.5	4704.25
15.	Aug-07	52	187	21	62	36.5	113	235.75	3654.125
16.	Sep-07	57	104	5	48	19	64	188.5	2827.5
17.	Oct-07	27	88	8	43	11	25	101	1565.5
18.	Nov-07	39	88	7	31	9	25	99.5	1492.5
19.	Dec-07	20	50	1	6	5	15	48.5	751.75
20.	Jan-08	14	44	0.25	5	4	16	41.625	645.187
21.	Feb-08	16	61	0.25	7	4	34	61.125	886.312
22.	Mar-08	16	62	2	29	4	33	73	1131.5
23.	Apr-08	16	60	2	25	4	30	68.5	1027.5
24.	May-08	16	50	0.5	10	4.5	30	55.5	860.25
	Total	438.75	1467.5	101.25	688	214.25	779.25	1844.5	28775.5615

society had made an agreement with Intellect Agri Product Private Company (IAPPC), Andheri East Mumbai (M.S.). According to the agreement, the company has to stock the fingerlings of Carps (Catla, Rohu, Mrigal and Cyprinus) and prawn juveniles in Yeldari reservoir and have to purchase all the harvested fish and prawn catch. Similarly the IAPPC have to pay Rs. 05/kg. of fish catch and Rs. 10/kg. of prawn catch as a commission to the society during purchase of catch from the fishermen. Due to this new policy adopted by the society for the reservoir fishery management, there was no financial burden faced by the society to conduct the fishery activities. Earlier the Yeldari reservoir fishery was not having the involvement of any effectively working private company.

2. On Yeldari reservoir there was 'no closed day for fishing' (Sakhare, 2003). It was observed that, the Purna Fish Co-operative Society helped the IAPPC to stock 2,54,13,199 fingerlings of Indian major carps (IMC) and 60,85,911 juveniles of prawn *Macrobrachium rosenbergii* in to the Yeldari reservoir from June 2006 to December 2006, in this duration of fish and prawn stocking, the society had banned the fishing activities to prevent the catch of young stages of fish and prawn, for this purpose the society had given Rs. 1500/month as compensation amount to about 200 full time active fishermen working at Yeldari reservoir. Only local fish species and cat fish species harvesting from the reservoir was permitted during banned period of fishing. Earlier there was no control of the other societies on fish harvesting period and use of small mesh nets.
3. The Purna Fish Co-operative Society has suggested the IAPP Company to stock the fish and prawn seed in to the reservoir within period of six months instead of one time stocking of all fish seed, it was an important suggestion by the society to maintain the age difference of the stock. Under the guidance of Purna Fish Co-operative Society the IAAP Company had stocked the fish fingerlings @ 4000/ha. And prawn juveniles @ 1000/ha. in to the reservoir, this rate of fish seed stocking was 20 times higher than earlier rate of fish seed stocking by any other fish co-operative society. Earlier, the other fish co-operative societies had stocked the IMC fish seed@ 211/ha.in 1998-99 and it was 44/ha. in 1999-2000, there was no any record of prawn seed stocking in to the reservoir (Sakhare,2002).
4. The society had strictly banned the use of small mesh sized (Mosquito net) traditional Drag-Bag net (Zorli or Wadap) and Gill net of mesh size less than 5 centimeter for the fishing in Yeldari reservoir. The studies on the earlier situation show that, there was common use of the mosquito-nets and cast-nets in fishing.
5. Purna Fish Co-operative Society agreed to cooperate the IAPPC to take necessary action against poaching of fish and prawn from the reservoir, regular checking of the mesh size of the nets used by the fishermen; for this purpose, the society had appointed about 50 employees as security guards on temporary basis in January 2007 onwards to patrol the fish poaching activity from Yeldari reservoir. A petrol engine speed boat, three jeeps, one luggage van and few motor-cycles were provided to the security guards. The security guards also work to collect the catch from different fish and prawn collection centers of the reservoir by using luggage vans. This special step of reservoir fishery management was found firstly undertaken by a fish co-operative society working in the reservoir fishery sector of Maharashtra.
6. Purna Fish Co-operative Society had organized the all fisher communities living in 34 different villages around the Yeldari reservoir and provided daily wages to about 964 fishermen; the society also provided temporary employment @ Rs. 1500/month as salary to about 130-150 active full time fishermen working during June 2006 in Yeldari reservoir for

Figure 20.1: Fish and Prawn Collection Centre Bamni, Yeldari Reservoir

Figure 20.2: Prawn Preservation in Crushed Ice in Thermocol Box

Figure 20.3: Young Team of Fishermen Involved in Prawn Catch

Figure 20.4: Use of Diesel Engine Boat for Effective Fishing

Figure 20.5: Office of the Purna Fish Cooperative Society

Figure 20.6: View of Yeldari Reservoir, District Parbhani (M.S.)

Figure 20.7: Petrol Engine Speed Boat for Patrolling the Fish Poaching

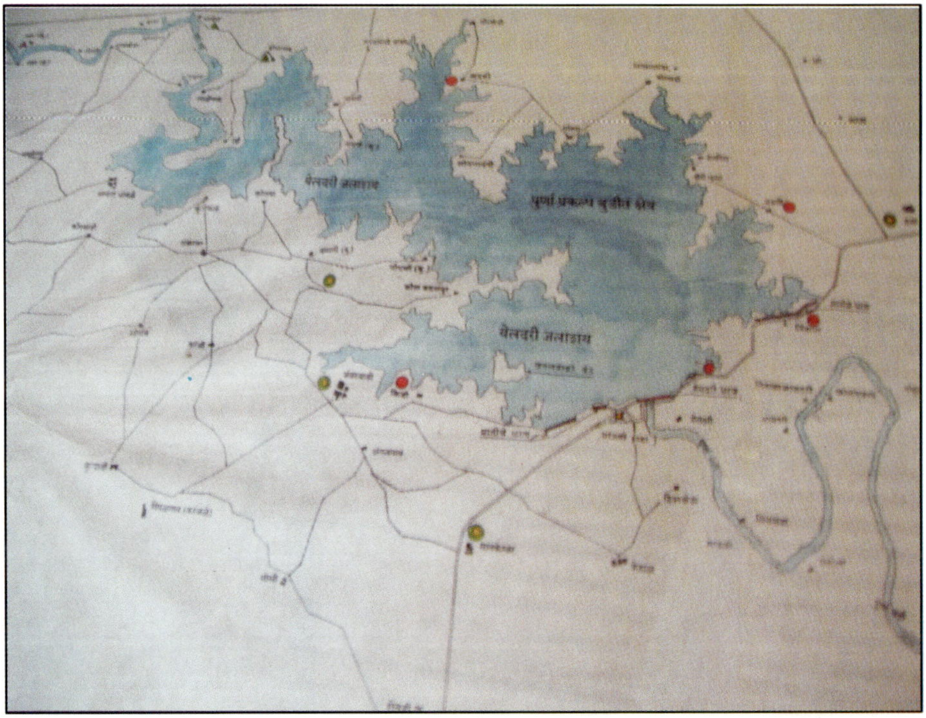

Figure 20.8: Map of Yeldari Reservoir showing Villages in Surrounding

Table 20.3: Fishermen Population Status of Yeldari Reservoir

Sl.No.	*Village Name*	*Taluka*	*Fishermen Population*	*Fisher Community Tribes and Caste and their population*
1.	Yeldari camp	Jintur	85	Bhoi-51, Boudh-30, Telgu-1, Muslim-03
2.	Sawangi Mahalsa	Jintur	46	Banjara-03.Bhoi-09, Boudh-20, Holar-02, Muslim12,
3.	Murumkheda	Jintur	08	Bhoi-02, Boudh-06,
4.	Kini	Jintur	69	Bhoi-12, Boudh-15, Chambar-36. Hatkar-01, Vanjari-05
5.	Ambarwadi	Jintur	56	Andh-2, Boudh-11, Chambar-11, Matang-23, Vanjari02, Waddar-07
6.	Kawtha	Jintur	76	Andh-48, Bhoi-03, Boudh-12, Hatkar-08, Matang-05,
7.	Badnapur	Jintur	07	Bhoi-07
8.	Chaudharni	Jintur	21	Boudh-17Hatkar-02, Muslim-02
9.	Bamni	Jintur	32	Bhoi-06, Baudh-03, Chambar02, Dhanger-02, Hatkar-01, Koli01, Matang10, Muslim7,
10.	Kolpa	Jintur	37	Bhoi-06, Boudh-22, Hatkar-07, Koli-02
11.	Kumbhephal	Jintur	09	Boudh-09
12.	SawangiBhambre	Jintur	43	Bhoi-09, Boudh-11, Koli-21, Muslim-02
13.	Umrad	Jintur	23	Boudh-23
14.	Bekheda	Jintur	03	Boudh-03
15.	Saikheda	Jintur	03	Boudh-03
16.	Wazar(Kh)	Jintur	08	Boudh-02, Rajput-05, Muslim-01
17.	Yeldari	Sengaon	10	Hatkar-10
18.	Limbala-Tanda	Sengaon	23	Banjara-23
19.	Bhandari	Sengaon	32	Andh-01, Banjara-05, Bhoi-03, Boudh-05, Hatkar-01, Muslim-17
20.	Khairi-Ghumat	Sengaon	13	Banjara-05, Hatkar-06, Muslim-02.
21.	Holgira	Sengaon	07	Banjara-05, Hatkar-02
22.	Borkhadi	Sengaon	39	Banjara-26, Boudh-12, Hatkar-01
23.	Dhotra	Sengaon	49	Boudh-41, Hatkar-07, Maratha-01
24.	Pathonda	Sengaon	26	Andh-11, Banjara-04, Boudh-10.Hatkar-01
25.	Son Sawangi	Sengaon	25	Banjara-22, Hatkar-03
26.	Khadki	Sengaon	43	Banjara-21, Boudh-21, Hatkar-01
27.	Bamni(Kh)	Sengaon	38	Banjara-31, Hatkar-03, Muslim-04
28.	Nansi	Sengaon	27	Banjara-02, Boudh-25
29.	Dongergaon	Sengaon	43	Banjara-21, Bhoi-16, Boudh-06
30.	Ooty(Purna)	Sengaon	20	Bhoi-08, Boudh-12,
31.	Salegaon	Sengaon	13	Boudh-11, Muslim-01, Vanjara-01
32.	Dhanora	Sengaon	20	Bhoi-08, Boudh-06, Hatkar-04, Maratha-02
33.	Pimpri	Sengaon	06	Boudh -06
34.	Barda	Sengaon	04	Boudh-04
	Total Fishermen Population =		**964**	

the different work processes like fish and prawn seed stocking, prevention of fish poaching, prevention of fish harvesting during monsoon, close watch on the mesh size of the nets used by the non-member fishermen of the society, collection and transportation of the fish and prawn catch from various fish collection centers around the reservoir etc.

7. For daily payments and convenience to the fishermen, the society had established three different collection centers for fish and prawn catch at villages Yeldari-camp, Bamni and Khadki; this has helped the fish co-operative society to increase the involvement of the fishermen in the fishing process, in breaking the channel of middleman fish marketing.
8. The total fish production from Yeldari reservoir by the application of improved new management techniques by Purna Fish Co-operative Society was 1,79,501.5 kg. @ 28.61/ha. in the year 2006-2007 and 4,88,534 kg.@ 77.89/ha. (Table 20.1.), this was greater than earlier fish production obtained by the fishery management of other co-operative societies working on Yeldari reservoir *i.e.* 7.60 kg./ha. in the year 1998-1999, 10.71 kg/ha in the year 1999-2000 and 6.78 kg./ha. in 2000-2001 (Sakhre, 2002; Sugunan, 1995).
9. The remarkable success achieved by the Purna fish Co-operative society in the production of prawn species *Macrobrachium rosenbergii* was 28,775 kg @ 4.58kg/ha. in the two year duration, year 2006-2008.(Table 20.2). There was no stocking of prawn seed in to the Yeldari reservoir prior to year 2006 by any other agency. Therefore, it can be suggested that Yeldari reservoir has good potential for prawn development.
10. The Purna Fish Co-operative Society had issued fishing license to members and non-members of the society @ Rs. 05/month from June, 2006 on words on the condition that, every fisherman have to carry the fishing license during fishing and the catch obtained should be submitted only towards the fish and prawn collection centers established by the society on the periphery of the Yeldari reservoir @ Rs. 24/kg. for fish catch and @Rs. 150/kg. for prawn catch. But from July 2007, the wages given by the Purna fish co-operative society to these fishermen for prawn catch were reduced to Rs. 70/kg. due to increase in the rate of prawn catch from the reservoir.

Conclusion

It can be concluded that, the Purna Fish Co-operative Society, Sawali working at Yeldari reservoir in Parbhani district of Maharashtra had set a new and modern trend of reservoir fishery management by ensuring daily wages to the fishermen, suggested the need of high rate of fish seed stocking in large size reservoirs, established a special squad for the prevention of fish poaching, managed all fishery activities to increase the fish and prawn production and also suggested the importance of capital investment by private companies in the fisheries sector for the flourishment of reservoir fishery in India.

Acknowledgements

The authors are grateful to the Chairman and the Secretary of Purna Fish Co-operative Society, Sawli, Jintur, and Parbhani for providing the data. The authors are also thankful to Assistant Registrar Co-operatives (Animal Husbandry, Dairy and Fishery) Parbhani.

References

Ahmad, S.H. and Singh, A.K., 1992. Present status potentialities and strategies for development of reservoir fisheries in Bihar. *Fishing Chimes*, 12(8): 49–57.

CIFRI, 1997. *Ecology and Fisheries of Bhatghar Reservoir* No. 73. CIFRI Golden Jubilee Celebration, Barrackpore, Kolkata, W.B.

Desai, S.S, 1980. Fisheries of Nathsagar Reservoir. *India: Today and Tomorrow,* 8(4): 161–181.

Diwan, A.D., 2003. Fish for nutritional and food security, *Sovenier, Nat. Con. On Urban lake environment: Economics and Management aspects,* April 5th–6th 2003 at Vivek Vardhini College Hyderabad, p. 66–76.

Due, A., 1993. Fisheries in Gobindsagar. *Fishing Chimes,* 13(9): 53–54.

Jhingran, V.G., 1982. *Fish and Fisheries of India.* Hindustan Publishing Corporation, New Delhi.

Lonkar, R.L., 1992. Case studies on selected reservoirs on Godavari basin. *Dissertation of Training of Extension,* CIFE, Kakinada, A.P., p. 75.

Lu, X., 1986. *Review of Reservoir Fisheries in China.* FAO, Fisheries Circular No. 803 Rome, 37.

Sakhare, V.B., 1999. Fisheries of Yeldari reservoir, Maharashtra. *Fishing Chimes,* 29(8): 17–19.

Sakhare, V.B., 2003. Socio-economic status of fishermen around Yeldari reservoir, Maharashtra. *Aqua Tech.,* 2(5): 77–78.

Sakhare, V.B. and Joshi, P.K., 2003. Reservoir fishery potential of Parbhani district of Maharashtra. *Fishing Chimes,* 23(5): 13–16.

Sugunan, V.V., 1995. *Reservoir Fisheries of India.* FAO Fisheries Technical Paper, 345, Rome, FAO, p. 423.

Valsankar, S.V., 1980. Economic rehabilation of fishermen in Yeldari Reservoir. *India: Today and Tomorrow,* 8(4): 162–163.

Chapter 21

Distribution of Bacterial Flora in Pearl Spot *Etroplus suratensis* (Bloch) during Captivity

☆ *J.L. Rathod and U.G. Naik*

ABSTRACT

The bacterial flora in pearl spot *Etroplus suratensis* maintained in the captivity was quantitatively and qualitatively analysed. Total heterotrophic bacterial (THB) population in sediment ranged from 2.2 X 10^6 to 4.26 X 10^7/g and was higher than that of water (7.61 X 10^4 to 9.46 X 10^5/ml) whereas in fish tissue it varied between 4.96 X 10^3/g and 3.91 X 10^5/g respectively. In pearl spot, gills harboured more number of bacteria than skin but gut showed comparatively higher bacterial population than skin and gills. In skin, *Bacillus* was found to be predominant flora followed by *Micrococcus* and *Alcaligenes*. *Alcaligenes* was found to be major flora in gills followed by *Pseudomonas* and *Micrococcus*. On an average, *Pseudomonas* was recorded to be the major flora in the gut of pearl spot. It concludes, to certain extent that the bacterial flora in cultured *Etroplus suratensis* depend upon the fluctuation in their environment and type of food ingested by the fish. The *Bacillus* being predominant micro flora associated with water and sediment and *Alcaligenes* being the second dominant flora.

Introduction

The heterotrophic bacteria play a major role in aquaculture systems. The important role played by them is in the maintenance of water quality and nutrient recycling. The microflora of cultured pearl spot, *Etroplus suratensis* during various development stages have been investigated by several workers (Surendran and Iyer, 1984 and Sreekumari *et al.*, 1988) of Indian seas but there is no work carried out in the Karwar waters in recent past years. As a result of the low nitrogen content of detritus (Dieckman, 1978) the heterotrophic microorganisms associated with detritus apparently constitute the major source of protein in various detritus animals (Seiderer *et al.*, 1984 and Moriarty, 1989).

The fish harbours bacteria on skin, gills and in gut regions. Certain bacteria are present in the gut for a relatively long term and they form the microflora specific to the animal. These bacteria help fish to digest their food and provide energy for growth. Realising the significance of bacteria in the culture system, in the present investigation an attempt was made to study the possible relationship between the microflora of cultured fish and its environment.

Materials and Methods

Fishes were stocked in outdoor concrete pond and allowed to grow to a period of 105 days. During growth period changes in the bacterial (micro) flora were investigated. The impact of physico-chemical parameters on bacterial flora and fish was also examined. The materials used and the methods followed in the present investigation are given hereunder.

Experimental Tank Conditions

The present work was carried out in the outdoor concrete pond (5.5 x 3 x 2 m). The pond was cleaned, dried properly prior to stocking. The pond bottom was covered with mud and planted some aquatic plants in it so as to provide natural habitat to fish. After stocking, the pond was fertilized weekly with groundnut oil cake, to increases the primary productivity.

Fish Species

The fish, *E. suratensis* (Bloch) was caught from Sunkeri backwaters by using cast net in the month of September. The salinity of the backwater ranged between 11.07 per cent and 14.53 per cent respectively. The fish were brought to the laboratory in plastic bucket and acclimatized to the laboratory conditions for a period of one week. The uniform size fishes were sorted; length was measured and stocked in the experimental pond where they were allowed to grow for a period of 105 days. No artificial feed was given to fish during the culture period, but the pond was fertilized weekly with groundnut oilcake. This in turn increased the primary productivity of the pond.

Hydrological Parameters

The water temperature of the pond was measured using scientific thermometer, where as the salinity was determined by Mohr Knudsen titrimetric method and Dissolved Oxygen of water by Winkler's Titration method (Strickland and Parson, 1975). The pH of the water was measured using a pH meter. Primary production was calculated by following the "Light and Dark bottle" method (Gaarder and Gran, 1927).

Bacteriological Sampling and Analysis

For bacteriological analysis, the water sample was collected in sterilized glass bottles whereas sediments were collected in sterilized steel container for further analysis in the laboratory. One ml of water sample was taken and diluted with 9.0 ml of sterilized distilled water, similar serial dilutions were prepared using the same diluents, whereas for sediment, a known weight was homogenized in 100ml of sterilized distilled water and serial dilutions were made in 9 ml of sterile distilled water (Anon, 1957; APHA, 1998).

Five test fishes (*E. suratensis*) were caught by using cast net and brought to laboratory in sterilized polytehene bag for determination of bacterial count. For analysis of the skin, swab was used from either side of the fish body. About 2-3g of gills tissue was cut aseptically to analyze the floral content in it. For screening bacteria from the gut, the gut portion of fish was removed by dissecting aseptically in three regions namely Foregut (FG-Oesophagus and stomach), Mid gut (MG-anterior half of the intestine) and Hind gut (HG-posterior of the intestine) (Shewan, 1961 and 1962).

The samples were homogenized in 100ml of sterilized distilled water and serial dilutions were made using 9 ml of sterilized distilled water. The appropriate dilutions were selected and 1 ml of homogenated sample was inoculated into petridishes in duplicate and the fish samples were plated using Tryptone Glucose Beef Agar (TGA) by pour-plate method, media was allowed to solidify. Plates were then incubated at room temperature for 24-48 hours.

Bacterial Enumeration

For total heterotrophic bacterial (THB) count, the dilutions were inoculated on TGA media and incubated at room temperature (28 ± 2°C) for 24-48 hours. After incubation period, plates showing 30-300 colonies were selected for counting and calculated using the following formula (APHA, 1998).

$$\text{Colony Forming Unit / ml} = \frac{\text{Colonies counted}}{\text{Actual volume of sample in dish, ml}}$$

Computed the count as above and reported as estimated CFU per milliliter.

Identification of Isolated Bacteria

The plates showing discrete colonies were selected for the identification of bacteria. Colonies showing different types of morphology were picked up and streaked on agar slants. Further, they are restreaked for purification and maintained on TGA media. The following tests were carried out for identification.

1. Motility; 2. Gram's staining; 3. Shape and arrangement of cells; 4. Catalase production; 5. Cytochrome oxidation; 6. Penicillin sensitivity; 7. Pigmentation; 8. Hugh-Liefson oxidative/ fermentation; 9. Sugar fermentation.

Statistical Analysis

The correlation co-efficient (r) between the total heterotrophic bacteria of fish (Skin, Gills and Gut regions), water, sediment and physico-chemical characteristics of the pond were worked out. The relation between THB of gut and feeding intensity of fish were also studied. Two ways ANOVA was carried out to see the variation in THB between samples and period.

Results and Discussion

Table 21.1 explains the variations in the physico-chemical parameters of experimental pond during the study period. The water temperature, dissolved oxygen and salinity were ranged from 28 to 31 ±1°C, 4.1 to 6.43 mg/l and 0.11 to 0.15 psu respectively. The variation in organic matter coincides with that of primary production, with peak during mid of December (2.0 per cent and 3. 122 gm C/m^2/ hr^{-3} respectively). The length-weight characteristics of fish and their feeding details are given in Table 21.2. The initial length of 5.9 cm increased to 9.4 cm during 105 days of growth period. The feeding intensity varied between 0.11 to 0.896 per cent.

The total heterotrophic bacteria in the pond water varied between 0.761×10^4/ml and 9.46×10^5/ ml during October and December (Table 21.3). Where as the THB population of sediment was always recorded higher than that of water (2.12×10^6 to 4.26×10^7/gm (Table 21.3). The findings of Okpokwasili and Alapili (1990) stated that THB of sediment is about 200 times more than that of water column.

The total heterotrophic bacteria in different regions of fish is given in Table 21.2. The bacterial population on skin showed an increase from 0.0496×10^3/g to 3.91×10^5/g (October to November) followed by steep fall in December (Table 21.3). The present findings are slightly different from

Venugopalan *et al.* (1985) who recorded maximum value of bacteria on *Chanos chanos* during September and minimum during December. This difference could be due to the differential ecological niche of fish *i.e.*, concrete pond of the present study.

Table 21.1: Different Hydrological Parameters of the Culture Pond

Sampling Date	*Water Temp. (°C)*	*Salinity (psu)*	*pH*	*Dissolved Oxygen (mg/L)*	*Organic Matter (per cent)*	*Primary Production ($gmC/m^2/hr^3$)*
15th Oct.02	28	0.145	7.5	4.57	1.38	2.381
1st Nov.02	28	0.155	7.5	4.70	1.15	2.01
15th Nov.02	28.5	0.140	7.0	4.10	1.56	1.916
1st Dec.02	28.5	0.123	7.5	6.13	1.42	2.096
15th Dec.02	29.5	0.115	7.5	6.43	2.00	3.122
1st Jan.03	30.5	0.110	7.0	6.09	1.81	3.026
15th Jan.03	31	0.110	7.5	5.20	1.29	2.083
Mean ± S.D	29.14 ±1.21	0.129 ±0.018	7.42 ±0.2	5.32 ±0.91	1.52 ±0.299	2.376 ±0.498

Table 21.2: Length-Weight Characteristics of Test Fishes

Sampling Date	*Length (cm)*	*Weight (gm)*	*Feeding Intensity (per cent)*
15th Oct.02	5.9	13.68	0.21
1st Nov.02	6.5	17.93	0.15
15th Nov.02	7.1	22.13	0.32
1st Dec.02	7.8	26.38	0.22
15th Dec.02	8.4	30.66	0.18
1st Jan.03	8.9	34.96	0.11
15th Jan.03	9.4	39.18	0.89

FG: Fore Gut; MG: Mid Gut and HG: Hind Gut.

The variation in the bacterial floral density on fish skin was also affected by the fluctuation in the environmental parameters. Thus, it showed insignificant positive correlation with Temperature (r = 0.310), Dissolved Oxygen (r = 0.063), Primary Production (r = 0.222) and Organic Matter (r = 0.488) whereas insignificant negative relation with that of pH (r = -0.343) and Salinity (r = -0.204) (Table 21.4).

The gills harboured more number of bacteria than skin during the study period. The density varied between 6.83×10^5/g 8.98×10^6/g. Bacterial population of gill showed insignificant positive correlation with Temperature, pH, Dissolved oxygen and Organic matter (r = 0.037, 0.385, 0.144 and 0.411 respectively) and insignificant negative relation with salinity and primary production (r= - 0.154 and - 0.574 respectively). Venugopal *et al.* (1985) recorded positive correlation of gills flora with temperature and dissolved oxygen. However, they recorded negative relation with pH and positive with salinity. This difference could be due to the factors other than environmental parameters like physiological stress, the production of mucous; secretary bactericidins and tissue integrity are likely to influence bacterial population of the gills (Trust, 1975).

Table 21.3: Total Heterotrophic Bacteria (THB-NoX10^5/g) of Water, Sediment and *Etroplus suratensis* from Culture Pond

Sampling Date	*Water (No/ml)*	*Sediment (mg/g)*	*Fish (No/gm)*				
			Skin	*Gills*	*Fore Gut*	*Mid Gut*	*Hind Gut*
15th Oct.02	0.761	2.12	0.0496	6.83	6.39	9.42	6.64
1st Nov.02	6.7	8.11	0.848	2.17	1.15	6.21	0.01
15th Nov.02	3.2	10.50	3.91	9.92	4.62	7.12	7.8
1st Dec.02	1.49	6.24	0.108	8.98	1.75	4.31	9.36
15th Dec.02	9.46	42.60	6.19	1.41	9.33	5.75	8.32
1st Jan.03	6.44	18.30	0.643	7.12	2.94	4.26	9.77
15th Jan.03	5.88	35.60	0.86	4.67	0.31	8.32	3.39

Table 21.4: Correlation Co-efficient (r) Showing the Relationship of THB and Physico-Chemical Characteristics of Pond

	Water	*Sediment*	*Fish*		
			Skin	*Gills*	*Intestine*
Temperature	0.447	0.783*	0.310	0.037	0.44
pH	0.385	0.158	–0.343	0.385	0.291
Salinity	0.146	0.704*	–0.204	–0.154	0.44
Dissolved Oxygen	0.102	0.486	0.063	0.144	0.219
Primary Production	0.206	0.411	0.222	–0.574	0.68**
Organic Matter	0.365	0.495	0.488	0.411	0.892**

Table 21.5: Analysis of Variance (ANOVA) Showing Variations in THB between Samples and Anatomical Sites

Item	*S.S*	*D.F*	*M.S.S*	*F*
Total	53.6268	34		
Samples	12.8312	6	2.1385	2.1536NS
Anatomical Sites	16.9635	4	4.2409	4.271*
Error	23.8321	24	0.993	

NS: Not Significant.

*: Significant at 1 per cent level ($p<0.01$).

The bacterial population in foregut showed high value during October (6.39×10^5/g) and abrupt fall in November (1.15×10^5/g) reaching maximum scale in mid of December (9.33×10^5/g). Present findings corroborate with the findings of Fatima *et al.* (1980) who observed high population in gut of *Rastrelliger kanagurta* in the month of August to October with a sudden decrease in November. Venugopal *et al.* (1985) recorded maximum bacterial density in gut of *Chanos chanos* from August to October followed by sudden decrease from November to January. In the present study, a low bacterial density was recorded in the beginning of November and mid of January. This could be impact of environmental factors. The bacterial population in foregut showed an increase during mid of November followed by sudden decrease and attaining peak (9.33×10^5/g) during mid of December followed by second fall (Table 21.3).

Similar trend in the bacterial density was noticed in mid-gut with the peak during mid of December (5.75×10^5/g). Mid-gut showed low bacterial density than foregut during mid of December. However, the population of Mid-gut was slightly more than the foregut. The fluctuation in the bacterial population of hindgut was comparatively more (0.0124×10^5/g - 9.77×10^5/g in January).

The bacterial density in the gut in pearl spot showed insignificant positive relation (r =0.233) with feeding intensity of the fish. The low bacterial density in well-fed fish could be the fact that the fish might have eaten the food having the less microbial load. This finding corroborates with that of Srinivasa and Floodgate (1966) in *Limanda limanda* and Venugopal *et al.* (1985) in *Chanos chanos*. The bacterial density in gut showed a significant positive correlation with primary production (r= 0.68) and Organic matter (r= 0.892). Marie (1991) also noticed similar type of positive correlation in *E. suratensis*. Sugita *et al.* (1981) suggested that the gut micro flora of fish may vary or changed by endogenous and exogenous factors, which include immune mechanisms, water temperature and diet as well as handling and environmental stress.

All the three micro-environment of gut (foregut, mid-gut and hind-gut) observed with higher bacterial population compared to skin and gills. Analysis of Variation showed that the difference was significant at 1 per cent level (F = 4.271) (Table 21.5).

The percentage distribution of different bacterial genus in pond water and sediment is given in Table 21.6. Genus *Bacillus* was predominant in both the samples and high incidence of this genus was recorded during December to January. This is in concurrence with that of Venugopalan *et al.*, (1985) who recorded high percentage of *Bacillus* in water and sediment of pond. Even Okpakwasili and Alapili (1990) recorded 100 per cent of Gram positive bacteria in pond sediment comprising *Bacillus* (31.3 per cent), *Lactobacillus* (25 per cent) and 93.3 per cent of Gram positive in water column including *Bacillus* (60 per cent). The second predominating genus was found to be *Alcaligenes* maximum percentage during the month of October. The incidence of *Alcaligenes* was more in sediment than in water except in November. *Pseudomonas* ranked 3rd whereas *Micrococcus* and *Vibrio* were sporadically recorded. According to Ogbondeminu *et al.* (1991), 52 per cent of flora of water column was constituted by *Pseudomonas*.

The generic composition of bacteria in cultured pearl spot is given in Table 21.7. In skin sample, *Bacillus* was found to be predominant flora and its peak (40 per cent) was noticed during January. The second predominant genus recorded in the skin was *Micrococcus* followed by *Alcaligenes*. *Alcaligenes* was found to be major flora in gills followed by *Pseudomonas* and *Micrococcus*. However, *Vibrio* was strikingly absent in the gills except in the first half of November. In the present study it was noticed that the *Vibrio* was absent in the gut of pearl spot and was in contrast with the observation made by Liston (1957). The *Pseudomonas* was predominant in midgut and hindgut whereas in the foregut, *Alcaligenes* formed major flora.

Table 21.6: Generic Composition of Bacteria (per cent) Isolated from Water (W) and Sediment (S) of Culture Pond

Sampling Date	*Sample*	*Micrococcus*	*Bacillus*	*Pseudomonas*	*Aeromonas*	*Vibrio*	*Flavobacter*	*Enterobacter*	*Alcaligenes*
15th Oct.02	W			12.5					87.5
	S		75	12.5			12.5		
1st Nov.02	W	8.3	33.3			16.7		8.3	33.4
	S	40	30	10			10		10
15th Nov.02	W	20	10	10			30		30
	S		20				40	10	30
1st Dec.02	W		40			10	20	30	10
	S		30	20		20		10	30
15th Dec.02	W		40				10	50	
	S	30	10	20	10				30
1st Jan.03	W		20	30	10				30
	S	40	30	10		10	10	20	
15th Jan.03	W	10	50						20
	S	10	20						50

Table 21.7: Generic Composition of Bacteria Isolated from *Etroplus suratensis*

Sampling Date	*Region*	*Micrococcus*	*Bacillus*	*Pseudomonas*	*Aeromonas*	*Vibrio*	*Alcaligenes*	*Flavobacter*	*Enterobacter*	*Arthobacter*
15th Oct.02	S	25	25	12.5	12.5		25			
	G	50	12.5							37.5
	FG	37.5		25	25					12.5
	MG		30	20	10		30			10
	HG		33.3	25			25			16.7
1st Nov.02	S		26.7	33.3		20	20			
	G	26.7	13.3	20		13.3	13.3	13.3	13.3	
	FG		25	25	8.3		25		16.7	
	MG		16.7	25	16.7	18.3	16.7		16.7	
	HG	16.7	25	33.3	8.3		8.3		8.3	
15th Nov.02	S	33.3	13.3	26.7				6.7	13.3	6.7
	G	25	25	16.7			25			8.3
	FG		20	20			40			20
	MG	20	20	10		30	20			
	HG		25	33.3	8.3			8.3	8.3	13.3
1st Dec.02	S	10	20	10		30			20	10
	G	20		30			20	20	10	
	FG	50	12.5				12.5		12.5	12.5
	MG	30	20	40			10			
	HG	16.7	25	25	16.7		16.7			
15th Dec.02	S	26.7		33.3		13.3	26.7			
	G	8.3	16.7	25			25		16.7	8.3
	FG		20	40			30			
	MG		30	40			10		20	
	HG		30	30			20			

Contd...

Table 21.7–*Contd...*

Sampling Date	*Region*	*Micrococcus*	*Bacillus*	*Pseudomonas*	*Aeromonas*	*Vibrio*	*Alcaligenes*	*Flavobacter*	*Enterobacter*	*Arthobacter*
1st Jan.03	S		30				30	10		
	G		30	30			30		10	
	FG	40		30	10		10		10	
	MG	25	13.3	25			25		8.3	
	HG	20	30	20			20	10		
15th Jan.03	S	10	40				30		20	
	G						30	20	50	
	FG						80	10	10	
	MG	20	30	10		20	20			
	HG	8.3	25	33.3	16.7		16.7			

S: Skin; G: Gut; FG: Fore Gut; MG: Mid Gut; HG: Hind Gut.

The average percentage of *Bacillus* was high in mid-gut and hindgut. This genus was found abundantly during October, December and middle of January (30 per cent) in mid-gut and October (33.3 per cent) in hindgut respectively. In the present study, the presence of *Bacillus* in gut region of the test species was in concurrence with the findings of Mary *et al.* (1975) and Mary (1977). *Micrococcus* occurred sporadically in the gut of pearl spot and comparatively more in foregut being maximum in December (50 per cent). The occurrence of *Aeromonas, Arthrobactor* and *Flavobactor* was rare in the fish.

The intestinal micro flora might have been derived from the food chain, diet or environment. This is in agreement with Sakata *et al.* (1980) and Sugita *et al.* (1985) who suggested that the intestinal micro flora are derived from diet or from the environment. Even Fang *et al.* (1989) showed influence of several environmental factors on the bacterial load from a high yielding pond.

The skin flora was predominated by starch hydrolyser followed by gelatin liquefier than nitrate reducers. The maximum starch hydrolytic activity was noticed among the isolates from gills (Figure 21.1). Colwell (1962) reported that the mucoid slime in the gills served as an ideal environment for polysaccharide splitting bacteria. Liston (1957) also reported similar observation. Relatively higher percentage of lipid hydrolysers (77 per cent) in the month of November in gut of pearl spot, which feed on plankton, some of which like calanoid copepods which are rich in lipids (Cowey and Sargent, 1979).

The percentage of gelatin liquefiers was low (30.6 from skin, 37.6 from gills and 41.5 from gut) in the pearl spot when compared to starch hydrolysers. This is in agreement with Marie and Pereira (1991) who recorded 53.3-62 per cent of isolates from sediment and 24-50 per cent isolates from gut of

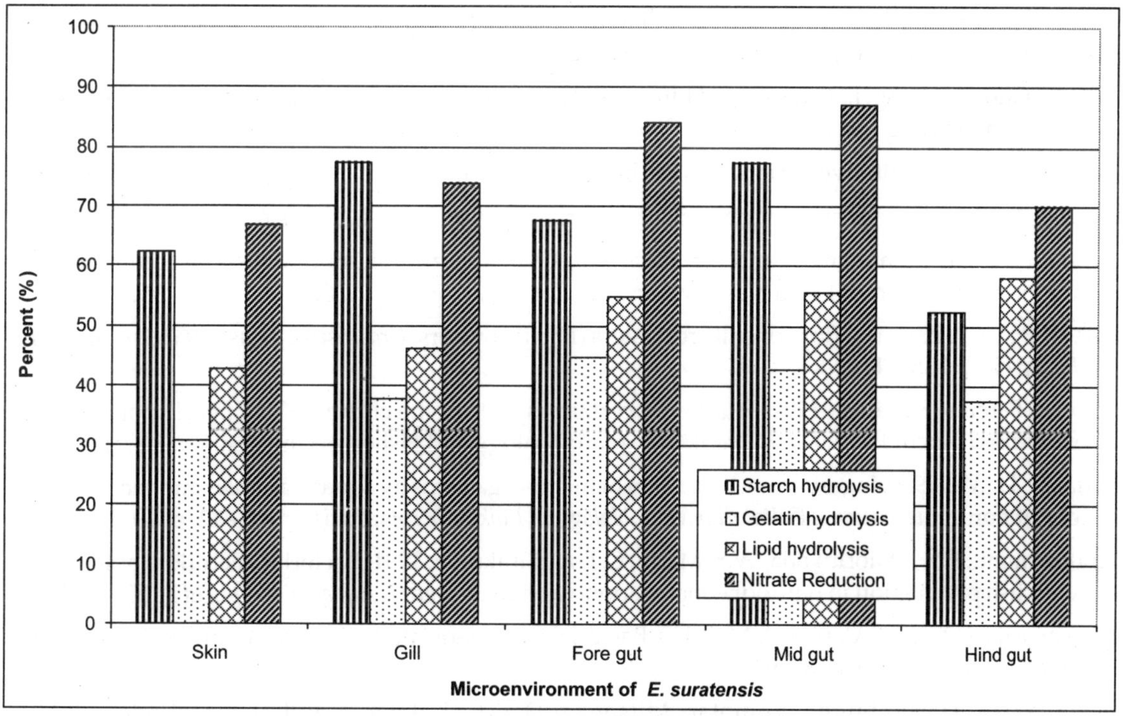

Figure 21.1: Biochemical and Physiological Characteristics of Bacterial Flora with Resepct to Different Microenvironment of *E. suratensis*

pearl spot to be gelatin liquefier. However, in present investigation the isolates from sediment were tested for its hydrolytic activity and differs from Surendran and Iyer (1984) findings, who recorded 85 per cent, 70 per cent and 96 per cent isolates from skin, gills and gut of *Etroplus suratensis* respectively, which were active gelatin liquefier. This difference could be due to the fact that in the present study, the fish is reared in a small concrete pond where in feeding was on a limited food available (hydrilla, detritus and phytoplankton etc.). Fatima *et al.* (1980) reported more proteolytic bacteria in oesophagus of *Rastrelliger kanagurta* and she claimed it is due to the preference of fish for live material. It is interesting to note that, all the hydrolytic activities of bacteria isolated from different anatomical parts of pearl spot decrease during December-January.

It is surprised from the present data that the settlement of the bacterial flora in cultured fish (*Etroplus suratensis*) depend upon the fluctuation in their environment in general and abundance of micro flora in the gut region of fish in particular is mostly derived from the food ingested by the fish.

References

APHA, 1980. *Standard Methods for the Examination of Water and Wastewater*, 15th edn. American Public Health Association, Washington D.C., pp.747.

Colwell, R.R., 1962. The bacterial flora of Puget sound fish. *J. Appl. Bacteriol.*, 25: 147–158.

Cowey, C.B. and Sargent, J.R., 1979. Nutrition. In: *Fishery Biology*, Vol. 8., (Eds.) W.S. Hoar, D.J. Randall and J.R. Brett. Academic Press, London, pp.1–69.

Dieckman, G.S.. 1978. Aspects of growth and production of *Laminaria pollida* (Grev.) off cape Peninsula. *M.Sc. Thesis*, University of Cape Town.

Fatima, E.J., Lakshmanaparumalsamy, P., Chandramohan, D. and Natarajan, R., 1980. Bacterial flora in the alimentary canal of *Rastrelliger kanagurta*.

Fang, X., Guo, X., Wong, J., Fang, Y. and Liu, Z., 1989. The preliminary study of heterotrophic bacteria in a high yielding fish pond. *J. Fish. China*, 13(2): 101–107.

Liston, J., 1956. Quantitative variation in the bacterial flora of flat fish. *J. Gen. Microbiol.*, 15: 304–314.

Liston, J., 1957. Occurance and distribution of bacterial types on flat fish. *J. Gen. Microbiol.*, 16: 205–216.

Mary, P.P., Chandramohan, D. and Natarajan, R., 1975. The gut microflora of some commercially important fishes from Porto Novo water. *Bull. Dept. Mar. Sci. Univ., Cochin*, 7: 185–199.

Mary, P.P., 1977. Bacterial flora of culture oysters (Pacific oysters, *Crassostrea gigas*). *Bull. Jap. Soc. Sci. Fish.*, 45(9): 1189–1194.

Marie, J., Pereira, Shiranee, 1991. Studies on certain aspects of the culture of the pearl spot *Etroplus suratensis* (Bloch) *Ph.D. Thesis*, Dept. of Aquatic Biology and Fisheries University of Kerala.

Moriarity, D.J.W., 1989. Ingestion and digestion of micro-organisms by aquatic animais and the function of gut flora. *International Colloguim Microbiology in Poikilotherms*, Paris, July 1–12, p. 49.

Ogbondeminu, F.S., Omorinkoba, W.S, Madu, C.T. and Ibikunle, 1991. Bacterial microflora associated with fish production in outdoor concrete pond and their feed. *J. Aquacul. Tropi.*, 6(1): 55.

Okpokwasiki, G.C. and Alepiki, A.M., 1990. Bacterial flora associated with a Nigerian freshwater fish culture. *J. Aquacul.*, 5(1): 87–90.

Sakata, Taizo, Haruo Sugita, Tomotari Mitsouka, Daiichi Kakimoto and Hajime Kadota 1980. Association of obligate Anaerobes from the intestinal tracts of freshwater fish. *Bull. Jpn. Soc. Sci. Fish.*, 46(4): 511.

Seiderer, L.J., Davis, C.L., Robb, F.T. and Nowell, R.C., 1984. Utilisation of bacteria as a nitrogen resource by the Kelphed mussel. *Chloromytilus meridionalis* krauss. *Mar. Ecol. P. Ser.*, p. 109–116.

Shewan, J.M., 1961. The microbiology of seawater fish. In: *Fish as Food,* (Ed.) G. Borgstrom. Academic Press, New York, 1: 487.

Shewan, J.M., 1962. The bacteriology of fresh and spoiling fish and some related chemical changes. In: *Recent Advances in Food Science,* (Eds.) Hawthron and Muil Leitch. Butter Worth, London, p. 167.

Shrinivasa, K.P. and Floodgate, G.D., 1966. Studies on the intestinal microflora of Dab, *Limanda limanda,* L. *J. Mar. Biol. Asso., India,* 8: 1–7.

Strickland, J.D. and Parsons, T.R., 1975. A practical handbook of sea water analysis, 2nd edn. *Bull. Fish Res. Ed. Canada.,* 167: 1–310.

Sugita, H., Sakata, T., Ishida, Y., Deguchi, Y. and Kadota, N., 1981. Aerobic and anaerobic bacteria in the intestine of ayu *Plecoglossus altivelis. Bull. Coll. Agric. Vet. Med. Nihon Univ.,* 38: 302–306.

Sugita, H., Tokuyama, K. and Deguchi, Y., 1985. The intestinal microflora of Carp, *Cyprinus carpio* Grass carp *Ctenopharanged idella* and Tilapia *Sarotherodon niloticus. Bull. Jpn.Soc. Sci. Fish.,* 51: 1325–1329.

Surendran, P.K. and Iyer, K.M., 1984. The bacterial flora of pearl spot, *Etroplus suratensis* caught from Cochin backwaters. *Proc. Symposium on Coastal Aquaculture,* Part 3: *Fish Culture,* Cochin, 12–18, January 1980, pp. 852–855.

Tanasomwang, V. and Muroga, K., 1989. Intestinal microflora of Rock fish, Tiger puffer and Red Grouper at their larval and juvenile stages Nippon Suisen Gakkaishi. *Bull. Jap. Soc. Sci. Fish.,* 55(8): 1371–1377.

Trust, T.J., 1975. Facultative anaerobic bacteria in the digestive tract of chum salmon (*Oneorhynchus keta*) maintained in freshwater under defined culture conditions. *Appl. Microbiol.,* 29: 663–668.

Venugopalan, V.K., Nandakumar, R. and Ramesh, A., 1985. Microflora of pond cultured milkfish, *Chanos chanos* (forskal). In: *Harvest and Postharvest Technology of Fish* (Ed.) pp. 680–688.

Chapter 22

Effect of Climatic Changes on Prawn Biodiversity at Mochemad Estuary of Vengurla, South Konkan, Maharashtra

☆ *V.M. Patole, S.G. Yeragi and S.S. Yeragi*

ABSTRACT

The mangrove is the silent and undisturbed environment highly suitable for juvenile prawns for various purposes like feeding, breeding and monitoring. The population density of juveniles was high due to less predation. If such environment is properly utilized for the cultivation of prawns, it would provide maximum yield with a short period of time. Present investigation was carried out during 2007 to 2009. It is recorded that the juveniles of *Peneas indicus, Peneaus monodon* and *Macrobrachium rosenbergii* were abundant at M_2 and M_3 regions of Mochemad estuary of Sindhudurg district.

Introduction

Mangrove is the pin drop silent and undisturbed ecosystem, highly suitable for the habitat of juvenile prawns for various activities like breeding, feeding and growth monitoring. The population density of juveniles was high due to less predation more protection. If such habitat is scientifically utilized for the cultivation of prawns, it would provide maximum yield within a short span of time. The catch of juvenile of mangrove habitat is always differ both qualitatively as well as quantitatively. It is also recorded that the growth of prawns in mangrove ecosystem is faster than that of the open areas because of the availability of organic debris and more stability in feeding.

The present investigation was carried out in Mochemad estuary during 2007 to 2009. It is recorded that the white prawn *Penaeus indicus, P. monodon, Metapenaeus dobsoni, M. monoceros* and *Macrobrachium rosenbergii* were plenty at M_1 and M_3 regions of Mochemad estuary of Vengurla. The maximum catch per unit time per unit area was at M_2 station only because of availability of lagoons with dense mangrove with mudflat floor, *Penaeus indicus* spend more time in estuary and mangroves other than four species.

In India during recent times, prawns have become a significant marine item being an export potential on a large scale due to which our country earns a heavy amount of foreign exchange from their export. They are highly nutritious and contain high values of glycogen and proteins (Pillai, 1978). Naturnal catch of prawns is rapidly dwindling because of the destruction of their breeding grounds due to over-exploitation, pollution and disposal of estuaries and environment. Not much knowledge has been gained about the prawn larval, juveniles and sub-adult abundance in estuaries, backwaters and mangroves. Their steady state of recruitment to adult stock of the adjoining neritic province is very much vital in order to balance their optimal population level. Additional information on juvenile distribution and species composition is very little apart from the work done by the Gopalkrishna and Rad (1968). The present study was undertaken at Vengula to further understand the pattern of juvenile distribution in relation to the environmental and physiological parameters.

Habitat also acts as an excellent breeding and spawning ground for prawns apart from favourable abiotic conditions and abundance of food. The mangrove vegetation is most favourable habitat for juvenile prawns. The population density of some species of prawn is much higher due to pressure of protection from predators.

Macrobrachium rosenbergii (giant prawn) has never-ending demand in the export section as well as markets coupled with the availability of enormous span of water resource for its aquaculture ventures in the country. The yearly yield production is much higher; the expansion is quite a lot as well as the gain is more. The most essential aspect is to maintain the hydrological parameter. In monsoon, they entered in the estuary. Their growth is very fast compared to *Penaeus* prawn. They show distinct sexual dimorphism. They have bluish colour prominent antennas. *Penaeus monodon* (tiger prawn) also breed in lagoons so that their youngones are found plenty in monsoon catch. The local people catch them with the help of filter net called "Yendi". The culture of both *Macrobrachium spp* and *Penaeus spp* are easy. The *Penaeus indicus* migrate in bulk during May-June in the estuary and settled in the lagoons. The lagoons are totally field with *Penaeus indicus*. Their survival rate without water is also high; therefore in market the demand for live specimen is more and costly. They are easy for digestion.

The potential of aquaculture of giant freshwater prawn has been widely recognized in the region and the species commands a lucrative market. In monoculture of *Macrobrachium rosenbergii*, the A. P. Agriculture University obtained the production of 509-712 kg.ha/6months. The prawns in the mangrove inlets are represented largely by juvenile Penaeids as had been observed at Mochemad estuary. Several species of prawns like *Penaeus indicus, Macrobrachium rosenbergii* use these inlets as nursery ground. These prawns get readymade food on the mud surface of mangrove belts.

Materials and Methods

Weekly sample collection was done from Feb.2007 to January 2009. Data was collected from three stations. Using the filter net called 'Yendi' and 'Aghali' were used for catching the juveniles present on mudflat of mangrove regions. The collected juvenile were analysed to study their distribution and relative abundance.

Results

Analysis of the monthly collection was done for distribution and relative abundance of juveniles belonging to different species. Five species of prawn i. e. *Penaeus indicus* (white prawn) *Penaeus monodon* (tiger prawn), *Metapenaeus dobsoni*, *M. monoceroses and Macrobrachium rosenbergii* (Giant) were identified. There percentage composition in total juveniles prawn population and monthly frequency polygon are presented in Figure 22.1. The study reveals that the juveniles of *P. indicus*, *P. monodon* and *M. rosenbergii* were found plenty in M_2 and M_3 regions while remaining two were available at M_1 zone only. *Metaenaeus* spp *is* not tolerating low salinity therefore they found near the mouth of the estuary. *P. indicus* is dominating at all stations. The density count was high in mangrove roots. *M. rosenbergii* is very sensitive to oxygen and salinity. Growth of this species was quite slow in high salinity. The growth is affected from November onwards due to increase in salinity.

Table 22.1: Percentage Composition of Juvenile Prawns of Mochemad Estuary

Name of the Species	*Pre-monsoon (%)*	*Monsoon (%)*	*Post-monsoon (%)*
Penaeus indicus	65	73	67
Penaeusmonodon	11	15	14
Metapenaeus dobsoni	10	08	05
Metapenaeus monoceros	04	05	02
Macrobrachium rosenbergii	10	09	12

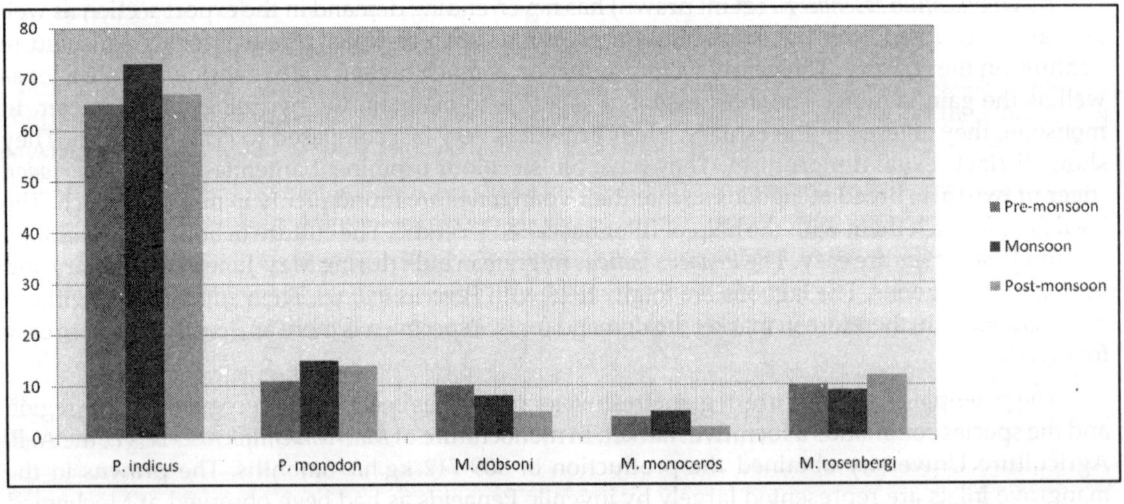

Figure 22.1: Percentage Composition of Juvenile Prawns of Mochemad Estuary

Crustacean larvae and adults come in from upstream (*Macrobrachiun*) and downstream (*Penaeids*), majority of them feed and develop apart from the decapods, mostly commerce their life cycle somewhere else.

It was displayed by the frequency polygon curve that all the five species were present throughout the study phase. The frequency of *P. indicus* was larger from May-September and then turned down. The highest frequencies were observed in September and lowest in February. The frequency of *P. monodon* was parallel to *P. indicus*. The frequency of *M. rosenbergii* hit the highest point in the month of October and least in February and March. In case of *M. monoceros* the frequency was observed highest in August and lowest in February. The freshwater species *Macrobrachiun rosenbergii* showed high density in September and low in March. It was observed that monsoon favours the growth and post-monsoon, pre-monsoon decline the density as well as growth. The growth is directly correlated to oxygen content of water.

Patwardhan (1990), Shet (1994), Yeragi (2004) have affirmed that the environmental condition may be the most important aspect in controlling the migrant population of the juveniles. In the current study, it was observed that, in monsoon due to heavy rainfall, there was a drastic change in the hydrobilogical factors, mainly the salinity. Patwardhan (1986), Yeragi (2004) in earlier studies has observed mass migration of prawns following heavy rainfall followed by flooding of the areas. Parulekar (1985) also observed a direct relation between rainfall and prawn catch in the Cochin backwaters.

The juveniles prefer soft muddy ground, where presence of mangrove vegetation provides abundant food in the form of decaying organic matter along with protection from predators. The juveniles were found clinging from predators. The juveniles were found clinging to vegetation. During the present study, it was noticed that abundance of prawn juveniles always associated with sea grasses. The abundance of variety of food items and the turbid waters appear to be the main contributing factors. Thus mangroves provide significant support to fish and prawn communities. Fish and prawns usually found in greater numbers and biomass in mangroves habitat than in adjacent near shore habitats.

Discussion

If the mangrove ecosystem were properly utilized for the cultivation of prawns would provide maximum yield within a short span of time. The catch of juveniles of mangroves areas constantly diverge both qualitatively as well as quantitatively because of availability of organic debris and stability for feeding. This always favour in fast growth of juveniles. It is evident that significant qualities of mangrove carbon are being assimilated through secondary sources (Yeragi,1997). The implementation of mangrove would positively reinstate the coastal ecological equilibrium and get a better environment and also the economy of the coastal population.

References

Patwardhan, D.D., 1990. A study of estuarine ecology of South Konkan. *Ph.D. Thesis*, University of Mumbai, Mumbai.

Parulekar, A.H., 1985. Aquaculture in mangrove ecosystem of India. *Proc. Nat. Symp.*, November.

Pillai, N., 1978. Summer Inst. Breeding and rearing Marine prawns. May and June, 1977. CMFRI, Cochin, Publ. pp. 75–85.

Sheth, P.B., 1994. Ecology and mangrove in relation to organisms. *M.Sc. Thesis*, University of Mumbai, Mumbai.

Yeragi, A.S., 2004. Species competition and distribution of prawn juveniles in mangroves habitat of Achara creek. *J. Comp. Toxical. Physiol.*, 1(1): 100–104.

Yeragi, S.S., 1997. Species competition and distribution of prawn juveniles in mangroves. *J. Aqua. Biol.*, 12(1& 2): 16–17.

Chapter 23

Impact of Climate Change on the Fish Farming

☆ *Meenakshi Jindal and Kavita Sharma*

ABSTRACT

There is no life without water and water forms the earth different from all other planets. Aquatic environments are known as homes for the diverse assemblage of organisms. Dashing through sparkling water, probing muddy depths, or weaving through a maze of corals, fish inhabit nearly every watery corner of the planet. With at least 27,000 known species living in oceans, lakes and rivers, fish are a cornerstone of global biodiversity, as well as an essential resource for humans. But fish are now increasingly threatened by climate change. All waters in the world are affected one way or the other by climate change. It is one of the most important issues confronting the global community.

Introduction

Fisheries and aquaculture plays very important roles for food security and income generation. Become a unusual role in Food supply, food security and income generation. 520 million people depend on the fisheries and aquaculture sectors for their livelihoods and 42 million people work directly in this sector. Aquatic products are among the most widely traded foods. Net earnings from aquatic products are greater than the combined earnings from the major agricultural commodities and increase at average global growth rate of 8.8 per cent per year.

Fish is the main source of animal protein for a billion of people worldwide. Common among the global poor, fish is an important source of essential vitamins and fatty acids. Fish provides 15 per cent of the average per capita animal protein intake for 3 billion people. Fish contributes 50 percent of total animal protein intake in some small island and other developing states.

Fisheries are a globally important economic activity, not the least from the perspective of human nutrition and underdeveloped societies. Fisheries, due to their primitive nature, are among the human activities most exposed to climate changes. The output of fisheries, as well as their costs and benefits, are directly and strongly affected by variations in natural conditions. Habitat conditions, which are the main determinants of the productivity and location of fish stocks, are strongly affected by ocean and atmospheric temperatures. The current prospect of substantial global warming, therefore, leads to concern about what this is likely to mean for the world's fisheries.

The build-up of carbon dioxide and other greenhouse gases in the atmosphere is known to be changing air and sea surface temperatures, rainfall and wind patterns, ocean acidity, sea levels and the intensity of tropical cyclones. Research has found that climate change is already modifying the distribution and productivity of marine and freshwater species, affecting biological processes, and altering food webs.

Aquatic ecosystems not only support fisheries by providing food, habitat and nursery grounds, the brief notes, but also protect communities from storms, which are predicted to become stronger and more frequent with climate change. Mangroves create barriers to destructive waves and hold sediments in place, reducing coastal erosion. Healthy coral reefs, sea grass beds and wetlands provide similar benefits.

What is Climate Change?

Climate is the result of the exchange of heat and mass between the land, ocean, atmosphere, polar regions (ice sheets) and space. Climate change is the change in the statistical distribution of weather over periods of time that range from decades to millions of years. This may be limited to a specific region or may occur across the earth. As we know that a major component of the global climate system is the oceans and hence they are adversely affected. Now it is being a major threat to sustainable growth and development of aquaculture. Global warming resulted into the increasing pressure on the world fisheries.

Major Environmental Challenges

1. Depletion of Natural Resources
    - ✰ Degradation of Land
    - ✰ Deforestation
    - ✰ Loss of Biodiversity
    - ✰ Overexploitation of Fisheries
2. Disposal of Solid and Hazardous Waste
3. Water Pollution
4. Air Pollution
5. Global Changes in the Chemical Composition of the Atmosphere
    - ✰ Greenhouse Effect
    - ✰ Stratospheric Ozone Depletion

Intergovernmental Panel on Climate Change (IPCC)

The Intergovernmental Panel on Climate Change (IPCC) is a scientific intergovernmental body tasked to evaluate the risk of climate change caused by human activity. The panel was established in

1988 by the World Meteorological Organization (WMO) and the United Nations Environment Programme (UNEP), two organizations of the United Nations. The IPCC shared the 2007 Nobel Peace Prize with former Vice President of the United States Al Gore.

The IPCC does not carry out its own original research, nor does it do the work of monitoring climate or related phenomena itself. A main activity of the IPCC is publishing special reports on topics relevant to the implementation of the UN Framework Convention on Climate Change (UNFCCC), an international treaty that acknowledges the possibility of harmful climate change; implementation of the UNFCCC led eventually to the Kyoto Protocol. The IPCC bases its assessment mainly on peer reviewed and published scientific literature. The IPCC is only open to member states of the WMO and UNEP. IPCC reports are widely cited in almost any debate related to climate change. National and international responses to climate change generally regard the UN climate panel as authoritative.

Climate Change 2007, the Fourth Assessment Report (AR4) of the United Nations Intergovernmental Panel on Climate Change (IPCC), is the fourth in a series of reports intended to assess scientific, technical and socio-economic information concerning climate change, its potential effects, and options for adaptation and mitigation. The report is the largest and most detailed summary of the climate change situation ever undertaken, involving thousands of authors from dozens of countries, and states in its summary,India, the seventh largest country in the world and the second largest in Asia, has a total geographical area of 329 Mha, of which only 305 Mha is the reporting area (the area as per the land records of villages and towns). The mainland stretches from 8°4' N to 37°6' N and 68°7' E to 97°25' E. It has a land frontier of 15,200 km and a coastline of 7,516 km.

In developing countries like India, climate change could represent an additional stress on ecological and socioeconomic systems that are already facing tremendous pressures due to rapid urbanization, industrialization and economic development. India is a major maritime state and an important aquaculture country in the world. With its huge and growing population, a 7500-km long densely populated and low-lying coastline, and an economy that is closely tied to its natural resource base, India is considerably vulnerable to the impacts of climate change. The various studies conducted in the country have shown that the surface air temperatures in India are going up at the rate of 0.4°C per hundred years, particularly during the post-monsoon and winter season. Using models, they predict that mean winter temperatures will increase by as much as 3.2°C in the 2050s and 4.5°C by 2080s, due to Greenhouse gases. Summer temperatures will increase by 2.2°C in the 2050s and 3.2°C in the 2080s. Extreme temperatures and heat spells have already become common over Northern India, often causing loss of human life. In 1998 alone, 650 deaths occurred in Orissa due to heat waves. Climate change has had an effect on the monsoons too. India is heavily dependent on the monsoon to meet its agricultural and water needs, and also for protecting and propagating its rich biodiversity. Subtle changes have already been noted in the monsoon rain patterns by scientists at IIT, Delhi. They also warn that India will experience a decline in summer rainfall by the 2050s, summer rainfall accounts for almost 70 per cent of the total annual rainfall over India and is crucial to Indian agriculture.

Apart from monsoon rains, India uses perennial rivers, which originate and depend on glacial melt-water in the Hindukush and Himalayan ranges. Since the melting season coincides with the summer monsoon season, any intensification of the monsoon is likely to contribute to flood disasters in the Himalayan catchment. Rising temperatures will also contribute to the raising of snowline, reducing the capacity of this natural reservoir, and increasing the risk of flash floods during the wet season. A trend of sea level rise of 1 cm per decade has been recorded along the Indian coast. Sea level rise due to thermal expansion of sea water in the Indian Ocean is expected to be about 25-040 cm by 2050. This could inundate low lying areas, down coastal marshes and wetlands, erode beaches, exacerbate flooding and increase the salinity of rivers, bays and aquifers. Deltas will be threatened by flooding, erosion and salt intrusion. Loss of coastal mangroves will have an impact on fisheries. The major delta area of the Ganga, Brahmaputra and Indus rivers, which have large populations reliant on riverine resources will be affected by changes in water regimes, salt water intrusions and land loss.

Biological Impact of Climate Change on the Aquaculture

- ☆ Decrease in primary production in the seas and oceans
- ☆ Affecting algae, plankton, fish and zooplankton
- ☆ Lower dry season and reduced water level.

Influence of Climate Change on Aquatic Animals

Fish are more sensitive to temperature than many animals because they cannot maintain a constant body temperature like we do–in most cases, their body is exactly the same temperature as the water they are swimming in. Different species can live in very cold or very hot water, but each species has a range of temperatures that it prefers, and fish can't survive in temperatures too far out of this range. When fish encounter water that is too cold for them, their metabolism–the chemical engine that drives their body–slows down and they become sluggish. As the surrounding water warms up, their metabolism speeds up–they digest food more rapidly, grow more quickly, and have more energy to reproduce. But fish need more food and more oxygen to support this higher metabolism. Warmer fish tend to mature more quickly, but the cost of this speedy lifestyle is often a smaller body size. Ninety percent of aquatic animals like fish raised in warm water end up smaller than their peers raised at cooler temperatures. Southern calamari, for example, grow more quickly at higher temperatures, but they also hatch much earlier (and smaller), and reach sexual maturity earlier, so they can't catch up in size to squid who have more time to grow in cool water. Many fish will also have less offspring as temperatures rise, and some may not be able to reproduce at all. Tropical fish like guppies produce smaller broods, and grass carp ovulate less frequently in warmer water. Temperate species like salmon, cat fish, and sturgeon cannot spawn at all if winter temperatures do not drop below a certain level. If there is not enough food, all of a fish's available energy goes to fuelling its high metabolism, and less energy is available for growth and reproduction. Rainbow trout grow significantly more slowly when their water temperature is raised only 2°C and food is limited, and fish such as salmon, whitefish, and perch are all expected to grow more slowly if food supply does not increase as temperatures rise.

Crucial Role of Oceans in Climate Change

Buffer to climate change and will likely bear the greatest burden of impacts. Oceans removed about 25 per cent of atmospheric carbon dioxide emitted by human activities. Oceans absorb more that 95 per cent of the sun's radiation, making air temperatures tolerable for life on land. Oceans provide 85 per cent of the water vapour in the atmosphere, and these clouds are key to regulating climate on land and sea. Ocean health influences the capacity of oceans to absorb carbon. Coastal circulation pattern affecting nutrient supply and increase acidity and coral bleaching. Reef building capacity of corals and also the spawning cycles of reef fishes and invertebrates.

- ☆ A modelling study showed significant large-scale changes of skipjack tuna habitat in the equatorial Pacific.
- ☆ The migration route and migration pattern and, hence, regional catch of principal marine fishery species, such as ribbon fish, small and large yellow croakers.
- ☆ Declines in fish larvae abundance in coastal waters of South and South-East Asia.

- Decline in fishery production in the coastal waters of East, South and South-EastAsia.
- Arctic marine fishery would also be greatly influenced by climate change.
- Northern shrimp will likely decrease with rise in sea-surface temperatures.
- Climate change may lead to a 30 to 70 per cent increase in catch potential in high-latitude regions and a drop of up to 40 per cent in the tropics, study by Sea around us project.
- On average, fish are likely to shift their distribution by more than 40km per decade.
- Increasing abundance of more southern species.
- Developing countries in the tropics will suffer the biggest loss in catch.
- Nordic countries such as Norway will gain with increased catch.
- In the North Sea, the northward shift of Atlantic Cod may reduce its abundance by more than 20 per cent.
- European plaice - a more southerly fish - may increase by more than 10 per cent.
- In the US, 50 per cent reduction in the Some species will face a high risk of extinction, including Striped number of some cod populations on the east coast by 2050.
- High risk of extinction of species Rock Cod in the Antarctic and St Paul Rock Lobster in the Southern Ocean.

Which Fish are Feeling Heat?

Polar Marine Fish

Polar species are uniquely adapted to narrow, cold temperature ranges and well-oxygenated water, making them vulnerable to even slight increases in temperature. Some species, such as the emerald rockcod and striped rockcod, are killed when temperatures climb only a few degrees above 0°C, and many Antarctic fish lack heat shock proteins–molecules that most animals have to repair cellular damage caused by heat. Arctic cod and other species associated with rich ice-edge communities already appear to be declining as polar ice melts and their habitats disappear.

Freshwater Fish that are Geographically Isolated

Freshwater species that cannot migrate to cooler waters as temperatures rise may be stuck in hot water. Migration is impossible from many isolated lakes and wetlands, and many major river systems worldwide run from east to west, making poleward migration impossible. Nearly all major river systems in the southern Great Plains and southwestern US for instance run from east to west, and native fishes are already living near their thermal tolerance limits in some of the hottest free-flowing water on earth. Increased warming could lead to the extinction of up to 20 species that are found nowhere else in the world.

Coral Reef Fish

Coral reefs support a huge diversity of fish and contribute about one quarter of the total fish catch in developing countries. Climate warming leads to coral bleaching–the loss of symbiotic bacteria that corals depend upon. In 1998, mass coral bleaching destroyed 16 per cent of the world's coral reefs. Significant changes in the abundance of some fish have been observed where intense bleaching has occurred, and fish that rely on live coral to survive have shown little recovery from these events.

Effect on Aquatic Communities

A recent study on the vulnerability of national economies and food systems to climate impacts on fisheries has revealed that African countries are most at risk. What makes them so vulnerable? The first reason is ecological — it is because many African countries are semi-arid with significant coastal or inland fisheries. This gives them high exposure to future increases in temperature and linked changes in rainfall, hydrology and coastal currents. The second reason is social — these countries also depend greatly on fish for protein, and have low capacity to adapt to change due to their comparatively small or weak economies and low human development indices. Countries in this category include Angola, Congo, Mauritania, Mali, Niger, Senegal and Sierra Leone. Other vulnerable African nations include Rift Valley countries such as Malawi, Mozambique and Uganda. Beyond Africa it is the Asian river dependent fishery nations including Bangladesh, Cambodia and Pakistan that are most at risk. The often overlooked links between fisheries and agriculture also make the semi-arid areas of Africa vulnerable. In these areas the higher-potential agricultural zones are around lakes, swamps and river-floodplains. Here fisheries often provide both safety nets and capital to invest in agricultural inputs and livestock. If the fishery system is under stress, the potential of the other components of the 'tri-economy' is reduced.

The system as a whole is resilient to local-scale perturbation, but with reduced rainfall stressing both fisheries and crop agriculture, that resilience could be threatened by climate change. So there is a case for not forgetting the fish in the wider discussion of adaptation and coping in these systems and particularly fisheries of inland and near-shore waters.

Some Studies on Climate Change Effect

- A substantial portion of the vast mangroves in South and South-East Asian regions has also been reportedly lost during the last 50 years of the 20th century, largely attributed to human activities.
- Salt water from the Bay of Bengal is reported to have penetrated 100 km or more inland along tributary channels during the dry season.
- Severe droughts and unregulated groundwater withdrawal have also resulted in sea-water intrusion in the coastal plains of China.
- Glaciers in Asia are melting faster in recent years than before, as reported in Central Asia, Western Mongolia and North-West China, particularly the Zerafshan glacier, the Abramov glacier and the glaciers on the Tibetan Plateau.
- Biodiversity in Asia is being lost as a result of development activities and land degradation (especially overgrazing and deforestation), pollution, over-fishing, hunting, infrastructure development, species invasion, land-use change, climate change and the overuse of freshwater.
- Wetlands in the major river deltas have been significantly altered in recent years due to large scale sedimentation, land-use conversion, logging and human settlement.
- Coastal erosion in Asia has led to loss of lands at rates dependent on varying regional tectonic activities, sediment supply and sea-level rise.
- Rapid melting of glaciers, glacial runoff and frequency of glacial lake outbursts causing mudflows and avalanches have increased.

- ☆ A recent study in northern Pakistan, however, suggests that glaciers in the Indus Valley region may be expanding, due to increases in winter precipitation over western Himalayas during the past 40 years.

Fish Help to Buffer Climate Change

Fish act as a buffer against carbon dioxide in the world's oceans because they secrete calcium carbonate that dissolves easily in water and lowers acidity. The ocean to maintain its optimal pH needs to have lots of healthy, young fish. Old fish are not as good at producing calcium carbonate. They're less active and drink less water. The ocean needs a constant supply of young fish that produce more calcium carbonate.

What Needs to be Done?

Role of Governance in Climate Change

Full implementation of code of conduct for responsible fisheries (CCRF). Building of institutional and legal frameworks and effective public, private and NGO partnership. Integration in research and management across the sector ensuring regulations and linking disaster management with development planning. Adoption environmentally friendly and fuel-efficient fishing and aquaculture practices. Eliminate subsidies that promote overfishing and excess fishing capacity. Undertake assessments of local vulnerability and risk. Build local-level ocean climate models. Strengthen knowledge of the dynamics of biogeochemical cycles in aquatic ecosystems, especially of carbon and nitrogen. Encourage sustainable, environmentally friendly biofuel production from algae and seaweed. Explore carbon sequestration in aquatic ecosystems. Implement comprehensive and integrated ecosystem approaches to managing oceans, coastal zones, fisheries and aquaculture; to adapting to climate change; and to reducing risk from natural disasters.

Carbon Sequential Cycle

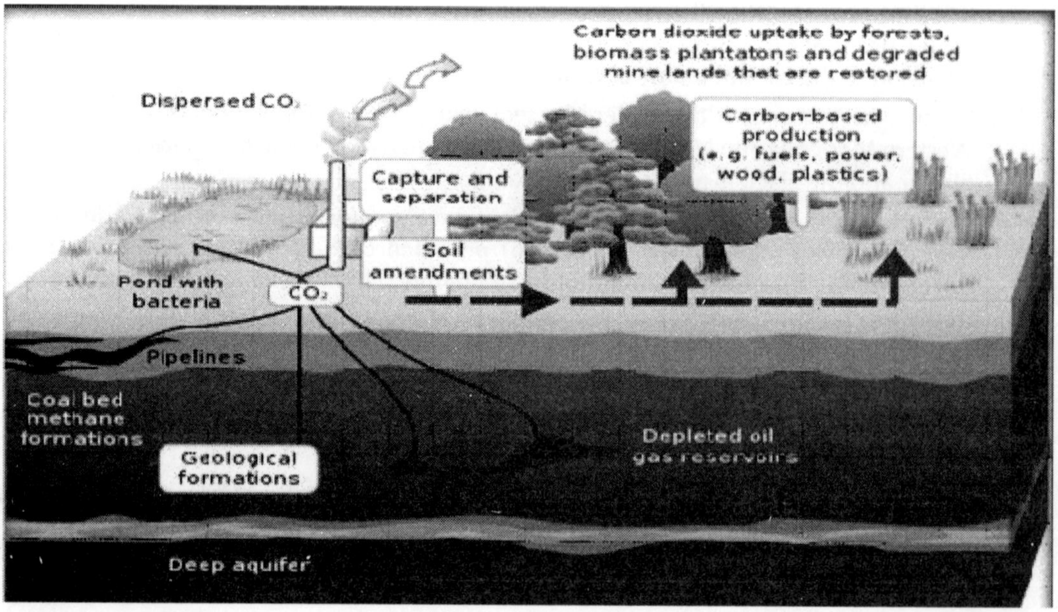

Industrialized countries need to cut their CO_2 emissions as obliged under the Kyoto Protocol, and all must agree to much more serious emission reductions in the next period, after 2012. To stay well below the 2°C danger threshold they must reduce their emissions by 60-80 per cent. The rapidly industrializing countries also need to lower their emissions while meeting their development goals by 'leapfrogging' into clean and efficient technologies. This will only be possible when developed economies–governments as well as the business and financial communities–engage in this endeavor.

- ☆ The single largest source of man-made CO_2 is electricity generation, accounting for 37 per cent of worldwide CO_2 emissions. The first step to move to a clean energy future is to clean up the power sector.
- ☆ Governmental and private aid agencies are starting to take climate-related impacts and catastrophes seriously. Comprehensive strategies to build resistance and resilience to climate change impacts need to be developed–for threatened communities as well as for nature reserves. All this must happen while curbing CO_2 emissions rapidly, as resilience building can only buy some time and becomes an insurmountable challenge if global temperatures are allowed to rise too high.

In order to preserve the diversity and abundance of fish–one of our most valuable biological, nutritional, and economic assets–we must keep global warming below dangerous levels. This is also crucial to help fish recover from threats like overfishing and the destruction of their habitats. WWF seeks to limit global warming of average global temperature to below 2°C (3.6°F) over pre-industrial levels.

Carbon dioxide (CO_2) is the main pollutant causing climate change. It rises through the atmosphere and captures heat, intensifying the effect of the greenhouse gases that keep the earth warm. This has dramatic consequences for the globe's climate system–more extreme weather like droughts, floods, and storms; rising sea levels and changes of large ocean currents, and changes of regional weather systems during events like elno.

Fisheries, Climate Change-Related Warming may Result in

- ☆ Longer growing seasons and increased rates of biological processes - and often of production;
- ☆ Greater risk of oxygen depletion;
- ☆ Species shift to more tolerant of warmer and perhaps less-oxygenated waters;
- ☆ Redeployment or re-design and relocation of coastal facilities;
- ☆ Coastal cultures may need to consider the impacts of sea-level rise on facilities and the freeing of contaminants from nearby waste sites;
- ☆ Changes in precipitation, freshwater flows, and lake levels;
- ☆ Introduction of new disease organisms or exotic or undesired species;
- ☆ Establishment of compensating mechanisms or intervention strategies;
- ☆ A longer season for production and maintenance.
- ☆ Modification of aquaculture systems, *e.g.* Keeping them indoors under controlled light, may be needed more often to protect larvae from solar UV-b. In addition, several of the above and other factors, such as competing demand for coastal areas, may argue for technological intensification in ponds and non-coastal facilities.

While the fisheries sector cannot do much to impede or seriously affect global climate change, it could contribute to its stabilization or reduction, and to mitigating its effects. Climate changes notwithstanding, there are several actions to consider. The most important strategies are those needed to promote sustainability and which are useful and practical, even in the absence of climate change. Further, when developing strategies, we need to consider both the problems and the opportunities that are being presented, in the following way:

- ✰ Active participation at global and regional level, to ongoing debate and collaboration, to obtain the best possible information of fisheries-related impacts;
- ✰ Allocating research funds to analyze local and regional potential changes in resource magnitude and composition and likely socio-economic impacts;
- ✰ Sharing information obtained with the sector on potential changes, their scale and possible effects on resources and fisheries;
- ✰ Establishing institutional mechanisms to enable or enhance the capacity of fishing interests (fleets and other infrastructures) to move within and across national boundaries as a consequence of changes in resources distribution. This implies developing bilateral agreements;
- ✰ Preparing contingency plans for segments of the sector that might not be able to move, particularly for disadvantaged areas and small-scale fishers lacking mobility and alternatives;
- ✰ Developing effective national and international scale resource management regimes and associated monitoring systems to facilitate adaptation of exploitation regimes in a shifting environment;
- ✰ Strengthening regional fisheries management organisation and other mechanisms to deal with cross-border stocks;
- ✰ Integrating fisheries management into coastal areas management to ensure that fisheries needs are taken up when dealing with protection of coastal areas from sea level rise, etc. For instance, to ensure that public works to protect coastal areas do not unnecessarily obliterate nursery areas important to fisheries;
- ✰ Analyzing aquaculture sustainability in an ecoregional context, forecasting changes in productivity or resistance and in required related changes in culture systems, cultured species or delocalisation of productive systems. Particular attention should be given to coastal investments;
- ✰ Fostering interdisciplinary research, with scientists meeting periodically to exchange information on observations and research results, and meeting with managers to ensure the proper interpretation of results and the relevance of research; and,
- ✰ Foreseeing and planning infrastructure adaptations. It could be expected that, in response to shifting populations and species, the industry will respond with faster, longer-range fishing craft, install on-board processing equipment to replace endangered coastal ones or use floating processors when feasible, and find alternative means of transport when coastal roads are flooded and relocation is not possible. Governments should also consider constructing and maintaining appropriate infrastructure for storm forecasting, signalling systems and safe refuges for dealing with possible rising sea level and increased storminess. There may be opportunities to take advantage of reduced need for ice strengthening of

vessels and infrastructures in a warmer climate, except perhaps for areas with increased icebergs.

The international activity already related to climate change is very intense as can be seen looking at the various Websites dealing with it. Most of the action refers, however to research and international agreements. Research focuses on tracking indicators of change, studying cause-effect relationships, modelling, assessing and forecasting impacts. International agreement such as the UN Framework Convention on Climate Change aim at mobilising attention and commitments of governments to reduce greenhouse gases. Little or no action has been taken by governments to mitigate the possible effects, and information on contingency plans is lacking.

In fisheries, while climate change has been addressed occasionally in scientific literature,the subject has not yet been formally addressed by most industry or fishery management administrations. However, the fishery sector and fisheries research are fairly advanced in this matter, through their dealing with the *El Niño*, decadal changes in ocean environments and other longer terms fluctuations in fisheries environments and resources. The observation programmes, scientific analyses, computer models, the experience gained and strategies developed by fishers, processors, fishfarmers, and management authorities confronted with the problem of medium-to-long-term natural fluctuations, is extremely useful for dealing with climate change. Many of the principles and strategies developed to deal with "unstable" stocks will be of use when having to deal with climate change. The Global Ocean Observing System (GOOS) has been established under the aegis of IOC-UNESCO. Changes remain uncertain and competing theories are still developing as to the reality of the change, its magnitude and its mechanisms. Progress in implementing the UN Framework Convention on Climate Change is slow and resistance from some of the major players to pledge reduction of gas emissions remains a stumbling block. It is not possible to forecast how the question will evolve. Fisheries will be able to move faster towards specific assessments and contingency plans when more precise and reliable predictions are available on projected climate change with at least a regional resolution. In the meantime, dealing effectively with medium-term natural changes offers a good "training field".

Chapter 24

Standardisation of Natural Colour in Surimi made Shrimp Analogs

☆ *Y.T. Patil, A.T. Tandale, S.S. Todkari, M.M. Girkar and S.B. Gore*

ABSTRACT

In seventies, a new generation product crabstick also known as crab leg was developed. New generation products are also called as shellfish analog or Surimi imitation products. Surimi analog means the product looks like that of the natural shellfish or fishes in terms of colour, appearance, taste etc. In this experiment five types of natural colours were used, namely paprika (water soluble), annatto seed (water soluble), annatto seed (oil soluble), caramel (oil soluble), beet root colour (water soluble). Among five colours annatto seed (water soluble) at 1:1.5 (colour: water) was selected as best colour for shrimp analog preparation.

Introduction

In 1960 heavy competition among the *kamaboko* producing industries led to the development of the variety of new products to compensate the surimi cost and to attract the consumer (Kanmuri and Fujita, 1985). This lead to the development of new products such as analogs or imitation products from surimi.

The new crab flavour analogs became more popular and its production increased from 1,300 to 71,000 tons during 1975 to 1984. Following the development of crab flavoured analogs, scallop and shrimp shape analogs were introduced in the market successfully. The crab sticks and other Surimi made imitation products became famous and became main item in restaurants, salad makers and super markets in USA (Zalke, 1992).

The product was given natural appearance by colouring it same as that of the natural colour of fin fishes and shell fishes. Surimi products can be coloured using artificial or natural colours. Now days, people are approaching towards the natural food rather than artificial foods. Therefore, in this study natural colours were used to standardise the colour for shrimp analogs.

Material and Methods

Sample Preparation

Shrimp analogs were prepared using standardized method (Patil, 2002). Surimi 60 per cent, salt 1 per cent, chilled ice water 11.2 per cent, egg white 4.00 per cent, starch 11 per cent, monosodium glutamate 0.40 per cent, vegetable oil 2.50 per cent, ginger garlic paste 3 per cent and shrimp extract powder 5 per cent.

Ingredients were mixed properly and dough was prepared. The shrimp shaped analogs were prepared by moulding the dough with the help of shrimp shaped die.

Natural Colour Preparation

Five natural colour namely paprika (water soluble), annatto seed (water soluble), annatto seed (oil soluble), caramel (oil soluble), beet root colour (water soluble) were purchased from Vinayak Co-operation, Mumbai.

Colour solutions were prepared at 1:1 ratio. Water soluble colours paprika, annatto seed colour and beet root colour were mixed in water at 1:1(colour: water) ratio. While oil soluble colours annatto seed and caramel were mixed in edible oil at 1:1 (colour: oil) ratio.

Each colour was separately given on the moulded shrimp analogs. The product was then steam cooked at temperature 85°C for 5 minutes and then cooled to room temperature.

Sensory Analysis

Ten trained panelists were chosen for further colour analysis. The coloured product of each colour was placed in white enamel dishes. This dish was subjected to organoleptic analysis. The results obtained are given in Table 24.1. The natural cooked shrimp was used as control.

Table 24.1: Organoleptic Evaluation of the Colour for Shrimp Analog

No. of Panelist	*Annatto Seed Colour*	*Annatto Seed Oil Soluble*	*Beet Root*	*Caramel Colour*	*Orange Red Colour*	*Paprika*
1	6	6	6	4	4	8
2	7	7	5	4	5	6
3	7	7	5	5	5	7
4	7	7	4	2	4	7
5	8	5	7	2	6	8
6	8	7	5	5	5	9
7	8	7	4	2	3	6
8	7	6	5	2	3	8
9	8	6	5	3	7	7
10	8	5	5	7	3	7

Results

The standardisation of colour was carried out in the two steps Phase 1 and Phase 2.

Phase 1

The shrimp analogs coloured using five different natural colours were subjected to sensory analysis. According to the opinion of the panelist two colour *viz.*, paprika (water soluble) and annatto seed (water soluble) colour gave the same colour *i.e.* orange red colour. Therefore it was suggested by the panelist to carry out further study between these two colours at higher concentration. Statistical analysis (ANOVA) showed that both the colours had the same scores.

Phase 2

In this phase, the above selected colours namely paprika and annatto seed were taken. Two higher concentration ratios 1:1.5 and 1:2 were made for both the colours. The various colour concentrations were applied on moulded shrimp analogs. The shrimp analogs were steam cooked at 85°C for 5 minutes. Further sensory analysis was carried out by the panelists. Statistical analysis for phase 2 is given in Table 24.2 and Figure 24.2. The ANOVA was carried out and indicated that treatment given were significantly different. During standardization annatto seed colour at 1:1.5 ratio was found to be significantly different from other.

Discussion

Boiled natural shrimp gave orange red colour. Therefore it was necessary to use natural colour giving same colour resemblance.

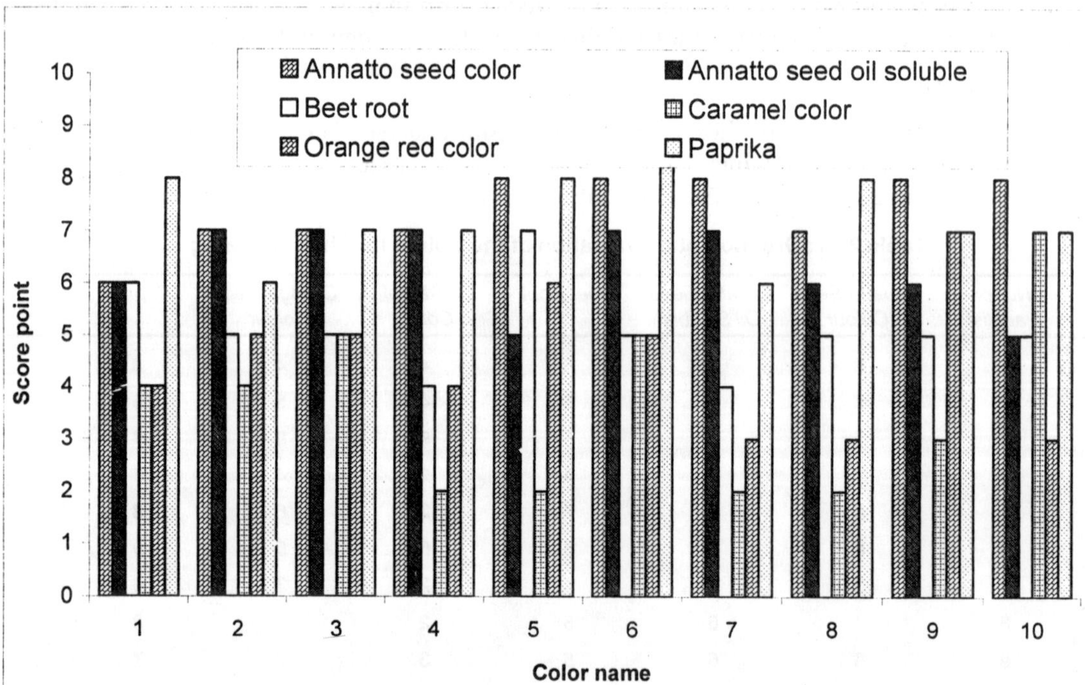

Figure 24.1: Standardisation of Different Natural Colours in Shrimp

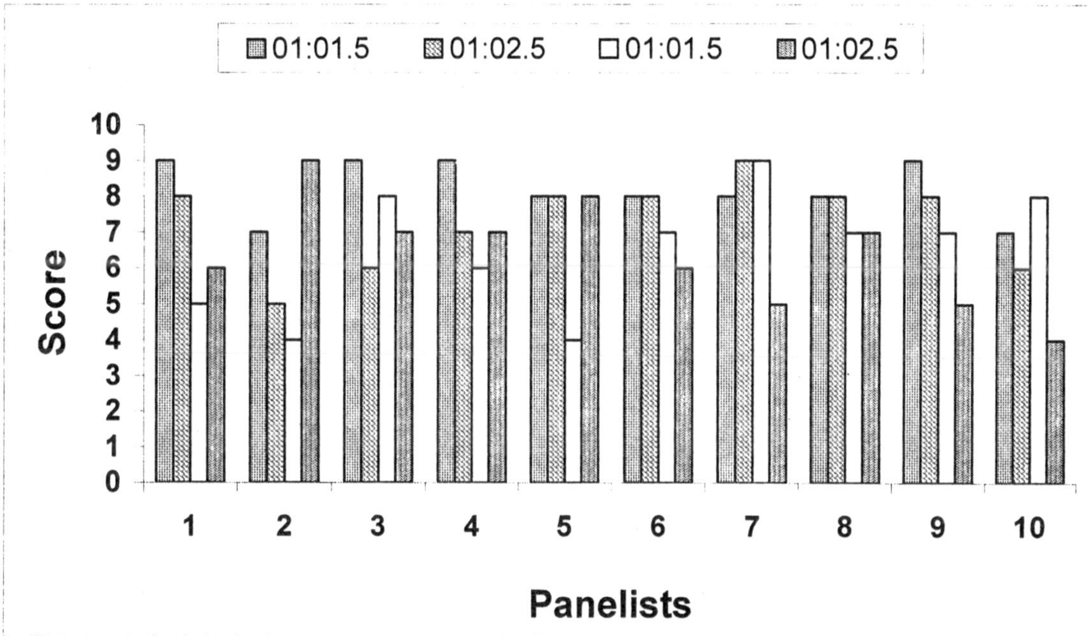

Figure 24.2: Standardisation for Two Selected Colours *viz.*, Annatto Seed and Paprika

Table 24.2: Sensory Analysis of Colour Selection for the Shrimp Analog

Sl.No.	*Annatto Seed (Water Soluble)*		*Paprika (Water Soluble)*	
	1:1.5	*1:2*	*1:1.5*	*1:2*
1	9	8	5	6
2	7	5	4	9
3	9	6	8	7
4	9	7	6	7
5	8	8	4	8
6	8	8	7	6
7	8	9	9	5
8	8	8	7	7
9	9	8	7	5
10	7	6	8	4

Five natural colours that were used had the properties to produce a colour hue (Table 24.3). All natural colours when applied on the moulded shrimp analog externally showed that they produce different shades.

Table 24.3: Colours Produced by Five Different Natural Colours

Sl.No.	Natural Colour	Colour Shade
1.	Annatto seed colour (water soluble)	Orange to a pale yellow colour
2.	Paprika (water soluble)	Yellow orange to the red orange colour
3.	Beet root colour (water soluble)	Bright red–purple coloured
4.	Annatto seed colour (oil soluble)	Yellow - orange to a red shade
5.	Caramel (oil soluble)	Dark brown colour

From all the shade, that were available paprika colour (water soluble) and annatto seed colour (water soluble) had the same shades *i.e.* orange to orange red. These shades resemble to the boiled natural prawn. But because of their ability to produce same colour shade, further concentrations of these colours at different ratios were made.

Finally, it was decided to choose the annatto seed colour (water soluble) to use for colouring the moulded shrimp analog externally. This was supported by the statistical analysis (ANOVA).

Lauro (2000) worked on the similar line for the natural colours effect on Surimi seafood. He had given different shades that were produced by different natural colours.

Paprika mixes easily with the surimi paste, giving yellow orange to red orange colour, depending on the hue of the starting material, the level of the usage, and the final pH of the surimi.

The starting material for annatto seed colour is a seed coat that is part of the fast growing shrub called *Bixa orellano* (Aparnathi *et al.*, 1990). This shrub is found typically in some parts of the United States, Africa and India.

Annatto seed colour (oil soluble) produces shade from yellow orange to red with decreasing pH. The colour stability is good when the surimi seafood is stored in dark, frozen state and is unaffected by pH. The annatto seed colour (water soluble) produces yellow orange to a pale yellow colour.

Beet juice extract easily colours surimi paste, producing a light pink to a bright cherry red colour depending on the amount of the colour used. Caramel is water soluble, viscous dark brown liquid made by controlled heating of number of food grade carbohydrate.

The results obtained also showed that annatto seed colour dose not change after steaming (at temperature 85°C for 5 minutes). It also adheres well to the Surimi mixture.

Lauro (2000) also explained that annatto seed colour (water soluble) known to react with protein to form strong bond. This colorant dissolves in water at the low concentration, when sprayed on the surimi paste before cooking and found to adhere well with minimal penetration. Surimi seafood coloured with the annatto seed colour (water soluble) found not to fade over a three months period in the frozen storage.

Acknowledgement

The Authors are thankful to the Associate Dean, College of Fisheries, Shirgaon, Ratnagiri to provide all necessary facilities during my research work.

References

Aparnathi, K.D., Lata, R. and Sharma, R.S., 1990. Annatto: It's cultivation, preparation and usage. *Intl. J. Tropical Agriculture*, 8(1): 80–86.

Kanmuri, Y. and Fujita, T., 1985. Surimi based products and fabrication process. In: *Proceedings of the International Symposium on Engineered Seafood including Surimi,* (Eds.) R. Martin and R. Collette. National Fish Institute, Washington D.C., p. 254–263.

Lauro, G.J., 2000. Natural colorants for surimi seafood. In: *Surimi and Surimi Seafood,* (Ed.) J.W. Park. Marcel Dekker Inc., New York, p. 417–443.

Patil, Y.T., 2002. Studies on preparation of shrimp analogs. *M.F.Sc. Thesis,* Dr. Balasaheb Sawant Konkan Krishi Vidyapeeth, Maharashtra, India, 156 p.

Zalke, J., 1992. Global market for surimi based products. In: *Pacific Whiting Harvesting, Processing, Marketing and Quality Assurance: A Workshop,* (Eds.) G. Sylvia and T. Morrissay. New Port, USA, pp. 67–72.

Chapter 25

Antimicrobial Activity of *Acanthus ilicifolius* Extracted from the Mangroves Forest of Karwar Coast

☆ *P.V. Khajure and J.L. Rathod*

ABSTRACT

The antimicrobial activity of n-hexane, chloroform and methanol extract extracts of leaves and roots of the plant *Acanthus ilicifolius* ware studied. Ampicillin and clotrimazole were used as standard antibacterial and antifungal agents respectively. The result of the study revealed that the n-hexane extract and chloroform extract of leaves exhibited strong inhibitory action against *Bacillus subtilis, Staphylococcus aureus, Candida albicans, Aspergillus fumigatus* and *Aspergillus niger* and moderate inhibitory action against *Pseudomonas aeruginosa* and *Proteus vulgaris.* The rest of the extracts showed good activity.

Introduction

Acanthus ilicifolius (family of Acanthaceae) is a valuable medicinal plant that is widespread in tropical Asia and Africa, through Malaya to Polynesia (Xie *et al.*, 2005). *Acanthus ilicifolius* extracts have been used in various folk medicines as remedies against rheumatism, neuralgia, poison arrow wounds, coughs, asthma and bacterial infections with subsequent scientific supports to these claims (Mastaller, 1997).These created an interest to test the possible antimicrobial activity of different parts of this plant, which has not been reported; hence, the present study was designed. The phytochemical literature reveals the presence of 2-benzoxazolinone, lignan glucosides, benzoxazinoide glucosides, flavone glycosides and phenylethanoid glycosides in this plant. (Kanchanapoom, *et al.*2001).The present study was aimed at the preliminary investigation of antibacterial and antifungal activity of n-hexane, chloroform and methanol extracts of leaves and roots of *A. ilicifolius.*

Acanthus ilicifolius (Taxonomic Authority: Lour, Family: Acanthaceae) was collected in mangroves forest of Karwar, west coast of India (Lat. 14° 47' 11.33N. Long. 74° 01' 48.38E), during December 2009 and identified by a systemic Botanist. (Figure 25.1)

The shade dried parts of the whole plant (Leaf and roots) were coarsely powdered (50–200 g) and extracted with n - hexane, chloroform and methanol respectively for 48 hours in Soxhlet apparatus. After evaporation of the solvent under reduced pressure, the respective extracts were obtained. Considering that the methanolic extract suits for better activity (Chatterjee, 2007), it was successively

Figure 25.1: Showing *Acanthus ilicifolius* Plant with Flower

partitioned with ethyl acetate and acetone affording 0.150 g and 0.200 g (residue dry) of each fraction respectively.

The *in vitro* antibacterial and antifungal studies of the n-hexane, chloroform and methanol extracts of the leaves and roots were carried out by the Agar disc diffussion method. (Barry,1976) All the extracts were separately dissolved in dimethylsulfoxide (DMSO) to get 10 mg/ml solutions. Ampicillin (1 mg/ml) and clotrimazole (1 mg/ml) were used as standard antibacterial and antifungal agents respectively. The antibacterial activity was evaluated by employing 24 h cultures of *Bacillus subtilis, Staphylococcus aureus, Pseudomonas aeruginosa* and *Proteus vulgaris* using Muller Hinton Agar medium. Antifungal activity was carried out against 24 h cultures of *Candida albicans, Aspergillus fumigatus* and *Aspergillus niger* using Sabouraud dextrose agar medium. Accurately 0.2 ml of the test and standard solutions were transferred to cups aseptically and labeled accordingly. The microorganism inoculated plates were then maintained at room temperature for 2 h to allow the diffusion of the solutions into the medium. The Petri dishes used for antibacterial screening were incubated at 37±1° for 24 h, while those used for antifungal activity were incubated at 28±1° for 48 h. The diameters of zone of inhibition surrounding each of the wells were recorded.

Table 25.1: Antimicrobial Activity of *Acanthus ilicifolius*

Test Organisms	*Zone of Inhibition in mm*					
	*n-Hexane Extract**		*Methanol Extract**		*Chloroform Extract**	
	Leaves	*Roots*	*Leaves*	*Roots*	*Leaves*	*Roots*
B. subtilis	21	18	17	15	20	15
S. aureus	20	17	12	14	22	10
P. aeruginosa	20	16	12	12	18	10
P. vulgaris	22	16	14	15	18	14
C. albicans	20	19	16	12	24	20
A. fumigatus	22	17	17	12	22	18
A. niger	22	20	15	14	22	18

*10mg/ml.

Table 25.1 enumerates the antibacterial and antifungal activity of the extracts of different parts of the *Acanthus ilicifolius*. The n-hexane, chloroform and methanol extracts of the different parts of the plant exhibited strong to moderate activity against the test microorganisms. The results revealed that, the n-hexane and chloroform extracts of leaves exhibited strong inhibitory action against *Bacillus subtilis, Staphylococcus aureus, Candida albicans, Aspergillus fumigatus* and *Aspergillus niger* and moderate inhibitory action against *Pseudomonas aeruginosa* and *Proteus vulgaris*. The rest of the extracts showed moderate activity.

References

Barry, A.L., 1976. *The Antimicrobial Susceptibility Test: Principle and Practices.* ELBS, London, pp. 180.

Geissberger, P. and Sequin, U., 1991. Constituents of *Acanthus ilicifolius* L.: Do the components found so far explain the use of this plant in traditional medicine? *Acta Tropica*, 48(4): 251–261.

Kanchanapoom, T., Kamel, M.S., Kasai, R., Yamasaki, K., Picheansoonthon, C. and Hiraga, Y., 2001. Lignan glucosides from *Acanthus ilicifolius. Phytochemistry*, 56: 369–72.

Mastaller, M., 1997. *Mangroves: The Forgotten Forest between Land and Sea.* Tropical Press, pp. 97.

Mclaughlu, J.L. *et al.*, 1988. The use of biological assays to evaluate botanicals. *Drug Information Journal*, (32): 513–524.

Xie, L.S. *et al.*, 2005. Pharmacognostic studies on mangrove *Acanthus ilicifolius. Zhongguo Zhong Yao Za Zhi*, 30: 1501–1503.

Chapter 26

Mangroves Diversity in Kali Estuary, Karwar

☆ *Pradeep V. Khajure and J.L. Rathod*

ABSTRACT

Mangrove ecosystems are most productive ecosystems which supports a rich species diversity of flora and fauna, but it is facing heavy human pressure, natural stresses leading loss of fish biodiversity, degradation and depletion of mangrove habitat. Altogether 20 species of mangroves and their associates have been reported from Kali estuary. Of these, 13 species are eumangroves belonging to 08 genera under 06 families and the other 07 their associates. In the recent past most of the mangrove fields are converted into aquaculture ponds, management of these extremely fragile ecosystems along with remedial measures to protect the mangroves and their habitat are also suggested.

Introduction

Mangroves are trees or shrubs that grow between low tide and high tide marks and beyond, where salt water reaches during spring tides. They form distinct communities commonly known as mangals or mangrove forest. Mangroves forest are the most productive and diverse wetlands found in the inter tidal zones of sheltered shores, estuaries, creeks, backwaters, lagoons, marshes and mud flats of tropical and subtropical region of world. These plants, with their many specialized adaptations such as pneumatophores, knee roots and stilt root for breathing, stilt roots and buttresses for supporting, specialized glands (in leaves and stems) for salt excretion and propagules for viviparous germination form a bridge between terrestrial and marine ecosystems. The total mangrove cover in the entire world is estimated to be around 181,000 km^2 and mangroves rich in density and diversity mainly occur in the Indo-Malayan region. A total of 90 mangroves and their associates are so far reported from around the world (Untawale A.G 2004). India has a total mangroves area of 4482 km^2 *i.e.*, 2.5 per cent of the world's cover. Of this, 59 per cent (2644 sq km) occurs along the east coast, 23 per cent (1031 sq km) on the west coast and the remaining 18 per cent (807 sq km) in the Bay of islands of Andaman and Nicobar.

Mangroves of the country exist in three major coastal settings, namely, deltaic, backwater-estuarine and insular environments (Kathiresan 2004). Deltaic type of mangroves mainly prevail along the east coast in vast deltas formed by the mighty rivers Brahmaputra, Ganga, Mahanadi, Godavari, Krishna and Kaveri. Backwater-estuarine types of mangroves are present along the characteristic funnel shaped estuaries (of west flowing rivers Indus, Narmada, Tapti, etc. devoid of any significant deltaic formations) or backwater creeks and neritic inlets. The other short estuarine rivers of west coast have fringing type of mangrove formations (Rao T.A *et al.*, 2002). Insular mangroves mainly exist in the Bay of Bengal islands interspersed by several rivulets, tidal estuaries, lagoons and neritic islets. As many as 69 mangroves and their associate taxa are reported to be present in India and among these 63 species occur in the east coast, 37 in the west coast and 44 along the coasts of Andaman and Nicobar group of islands (Kathiresan 2004). Mangrove forests are one of the most productive and biodiversity rich ecosystems on the earth. Yet, these unique costal tropical forests are among the most threatened habitats in the world. They are disappearing more quickly than inland tropical rain forest and so far with hardly any public notice.

Some people don't like mangroves, regarding them as muddy, mosquito and crocodile infested swamps, there fore, these have been classified as "waste land". In the past removal of the mangroves was seen as a sign of progress. But with rapid decline in the quantum of mangroves and scientific interpretation of the importance of mangroves to the society, various organizations, NGO'S,academicians all over the worlds are taking lead in the biodiversity study and conservation of these mangroves.

An estimated 75 per cent fishes caught commercially, spend some time in the mangroves or are dependent on food chains which can be traced back to these costal forests. Mangroves protect the coast by absorbing the energy and storm driven waves and wind, while providing a buffer for the land on one side and interact with the sea on the other.

Sediments trapped by mangrove roots prevent silting of adjacent marine habitats where clouding water might cause corals to die. In addition, mangrove plants and sediments have been shown to absorb pollution, including heavy metals. In many areas of the world, mangrove deforestation is contributing to fisheries decline, degradation of clean water supplies, salinisation of coastal soils, erosion and land subsidence as well as the release of carbon dioxide into the atmosphere. It is estimated that mangrove forests fix more carbon dioxide per unit area than phytoplankton in tropical oceans.

Mangroves have long functioned as a store house of materials providing food, medicines, shelter and tools. The fruit of certain species are eaten, the best honey is considered to be that produced from mangroves, particularly from river mangrove *Aegiceras corniculatum.* Numerous medicines are derived from mangroves. Ashes or bark infusions of certain species can be applied to skin disorders and sores including leprosy, headache, rheumatism, snake bites, boils, ulcers, diarrhoea, haemorrhages and many more conditions are traditionally treated with mangrove plants. The latex from the leaf of *Excoecaria agallocha* can be used on sores and to treat marine stings, the leaves are also used to fishing when crushed and dropped in water, fish are stupefied and float to the surface. Certain mangrove trees are prized for their hard wood and used for boat building and cabinet timber, as well as for tools such as digging sticks, spears, thatching, basket weaving etc, various barks are used for tanning, pneumatophores make good fishing floats and so on.

A tea spoon of mud from mangrove forest contains billions of bacteria. These densities are among the highest to be found in marine mud indicating high productivity of coastal forest habitat. Mangrove plants produce about 1 kg of litter per square meter a year. These form a part of the productive food chain, part of which is utilized and part is carried out to the sea. Mangroves in most of the places are in a degraded state today, owing to indiscriminate felling for domestic purposes, recreational and developmental activities of human being.

Many factors contribute to mangrove forest loss, including the charcoal and timber industries, urban growth pressures and mounting pollution problems. However, one of the most recent and significant causes of mangrove forest loss in the past decade has been the consumer demand for luxury food item shrimps or 'prawns' and the corresponding expansion of mangrove destructive production methods of export oriented industrial shrimp aquaculture. Vast tracts of mangrove forest have been cleared to make way for the establishment of coastal shrimp farms facilities. The failure of union government to adequately regulate the shrimp industry, and the head long rush of multi lateral lending agencies to fund aquaculture development, without meeting their own stated ecological and social criteria, are other important reasons for this unfortunate degradation.

Mangroves of the Kali Estuary Karwar, Karnataka

River Kali or Karihole is a major river of the northern costal Karnataka at lat. 14°48′ N and Long. 74° 07′ E. It has an estuary extending from Kodibag (S) and Devbagh (N) at the river mouth up to Kunnipet (S) and Kadra (N), stretching to a distance of about 30 kms. Another small river called Mavinahalla also joins this estuary, converting it into an estuarine complex.

This estuarine complex formed by rivers Kali and Mavinahalla; about 3 km north of Karwar (Uttar Kannada district of Karnataka) along the west coast, supports mangrove vegetation along its shores and mud flats. Floristic studies have reviled that the isolated and remnant patches of mangrove forests of this area are rich in species diversity of both eumangrove and mangrove associate plants. A total of 130 species representing 106 genera and 50 families of plants in Kali estuary which involves true mangroves species, associates and accidental mangroves have been identified (Nayak, *et al*.,.2010).

Of the 15 species of eumangroves reported from Karnataka, as many as 13 species were found growing here. This includes the major mangrove genera such as *Avicennia* (3 Species), *Bruguiera* (2 species), *Rhizophora* (2 species), *Sonneratia* (2 species) and one species each of *Aegiceras, Exoecaria, Lummitzera* and *Kandelia*. Among these *Sonneratia alba, Rhizophora apiculata* and *Avicennia officinalis* are the most dominant species

Materials and Methods

Several field trips were conducted to study mangrove areas along the Kali estuary from February 2009 to June 2010. Plants specimens were collected, poisoned, documented and made into herbarium as per standard methods for further studies. All the specimens were identified and deposited at Department of Marine Biology, Karnatak University Post Graduate Centre, Karwar, India.

Results

Literature published on the status, diversify and distribution of mangroves reported from Kali estuary, Karwar are quite inconsistent differing from one researcher to the other. According to compiled data from all such publications, a total of 20 species of mangroves belonging to 13 genera following under 9 families occur in the Kali estuary (Table 26.1), all the specimens were collected and deposited by the authors.

Figure 26.1: *R. apiculata*

Figure 26.2: *B. gymnorrhiza*

Figure 26.3: *Avicennia alba*

Figure 26.4: *R. mucronata*

Figure 26.5: *B. cylindrica*

Figure 26.6: *Kandelia candel*

Figure 26.7: *Avicennia officinalis*

Figure 26.8: *Avicennia marina*

Figure 26.9: *Sonneratia caseolaris*

Figure 26.10: *Sonneratia alba*

Figure 26.11: *Lumnitzera racemosa*

Figure 26.12: *Exoecaria agallocha*

Figure 26.13: ***Aegiceras corniculatum***

Figure 26.14: ***Acanthus ilicifloius***

Figure 26.15: Caesalpinea *crista*

Figure 26.16: ***Caesalpinea bonduc***

Figure 26.17:*Derris trifoliate*

Figure 26.18:*Acrostichum aureum*

Figure 26.19:*Dalbergia spinosa*

Figure 26.20: ***Derris scandens***

Table 26.1: Common True Mangroves and Associates found in Kali Estuary

Sl.No	Family	Genus	Species	Type of Mangrove
1.	Rhizophoraceae	Rhizophora	*R. apiculata*	True Mangrove
			R. Mucronata	True Mangrove
		Bruguiera	*B. gymnorrhiza*	True Mangrove
			B. cylindrica	True Mangrove
		Kandelia	*K. candel*	True Mangrove
2.	Avicenniaceae	Avicennia	*A. officinalis*	True Mangrove
			A. alba	True Mangrove
			A. marina	True Mangrove
3.	Sonneratiaceae	Sonneratia	*S. caseolaris*	True Mangrove
			S. alba	True Mangrove
4.	Combretaceae	Lumnitzera	*L. racemosa*	True Mangrove
5.	Euphorbiaceae	Exoecaria	*E. agallocha*	True Mangrove
6.	Myrsinaceae	Aegiceras	*A. corniculatum*	True Mangrove
7.	Acanthaceae	Acanthus	*A. ilicifloius*	Associate Mangrove
8.	Pteridaceae	Acrostichum	*A. aureum*	Associate Mangrove
9.	Fabaceae	Caesalpinea	*C. crista*	Associate Mangrove
			C. bonduc	Associate Mangrove
		Dalbergia	*D. spinosa*	Associate Mangrove
		Derris	*D. trifoliata*	Associate Mangrove
			D. scandens	Associate Mangrove

Discussion

Mangrove taxonomy needs to be paid grater attention, as the same has not been worked out in detail despite considerable number of publications. For example, it was pointed out that *Xylocarpus grantum* Koenig and *X. makongensis* (Prains) pierre can only be distinguished in situ but not from herbarium specimens (Raju, 2003). Also, ecological varieties of species like *Avicennia marina* (Forssk) Vierh. need to be clearly distinguished. Further, natural hybrids and their parental species need to be properly understood, especially in the case of *Rhizophora* species. The quandary between true mangroves species and associate species too needs to be resolved. Some authors consider *Acanthus ilicifolius* as an eumangroves while others don't. Discontinuous distribution of species like *Scyphiphora hydrophyllacea Gaertn.* (Raju, 1990) in distant geographical entities like Coringa mangroves of Andra Pradesh and Andaman islands of Bay of Bengal and *Aegialitis rotundifolia* Roxb.(Naithani *et al.*, 2004) In Krishna delta only are other interesting phenomena to be comprehended. Also more attention should be given to identify the mangrove wetlands and conserve the dwindling and fragile mangrove flora of the Kali estuary.

Mangrove research in Karnataka has been so far been concentrated in the major basins of Kadvad, Kinnar, Ulage, Devbagh and Hankon along the Kali River. (Anand Rao, 1989) In fact, mangroves are present at several other areas all along the coast. Considerable mangrove vegetation does exist along the study area of Siddar, Madhevada, Kache, Karge Jug and Ambe jug. *Avicennia marina* was observed to be a dominant species at all most all places of survey in Kali estuary along with it, *Exoecaria agallocha* L. was equally distributed. *Acanthus ilicifolius* L. the spiny mangrove was found forming extensive thickets on the banks of many study sites. The mud flats along the estuary were also seen harboring small bushes of *Avecinnia* and *Lumnitzera* species in a number of places. Therefore, actual areas of mangroves distribution cited in all most all earlier works as well as total extent of mangrove cover in the Kali estuary are subjected to modification.

As regards mangroves in other places, five mangrove species were reported from the Kinnar estuarine complex at Kali estuary in Karwar district. (Anand Rao, 1989) In the flora of Kodibag study site, it was stated that the "sea coast was completely devoid of mangrove forest as the estuarine region does not develop typical muddy flats essentials for mangrove species". Also, no information was found about the mangroves of these two study stations at Kali estuary. But intensive survey conducted by the authors revealed the presence of several mangrove wetlands in Kali estuary. Further present studies revealed in the occurrence of 13 eumangroves in Kali estuary instead as against a meager less reported earlier. (Anand Rao, 1989) Thus, these observations underline the importance of extensive exploration all along the Kali estuary of Karwar district.

Threats and Issues Affecting Mangroves

At present, mangroves like many other key ecosystems are under severe anthropogenic as well as ecological pressures. Therefore, they continue to diminish in size and composition that may ultimately threaten the very existence of these unique and important biological formations. In order to overcome this plight, conservation and management of these magnificent ecosystems is highly essential. Major threats affecting the mangrove ecosystems, especially in Kali estuary are identified as:

1. Indiscriminate felling for fire wood
2. Overgrazing by livestock
3. Conversion for agriculture, aquaculture and salt pans
4. Reclamation for human habitation and industrial enterprise

5. Bioinvasion of exotic species of plants
6. Sewage and effluent discharges of varied genesis
7. Innocence and ignorance of local populace on the importance and role of mangroves

In addition, due to frequent monsoon failure the much-needed silt inflow to mangrove niches from major rivers diverted for irrigation by way of dam construction. Further, the latest national issue of " Interlinking all major rivers in the country" to ensure a dependable agriculture calendar while enhancing the irrigable land will drastically affect the coastal ecosystems as a whole and mangroves in particular. During such an extreme exercise, unless euthenics is taken proper care of and appropriate caution exercised to secure the due share of mangroves, many wet lands all over the nation would perish, in all probability. Mangroves have often been cleared for the inhabitation of ever exploding population besides industrial development especially in coastal mega cities like Kolkatta and Mumbai. Smaller cities follow suit without any forethought and jurisprudence. Even in a moderate city like Karwar, which is blessed with a considerable stretch of lush mangroves along the Kali estuary amidst the city, any concern is hardly attested to shield them. Many plants in this lovely, lull and luxuriant mangrove cover were indiscriminately laid down for laying a road recently. Due to resultant cut off of water circulation, a major patch of these mangroves faces serious threat. The fate of adjoining mangrove area is also under stake due to ongoing development of an international airport in the vicinity. Similarly, a major mangrove cover at Karwar region is going to be sacrificed for the promotion of a sea port a. Much the same way, an extensive stretch of mangroves in the Karwar and Kali estuary- region is under consideration due to ensuing naval establishment.

Remedial Measures

In order to up keep the existing status of mangroves cover in the Kali estuary and further improve it, the action cores that requires to be sincerely attempted and strictly followed is:

1. Curtailing of human settlement in an around the mangrove wet lands.
2. Providing alternates to fire wood and other timber obtained from mangroves.
3. Banning live stock from entering into mangrove forest.
4. Arresting conversion of wet lands for agriculture, aquaculture and salt pans.
5. Secluding all mangrove areas from habitation and industrial enterprise.
6. Excluding mangrove wet lands while developing ports, Harbors naval establishments etc. By transforming prevailing areas into mangrove sanctuaries, park and landscape.
7. Eradicating establishment and growth of *Prosopis chilensis* (Molina) stuntz, *Parthenium hysterophorus* L. and other exotic plants into mangrove reserves.
8. Directing away of sewage and effluents discharges from mangrove sites.
9. Creating awareness among local residents about the numerous tangible and intangible benefits bestowed by the mangroves on mankind.

In addition, mangrove wet lands that are still under revenue custody without ant attention for improvement should be transferred immediately to the forest department for better care, nurture and propagation. Thorough participation and committed involvement should be generated among adjoining inhabitance, industries, establishments and entrepreneurs for proper conservation and sustainable development of mangroves in each place. Skills of administrative machinery should periodically be geared up to rise to the occasion in planning and implementing various community participation programmes and silvicultur practices. Degraded wet lands, cyclone prone estuarine regions and allied suitable areas should be afforested on a war footing for improving mangrove cover as buffer zone so that all accruing benefits are reaped forever from these ecosystems in the hinterlands.

Conclusion

The entire Karwar coast should be thoroughly explored and evaluated for the exact status, distribution and diversity of mangroves and their wetlands giving due importance to taxonomy and ecology. All converted, degraded and encroached wetlands should be recognized and steps to afforest the areas be worked out in detail and implemented. Special measures to ensure proper conservation, preservations and sustainable development of mangroves in the state should be adopted. Local communities should be involved in promoting all such steps along with government machinery. In order to perpetuate these activities, needed resources may be augmented through development of ecotourism spots and introduction of "mangrove cess" on brackish water entrepreneurs, salt manufactures and coastal industries and around mangrove vicinities.

References

Kathiresan, K., 2004a. Issue of mangrove conservation in India. In: *Proc. National Workshop on Conservation, Restoration and Sustainable Management of Mangrove Forests in India*, Vishakhapattanam, p. 1–17.

Kathiresan, K., 2004b. Biodiversity in mangrove ecosystems of India: Status, challenges and strategies. *ENVIS Forestry Bulletin*, 4: 11–23.

Naithani, H.B. *et al.*, 2004. Vegetation analysis of mangrove forest of Krishna wildlife santury, Andra Pradesh, India. *Indian Forester*, 130: 841–857.

Naskar, K.R. and Mandal, R., 1999. *Ecology and Biodiversity of Indian Mangroves*, Vols. 1 and 2. Daya Publishing House, New Delhi.

Nayak, V.N. and Bhandari, J.M., 2010. *A Field Guide to the Mangroves of Kali Estuary, Uttara Kannada.* Karnataka Forest Department.

Raju, J.S.S.N., 1990. *Scyphiphora hydrophyllacea* Gaertn (Rubiaceae): A rare and interesting mangrove taxon in peninsular India. *Journal of Indian Botanical Society*, 69: 207–208.

Raju, J.S.S.N., 2003. Xylocarpus (Melivaceae): A less-known mangrove taxon of the Godavari estuary, India. *Current Science*, 87(7): 199–209.

Rao, T.A. *et al.*, 2001a. Costal ecosystems of the Karnataka state, India. I. Mangroves Karnataka association for the advancement of science, Central College, Bangalore, pp. 319.

Rao, T.A. *et al.*, 2002b. Costal ecosystems of the Karnataka state, India. II. Beaches. Karnataka association for the advancement of science, Central College, Bangalore, pp. 192.

Selvam, V. *et al.*, 2006. Toolkit for establishing coastal Bioshield. MS Swaminathan Research Foundation, Chennai.

Untawale, A.G., 2004. Mangrove diversity and conservation needs for India. In: *Proceedings of National Workshop on Conservation, Restoration and Sustainable Management of Mangrove Forest in India*, Vishakapattanam, p. 1–13.

Chapter 27

Assessment of Water Quality of Byramangala Lake in Bangalore City

☆ *H. Krishnaram, M. Ramachandra Mohan Shivabasavaiah and Rmanjnath*

ABSTRACT

Aquatic ecosystem is one of the most productive ecosystems providing many critical services to humans, such as plants carry out photosynthesis and produce the oxygen; and bacteria process the organic waste products, and maintain good water quality. The objective of this endeavor was to investigate the ecological status of the Byramangala lake, the physico-chemical analysis of water shows with high values of pH, conductivity, TDS, nitrate, phosphates, BOD, COD, and so on. The nature of the lake is polluted, which is mainly due to the domestic wastes combined with sewage, which causes organic pollution because of higher concentration of nitrates and phosphates. This encourages the growth of obnoxious weeds, such as water hyacinths. These floating weeds prevent penetration of sunlight, which is essential for all life forms. This adversely affects aquatic life and upsets, the delicate the equilibrium of the wetland ecosystem and concentration of chemical constituents in partially dried up aquatic environment, which are the causes for the deterioration of water quality of lake.

Introduction

During the past centuries, increased urbanization and sewage disposal, regulation of wetlands and streams and more intensive farming practices have increased the nutrient loading to many shallow lakes world-wide, not least in the industrialized part of the world. This has resulted in major changes in the biological structure and dynamics of the lakes and often in a shift from a clear to a turbid state. Lakes are transitional zones between land and water, a collective term for marshes, swamps, bogs and similar areas. They have been described as the 'kidneys' of the landscape as they filter sediments and nutrients from surface water. Lakes are often referred to as biological supermarkets because they support all life forms through extensive food webs and biodiversity (Mitch and Gosselink, 1993). They help regulate water levels with in watersheds, improve water quality, reduce flood and storm damages, provide habitat for important fish and wild life, support hunting, fishing, other recreation activities and perform some useful functions in the maintenance of ecological balance.

Lakes have played a major role in the history of Bangalore serving as an important drinking and irrigation source. They occupy about 4.8 per cent of the city's geographical area (640 sq. km) covering both urban and non-urban areas. Bangalore has many man made wetlands but has no natural wetlands. They were built for various hydrological purposes and mainly to serve the needs of irrigated agriculture. The spatial mapping of water bodies in the district revealed the number of water bodies to have decreased from 379 (138 in north and 241 in south) in 1973 to 246 (96 north and 150 in south) in 1996 and 81 at present. The overall decrease of 35 per cent was attributed to urbanization and industrialization (Deepa *et al.*, 1998).

The ecology of lakes are under stress condition due to the fast pace of development, deforestation, cultural practices and agriculture. These activities trigger the rate of sedimentation of the lakes bed characterized by silt and organic suspended solids, which initiate the process of eutrophication at a very early stage and show a deterioration of habitat quality. It has become necessary to pay proper attention to find out the extent of possibilities of impounded water for raising the fishery wealth. Productivity of the lakes is greatly influenced by its morphometric and hydrological features. Dense human population in catchments, urbanization and varies anthropogenic activates has resulted in overexploitation of wetland resources, leading to degradation in their quality and quantity. New there is increasing concern to conserve and restore perishing lakes and endangered habitats to achieve ecological sustainability. This highlights the need for appropriate conservation, restoration and management measures. The aim of this paper was to assess the status of Byramangala Lake by analyzing the physico-chemical characterization of the lakes in the Bangalore city.

Materials and Methods

Background

The Byramangala lake, 12° 45′ 33.6″ NL and 077° 25′ 05-0″ E L is created on the river Vrishabhavati, 38 km away from the city of Bangalore. The 437 ha manmade lake has been receiving a steady inflow of treated sewage from the city and wastes from many industrial establishments for the past two decades. It is located near Bidadi industrial area, Ramanagaram district. The lake is highly polluted in nature.

Climatic Conditions

The climatic factors taken into consideration during present investigation are atmospheric temperature and rainfall. The atmospheric temperature was recorded with the help of thermometer and rainfall was measured by a rain gauge.

Atmospheric Temperature

The seasonal variation in mean atmospheric temperature. The results revealed that the mean monthly atmospheric temperature ranges from 26.0° to 36.3°C (recorded 2003), the highest temperature 36.3°C was recorded during May and the lowest temperature 26.0°C was recorded during August. In 2004 atmospheric temperature ranges from 26.0°C to 36.0°C. The highest temperature 36.0°C was recorded during May and the lowest temperature 26.01°C was recorded in July. In 2005, atmospheric temperature ranges from 25.65ºC to 35.3ºC with the highest temperature 35.3ºC recorded during May and the lowest 25.65ºC recorded during November. The results revealed that the highest atmospheric temperature was recorded during 2003 with a gradual decrease in 2004 and 2005 (Figure 27.1).

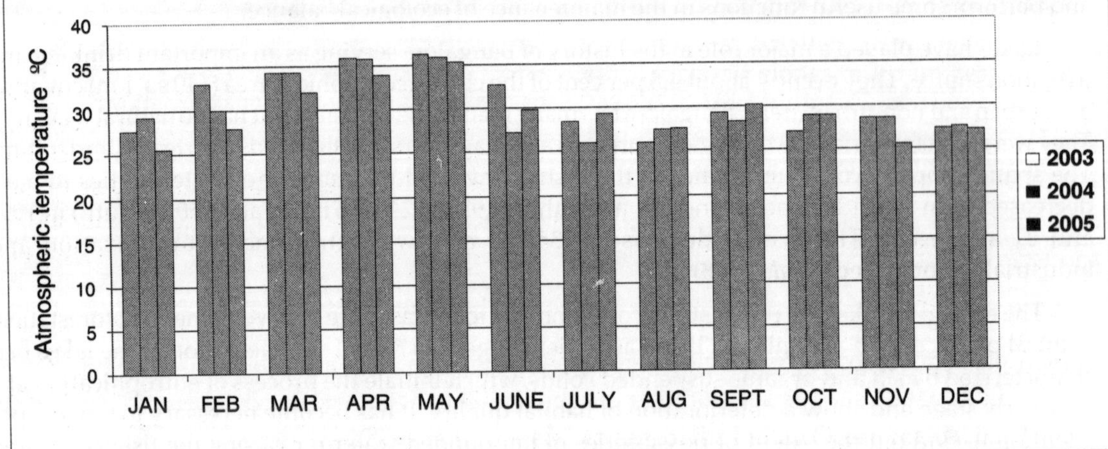

Figure 27.1: Atmospheric Temperature at Study Sites during 2003, 2004 and 2005

Rainfall

Rainfall plays an important role in the regularity of the biological activities of water bodies, the lakes being predominantly rainfed. In the Byramangala R.G. the highest precipitation (196.4 mm) was recorded during the month of October, and not recorded in the months of January, February and December of 2003. In 2004 the highest precipitation (200.6 mm) was recorded during the month of September, and not recorded in the months of January, February and December. In 2005, highest precipitation (262.6 mm) was in the month of October and not recorded in the months of January and February (Figure 27.2).

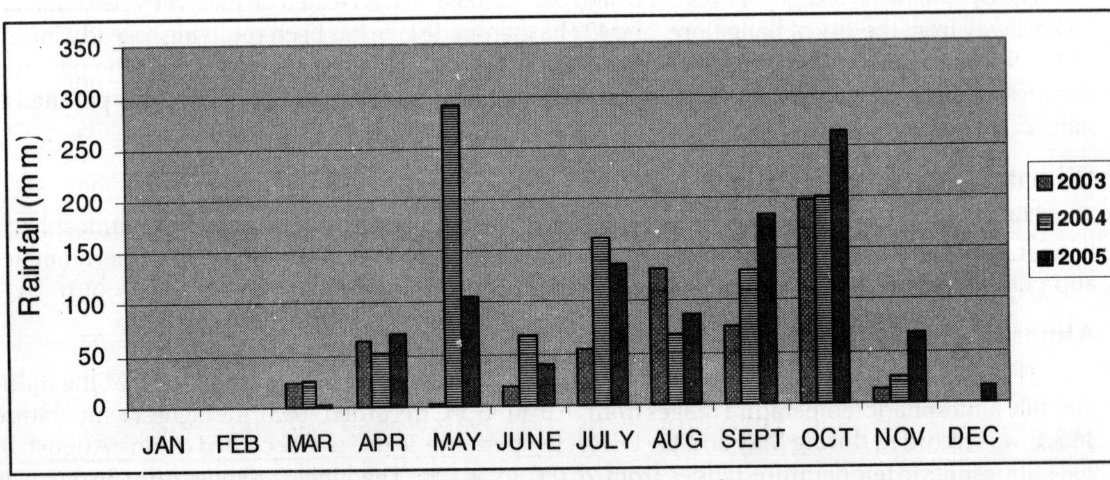

Figure 27.2: Rainfall in 2003, 2004 and 2005 at Brramangala RG

Sampling

Water samples from the lakes water from (Byramangala lake) were collected from January 2003 to December 2005. At each collection time, the analysis on water samples was performed twice. All the collections and field observations were made between 07.00 to 11.00 hours throughout the study period. The water samples were collected from the lakes in bottles and brought to the laboratory for analysis of parameters were studied as per the standard methods described by APHA (1995), Trivedy and Goel (1986) and Mathur (1999).

Results and Discussion

The results of the physico-chemical analysis of water samples from, Byramangala lake are given in Tables 27.1–27.3.

pH

pH is an important parameter in measuring water quality as every aspect of water, such as acid and base neutralization, water softening, precipitation, coagulation, acid disinfections is pH dependent. In the samples of Byramangala lake exhibited during 2003, high pH value were recorded as 7.9 in April and a lower pH was recorded as 7.0 in January. In the 2004, the high pH was recorded as 8.4 in March and lower pH was recorded as 7.3 in January, July and October. In 2005, highest pH was recorded as 8.3 in May and a lower pH was recorded as 7.6 in February. The pH values were mostly within desirable limits prescribed by ICMR (WHO). A pH range of 6.7 and 8.4 is considered to be safe for aquatic life and to maintain productivity. However, pH below 4.0 and above 9.6 is hazardous to most life forms. pH gives an idea to the type and intensity of pollution (Verma *et al.*, 1987; Mishra and Seksena, 1991) and it is also considered a very important single factor that influences aquatic production (Swingle, 1967).

Table 27.1: Monthly Variation of Different Physico-chemical Parameters of Byramangala Lake-2003

Parameters	*Units*	*Jan*	*Feb*	*Mar*	*Apr*	*May*	*Jun*	*July*	*Aug*	*Sept*	*Oct*	*Nov*	*Dec*
pH	–	7.0	7.2	7.8	7.9	7.8	7.6	7.4	7.5	7.3	7.4	7.6	7.5
Temperature	°C	23.3	24.0	28.0	27.0	29.0	24.0	26.0	28.5	25.8	24.0	25.2	23.8
TDS	mg/L	500.	540	580	630	650	650	780	760	700	580	650	670
Turbidity	FAU	36	32	24	28	29	21	25	26	30	22	30	32
Conductivity	μmhos/cm	1238	1278	1083	1183	1133	1316	1333	1300	1278	1238	1100	1008
Total Hardness	mg/L	310	315	390	370	385	330	350	370	338	280	315	310
Alkalinity	mg/L	440	468	486	520	457	467	434	456	459	450	460	434
DO	mg/L	3.6	3.7	2.3	2.2	2.0	3.6	3.7	3.5	3.2	3.0	2.6	3.0
BOD	mg/L	8.9	9.1	11.2	10.8	11.3	10.0	9.8	10.0	10.9	9.8	9.7	10.0
COD	mg/L	59	58	70	67	61	70	64	62	62	61	65	60
Phosphates	mg/L	9.8	9.6	10.8	9.8	9.0	12.6	12.2	12.7	12.3	11.3	12.0	11.0
Sulfates	mg/L	53.0	53.4	54.0	55.4	53.0	60.0	57.0	57.0	55.4	54.0	53.4	53.0
Nitrates	mg/L	1.8	2.0	1.6	1.3	1.8	2.0	2.1	2.8	3.0	2.0	2.1	1.3
Potassium	mg/L	17.0	18.3	17.0	18.3	17.3	19.1	18.5	18.0	18.6	18.3	16.3	18.0

Temperature

Temperature is an important parameter required to get an idea of self–purification of lakes and plays an important role in aquatic ecosystem. In Byramangala lake during 2003, high temperature value was recorded as 29.0°C in May and lower temperature was recorded as 23.3°C in January. During the year 2004, the high temperature was recorded as 27.0°C in May and lower temperature was recorded as 23.0°C in January. In 2005, high temperature was recorded as 29.0°C in April and lower temperature was recorded as 23.0°C in February and November. Increased temperature not only reduces oxygen availability, but also increases oxygen demand, a situation that would add to physiological stress of organisms (Giller and Matmquist, 1998).

Table 27.2: Monthly Variation of Different Physico-chemical Parameters of Byramangala Lake–2004

Parameters	Units	Jan	Feb	Mar	Apr	May	Jun	July	Aug	Sept	Oct	Nov	Dec
pH	–	7.3	7.5	8.4	8.3	8.0	7.8	7.3	7.5	8.0	7.3	7.8	8.1
Temperature	°C	23.0	24.3	27.0	26.8	27.0	24.9	24.0	25.1	24.2	24.7	25.0	25.1
TDS	mg/L	780	810	792	780	760	980	987	910	1000	792	880	805
Turbidity	FAU	31	33	28	29	28	25	21	27	26	27	29	30
Conductivity	µmhos/cm	1238	1316	1188	1316	1230	1516	1320	1420	1645	1300	1342	1332
Total Hardness	mg/L	227	176	344	360	376	339	327	242	278	178	162	180
Alkalinity	mg/L	464	928	860	816	498	234	478	418	424	464	242	268
DO	mg/L	4.8	3.6	4.5	3.2	3.3	3.4	2.9	4.0	3.7	3.8	3.9	4.6
BOD	mg/L	29.1	32.5	37.0	46.0	53.0	34.0	33.0	39.0	32.5	37.0	28.5	32.5
COD	mg/L	31.0	37.0	48.0	52.0	64.0	51.0	49.0	38.0	42.0	39.5	37.0	42.0
Phosphates	mg/L	9.9	6.5	8.5	8.3	6.5	10.3	10.6	11.5	10.3	10.0	8.5	9.0
Sulfates	mg/L	17.7	17.5	18.0	18.6	19.8	24.3	26.2	20.0	23.0	20.0	17.0	19.3
Nitrates	mg/L	3.9	4.0	3.6	3.9	3.4	5.5	5.2	5.7	6.9	3.8	3.0	3.6
Potassium	mg/L	18.1	16.4	21.5	23.0	19.0	18.9	20.3	22.3	19.4	19.8	19.0	20.2

Total Dissolved Solids (TDS)

Total dissolved solids are the infilterable solids that remain as residue upon evaporation and subsequent drying at defined temperature. It gives the measure of ions dissolved in the water. In Byramangala lake during 2003, high TDS value was recorded as 780 mg/l in July and lower TDS was recorded as 500 mg/l in January. In the year 2004, the high TDS was recorded as 1000 mg/l in September and lower TDS was recorded as 760 mg/l in May. In 2005, high TDS was recorded as 920 mg/l in August and lower TDS was recorded as 748 mg/l in December. A certain level of these ions in water is necessary for aquatic life; changes in TDS concentration can be harmful because of entry and exit of water into and out of an organism's cell (Mitchell and Stapp, 1992).

Turbidity

Turbidity is suspension of particles such as clay, silt and organic matter in water interfering with the passage of light. It is a measure of the cloudiness of water. In Byramangala lake during 2003, high turbidity value was recorded as 36.0 FAU in January and lower turbidity was recorded as 21.0 FAU in June. In 2004, the high turbidity was recorded as 33.0 FAU in February and lower turbidity was recorded as 21.0 FAU in July. In 2005, high turbidity was recorded as 34.0 FAU in November and lower turbidity was recorded as 26.0 FAU in March. All values recorded above fall within the permissible limit prescribed by ICMR. Turbidity has been considered as a limiting factor for biological productivity in freshwaters (Kaushik and Saxena, 1999). The sources are mainly the storm water, agricultural runoff and effluents from industrial and domestic sectors, which in turn restrict the penetration of light giving rise to reduced photosynthesis and aesthetically unsatisfactory odours (Kiran and Ramachandra, 1999).

Table 27.2: Monthly Variation of Different Physico-chemical Parameters of Byramangala Lake–2006

Parameters	*Units*	*Jan*	*Feb*	*Mar*	*Apr*	*May*	*Jun*	*July*	*Aug*	*Sept*	*Oct*	*Nov*	*Dec*
pH	–	7.8	7.6	8.0	8.2	8.3	8.0	7.9	7.9	8.0	7.7	7.8	7.7
Temperature	°C	23.3	23.0	28.0	29.0	30.0	28.0	27.2	26.0	25.2	24.0	23.0	23.4
TDS	mg/L	755	748.00	760	782	783	753	820	920	880	750	753	748
Turbidity	FAU	31	32	26	27	28	28	29	30	27	33	34	32
Conductivity	μmhos/cm	1263	1216	1238	1200	1226	1360	1420	1380	1460	1310	1260	1200
Total Hardness	mg/L	373	360	530	475	460	400	360	310	303	268	192	220
Alkalinity	mg/L	704	690	670	774	810	760	780	810	780	560	670	610
DO	mg/L	4.0	3.8	2.6	2.7	2.3	2.8	3.2	3.8	2.7	3.0	3.0	3.4
BOD	mg/L	11.3	12.3	20.8	21.3	28.7	17.0	18.2	18.0	19.0	12.0	17.0	19.2
COD	mg/L	23.0	28.0	61.0	63.0	58.0	48.0	49.0	50.0	52.0	38.0	42.0	40.0
Phosphates	mg/L	6.8	7.2	6.0	5.8	5.9	10.5	13.0	14.0	14.5	8.2	8.0	7.8
Sulfates	mg/L	26.0	23.0	22.0	23.0	24.0	38.0	32.0	40.0	29.0	27.0	26.0	25.0
Nitrates	mg/L	4.2	4.8	3.2	2.8	2.0	5.7	6.8	5.0	5.2	4.2	4.0	4.4
Potassium	mg/L	18.2	17.8	16.0	17.0	16.0	19.0	23.0	24.0	21.0	20.0	18.0	19.0

Conductivity

Conductivity is a numerical expression of the ability of an aqueous solution to carry electric current. This ability depends on the presence of ions, their total concentration, mobility, valence and relative concentration. In Byramangala lake during 2003, high conductivity value was recorded as 1333.0 μmhos/cm in July and lower conductivity was recorded as 1008.0 μmhos/cm in December. In the 2004, high conductivity was recorded as 1645.0 μmhos/cm in September and lower conductivity was recorded as 1188.0 μmhos/cm in March. In 2005, high conductivity was recorded as 1460.0 μmhos/cm in September and lower conductivity was recorded as 1200.0 μmhos/cm in April and December. Conductivity increases with increasing amount and mobility of ions and also can be used as an indicator of water pollution. Higher values of dissolved solids, greater the amount of ions in water (Bhatt *et al.*, 1999).

Total Hardness (TH)

Hardness is a measure of polyvalent cations in water; it is generally represents the concentration of calcium and magnesium ions, because these are the most common polyvalent cations. In Byramangala lake during 2003, high TH was recorded as 385.0 mg/l in May and lower TH was recorded as 280.0 mg/l in October. In the year 2004, high TH was recorded as 376.0 mg/l in May and lower TH was recorded as 162.0 mg/l in November. In 2005, high TH was recorded as 530.0 mg/l in March and lower TH was recorded as 192.0 mg/l in November. Water is classified according to its hardness mg/l, as Hem (1970), Water with a TH in the range of 0 to 60 mg/l are termed soft; 60-120 mg/l moderately hard; from 120 to 180 mg/l hard and above 180 mg/l very hard. High values of TH are probably due to the regular addition of large quantities of sewage, detergents and large-scale human use in lake.

Alkalinity

Alkalinity of water is its acid neutralizing capacity. Alkalinity of surface water is primarily a function of carbonate and hydroxide content. In Byramangala lake during 2003, high alkalinity value was recorded as 520.0 mg/l in April and lower alkalinity was recorded as 434.0 mg/l in July and December and lower alkalinity was recorded as 234.0 mg/l in June. In 2005, high alkalinity was recorded as 810.0 mg/l in August and lower alkalinity was recorded as 560.0 mg/l in October. In the 2004, high alkalinity was recorded as 928.0 mg/l in February. The variation in total alkalinity may be due to the seasonal effect, planktonic population, and bottom deposits. According to Kaur *et al.* (1996), high alkalinity values are indicative of the eutrophic nature of the water body. The conditions of the lake when tallied with the higher values of total alkalinity point towards the polluted condition of lakes. The increase of alkalinity content might be due to the fact that the accidental mixing of amount of industrial substances in low water quality and high evaporation rates.

Dissolved Oxygen (DO)

Dissolved oxygen is one of the important parameter in water quality assessment, its presence is essential to maintain to variety of forms of biological life in water. The oxygen balance of the system largely determines the effects of waste discharges in a water bodies. In Byramangala lake during 2003, high dissolved oxygen value was recorded as 3.7 mg/l in July and lower dissolved oxygen was recorded as 2.0 mg/l in May. In the year 2004, high dissolved oxygen was recorded as 4.8 mg/l in January and lower dissolved oxygen was recorded as 2.9 mg/l in July and October. In 2005, high dissolved oxygen was recorded as 4.0 mg/l in January and lower dissolved oxygen was recorded as 2.3 mg/l in May. This change may be due to fluctuation in temperature. Increase in dissolved oxygen is related to decrease in temperature or decrease in dissolved oxygen is possibly because of higher temperature as solubility of O_2 decreases with increase in temperature. The highest D.O. recorded at during monsoon and post-monsoon months may be due to the impact of rainwater resulting in aeration. Hannan (1979), Mahadevan and Krishna Murthy (1983), Jebanesan *et al.* (1987), Abbasi *et al.* (1997) and Mathew Koshy and Nayar (1999).

Biological Oxygen Demand (BOD)

Organic matter present in water, utilize the dissolved oxygen of the water for its decomposition, depletes the oxygen and makes difficult for biota to live in heavily loaded water bodies. In Byramangala lake during 2003, high biological oxygen demand value was recorded as 11.2 mg/l in March and lower biological oxygen demand was recorded as 8.9 mg/l in January. In the year 2004, the high biological oxygen demand was recorded as 53.0 mg/l in May and lower biological oxygen demand was recorded as 28.5 mg/l in November. In 2005, high biological oxygen demand was recorded as 28.7 mg/l in May and lower biological oxygen demand was recorded as 11.3 mg/l in January. BOD, a relative oxygen demand is the amount of oxygen required for the biochemical degradation of organic material present in the water and the oxygen used to oxidize in organic material such as sulfides and ferrous ions (APHA 1985).

Chemical Oxygen Demand (COD)

Chemical oxygen demand measures the oxygen equivalent of the organic and inorganic matter in a water sample that is susceptible to oxidation. In Byramangala lake during 2003, high chemical oxygen demand value was recorded as 70.0 mg/l in March and June and lower chemical oxygen demand was recorded as 58.0 mg/l in February. In the year 2004, high chemical oxygen demand was recorded as 64.0 mg/l in May and lower chemical oxygen demand was recorded as 31.0 mg/l in January. In 2005, high chemical oxygen demand was recorded as 63.0 mg/l in April and lower chemical oxygen demand was recorded as 23.0 mg/l in January. As a result the high value of COD values higher even then a continuous monitoring is necessary so that the direct disposal of organic matter into the lakes should be strictly prohibited. Pollution that is largely determined by the various organic and inorganic materials (Kiran and Ramachandra, 1999).

Phosphate

Phosphate is a nutrient required by all organisms for the basic process of life. In Byramangala lake during 2003, high phosphate value was recorded as 12.7 mg/l in August and lower phosphate was recorded as 9.0 mg/l in May. In the year 2004, the high phosphate was recorded as 11.50 mg/l in August and lower phosphate was recorded as 6.50 mg/l in February, May and October. In 2005, high phosphate was recorded as 14.5 mg/l in September and lower phosphate was recorded as 5.8 mg/l in April. Phosphate emphasized that weathering of phosphorus bearing rocks, leaching of the soils of the catchments area by rain, cattle dung and might soils are the main sources of phosphorus to natural waters (Jhingran 1982). The classification proposed on the basis of phosphorus content, the Byramangala lake under eutrophic condition. Verma (1981) and Paul (1992) established the importance of phosphate as a nutrient element in the growth of microscopic algae and is responsible for the maintains of the lakes productivity.

Sulfate

Sulfate is ions usually occur in natural waters. It contribute to the total solids content and in a reduce and anaerobic condition. In Byramangala lake during 2003, high sulfate value was recorded as 60.0 mg/l in June and lower sulfate was recorded as 53.0 mg/l in January, May and December. In the year 2004, high sulfate was recorded as 26.0 mg/l in July and lower sulfate was recorded as 17.0 mg/l in November. In 2005, high sulfate was recorded as 40.0 mg/l in August and lower sulfate was recorded as 22.0 mg/l in March. The high concentrations of sulfates have been reported from the lake waters and high concentration of sulfates stimulus the action of sulphur reducing bacteria that produced hydrogen sulphide gas highly toxic to fish life. Higher concentrations of sulfates were observed and could be attributed due to run off from the agriculture land during flood in the monsoon season and sulfate enters into the lake water body from the catchments area through surface run-off and domestic wastewater.

Nitrates

Nitrates is the most highly oxidized from of nitrogen compounds commonly present in natural waters, because it is a product of aerobic decomposition of organic nitrogenous matter. In Byramangala lake during 2003, high Nitrates value was recorded as 3.0 mg/l in September and lower nitrates were recorded as 1.3 mg/l in April. In the year 2004, a high nitrate was recorded as 6.9 mg/l in September and a lower nitrate was recorded as 3.0 mg/l in November and October. In 2005, high nitrates were recorded as 6.8 mg/l in July and lower nitrates were recorded as 2.0 mg/l in May. All the values with in the desirable level of (45 mg/l.) prescribed by ISI and 20 mg/l. (ICMR.) The higher concentration of nitrate is an indicator of organic pollution and eutrophication. Wetzel (1983) stated that nitrate was generated by heterotrophic microbes as a primary end product of decomposition of oxygenic matter either directly from protein or organic compound. Zutshi and Khan (1998) stated that presence of excessive nitrate in water is due to man made domestic activities and fertilizers from fields.

Potassium

Potassium is main cation in natural freshwater; it is an important macro-nutrient and plays a vital role in metabolism of fish. In Byramangala lake during 2003, high potassium value was recorded as 18.5 mg/l in July and lower potassium was recorded as 16.32 mg/l in November. In the year 2004, high potassium was recorded as 23.0 mg/l in April and lower potassium was recorded as 18.10 mg/l in January. In 2005, high potassium was recorded as 24.0 mg/l in August and lower potassium was recorded as 16.0 mg/l in March. Highest value of potassium was observed in monsoon months. The rainfall and sedimentary rock strata form almost the entire source of potassium in water body.

Conclusion

Lake ecosystems are interconnected and interactive with in a watershed. In Bangalore the environmental pressure of unplanned urbanization and growing population has taken its toll of lakes. The revealed about 35 per cent decline in the number and loss in the interconnectivity among lakes disrupting the drainage work and the hydrological regime leading to irreversible (some times) changes in wet land quality. The exploration survey and physico-chemical characterization of lakes located all over the city show that lake are polluted mainly due to sewage from domestic and industrial sectors. Detailed quantitative investigation of Byramangala lake involving physical and chemical parameters and statistical analysis of has higher degree of pollution. Any restoration efforts for lake must address the interconnected nature of these sources contaminating the lake.

References

Abbasi, S.A., Abbasi, Naseema and Bhatia, K.K.S., 1997. The Kuttiadi river basin. In: *Wetlands of India: Ecology and Threats*, Vol. 3. Discovery Publishing House, New Delhi, p. 65–143.

APHA (American Public Health Association), 1985. *Standard Methods for Examination of Water and Wastewater*. American Public Health Association, Washington, New York.

APHA (American Public Health Association), 1995. *Standard Methods for the Examinstion of Water and Wastewater*, 19th edn. American Water Works Association, Water Environment Federation, Washington D.C.

Bhatt, L.R., Lacoul, P., Lekhak, H.D. and Jha, P.K., 1999. Physico-chemical characteristics and phytoplanktons of Taudaha lake, Kathmandu. *Poll. Res.*, 18(4): 353–358.

Deepa, R.S., Ramachandra, T.V. and Kiran, R., 1998. Anthropogenic stress on wetlands of Bangalore. In: *Proceeding of the National Seminar on Environmental Pollution: Causes and Remedies*. PES Institute of Technology, Bangalore, pp. 166–182.

Giller, Paul and Matmqvist, Bjorn, 1998. *The Biology of Streams and Rivers*. Oxford University Press.

Hannan, H.H., 1979. Chemical modification in reservoir regulated streams. In: *The Ecology of Regulated Streams*, (Eds.) J.W. Ward and J.A. Standford. Plenam Corporation Publication, p. 75–94.

Jebanesan, A., Selvanayagam, M. and Thatlegur, A., 1987. Distributory pattern of dissolved oxygen in the selected stations of Cooam river and its effects on the aquatic fauna Proc-Symp. *Environ. Biol.*, p. 303–311.

Jhingran, V.G., 1982. *Fish and Fisheries of India*, 2nd edn. Hindustan Publishing Corporation (India), Delhi.

Kaur, H., Dhillon, S.S., Bath, K.S. and Mander, G., 1996. Abiotic and biotic components of freshwater ponds of Patiala (Punjab). *Poll. Res.*, 15(3): 253–256.

Kaushik, S. and Saxena, D.N., 1999. Physico-chemical limnology of certain water bodies of central India. In: *Freshwater Ecosystems of India*, (Ed.) K Vijaykumar. Daya Publishing House, New Delhi, p. 1–58.

Kiran, R. and Ramachandra, T.V., 1999. Status of wetlands in Bangalore and its conservation aspects. *ENVIS J. of Human Settlements*, p. 16–24.

Mahadevan, A. and Krishnaswamy, S., 1993. A quality profile of river vague. (Shindig). *Ind. J. Environ. Hlth.*, 25(4): 288–299.

Mathew Koshy and Vasudevan Nayar, J., 1999. Water quality aspects of river Pamba. *Poll. Res.*, 18(14): 501–510.

Mathur, R.P., 1999. *Water and Wastewater Testing: Laboratory Manual.* Nemchand and Brothers, Roorkee, India.

Mishra, S.R. and Saksena, D.N., 1991. Pollutional ecology with reference to physico-chemical characteristics of Morar (Kalpi) river, Gwalior, M.P.. *Current Trends in Limnology*, 1: 159–184.

Mitchell and Stapp, 1992. *Field Manual for Water Quality Monitoring*.

Paul, D.K., 1992. Limnological investigation of certain hill-streams of Godda district, Santhal Pargana. *Ph.D. Thesis*, Bhag. Univ.

Swingle, H.S., 1967. Standardization of chemical analysis for water and muds. *FAO. Fish. Rep.*, 4(4): 397–421.

Trivedy and Goel, P.K., 1986. *Chemical and Biological Method for Water Pollution Studies.* Env. Publication, Karad.

Verma, P.K., 1981. The Limnobiological survey of Badua Reservoir, Bhagalpur (Bihar). *Ph.D. Thesis*, Bhag. Univ.

Wetzel, R.G., 1983. *Limnology*, 2nd edn. Saunders. Coll. Publ., p. 267.

Zutshi, D.P. and Khan, A.V., 1998. Eutrophic gradient in the Dal lake, Kashmir. *Indian J. Env. Health*, 30(4): 348–354.

Chapter 28

Biological Index of Pollution in Ponds and Lakes of Mysore District, Karnataka

☆ *T.B. Mruthunjaya and S.P. Hosmani*

ABSTRACT

Biological index of pollution was calculated using the formula bB/A+B x 100 (A = producers, B = consumers) for the ponds and lakes. Kukkarahally lake fluctuates between septic zone to zone of active decomposition. Sewage oxidation and through out all the seasons full under the zone and moderate decomposition to septic zone. Dalvoi lake also structures between the zone of moderate decomposition and the septic zone.

Introduction

Some organisms may play a significant role in the intervention between the primary and secondary producers. Zooplankton do not engulf whole plants but enter into the plant body and devour the cell from within (Cantar and Lund, 1968). The work of Canter and Lund (1968) reveals that certain producers do not engulf whole plants but enter into the plant body and devour it cell by cell from within. The protozoans may add directly or indirectly to the silter seedling zooplankton and apart from this the undigested remains of the algae and their mucilage may support the growth of bacteria while the algal fragments may be small enough to be digested by the rotifers and crustaceans. These observations indicate that zooplankton have certain relationship to the phytoplankton. With a view of understanding such a relationship the present work was undertaken and a biological index of pollution was calculated. The present work is based on the data collected over a period of one year on the distributions of phytoplankton and zooplankton in sewage and ponds and freshwater bodies of Mysore.

Materials and Methods

Mysore city situated 12.50 N of equator in the northern subtropic region and 76.390 East. It lies in the semiarid region of the Indian subcontinent on the Deccan plateau at a height of about 2250 feet above ocean the sea level. It is an area of about 32 sq kms. The land scape is marked by Chamundi Hills. The river cauvery is in its northern side with its tributary Kabini on the southern side. Mysore mainly receives rains from both southwest and north-west; the south-west monsoon contributes to the bulk of the annual rainfall.

The seasonal distribution of phytoplankton and zooplankton in each water body was accounted. The collection, preservation and enumeration of the plankton forms has been given in details by Hosmani and Bharathi (1980a). for the purpose an index called Biological Index of Pollution (BIP) was obtained by using the formula BIP = B/A+B x 100 where A are producers (phytoplankton) and B are consumers (zooplankton). The results of these calculations and the details of the water bodies studied have been presented in Tables 28.1–28.3.

Table 28.2 represents the seasonal occurrence of phytoplankton and zooplankton in the ponds and lakes.

Table 28.1: Details of the Three Water Bodies Studied

Description	*Kukkarahally Lake*	*Dalvoi Lake*	*Sewage Oxidation Pond*
Area cubic meters	15 acres	10 acres	4 cubic meter
Maximum depth (meters)	10	08	10 feet
Minimum depth (meters)	06	06	10 feet
Inflow of water	Earlier received sewage (later stopped)	City sewage	City/campus sewage
Detention period of water	Until it reaches a maximum mark	Until it reaches a maximum mark	30 days (average)
Surface	Permanently covered with bloom of *M. aeruginosa*	Temporary blooms of chlorococcales	Solid/semisolid organic matter (often removed) (occasional appearance of Euglenaceae)
Usage	Recreation	Primary sewage treatment	Oxidation pond
Hydrophytes	Many	Nil	Nil
Type of pollutants	Domestic	Domestic	Domestic

Table 28.2: Seasonal Occurrence of Phytoplankton and Zooplankton in the Three Water Bodies (Total count of plankton)

Seasons		*Kukkarahally Lake*	*Dalvoi Lake*	*Sewage Oxidation Pond*
Monsoon	Phytoplankton	5000	8756	14654
	Zooplankton	158	110	359
Winter	Phytoplankton	5441	3046	15429
	Zooplankton	514	63	147
Summer	Phytoplankton	33794	24669	40994
	Zooplankton	179	274	490

Table 28.3: Biological Index of Pollution

Seasons	*Kukkarahally Lake*	*Dalvoi Lake*	*Sewage Oxidation Pond*
Monsoon	3	1	3
Winter	8	2	1
Summer	5.0	10	10.6

BIP values: Less than 1 = zone of clean water; 1 to 3 = zone of moderate decomposition; 3 to 6 = zone of active decomposition; 6 and above septic zone.

Results and Discussion

Kukkarahally lake supports larger number of cyanophyceae and are mainly represented by blooms and *Mocrocystis, Aerginosa* and *Spirulina nurdesteddti* while other from that occur in smaller numbers are *Euglena gracilis* and *Euglena elastica*. Rohdess among the zooplankton are quite abundant. Seasonal BIP values in this lake indicates that during the monsoon months the water is in a state of moderate decomposition. The active decomposition during summer due to coming waters from the surroundings into the lake. There is a greater human disturbance around the lake. Dalvoi lake is in a state of moderate decomposition. It is quite interesting to note that it supports a number of zooplanktons like rotifera, nemata, copepoda. During winter and summer when the disturbances around the lake are much less and when the level of the water goes down, it becomes a septic zone as evinced by the BIP values (5 10 1.6); but as monsoon approaches it becomes a zone of moderate decomposition, while the sewage oxidation and ranges between the zone of moderate decomposition to the septic zone. It must also be mentioned that a small amount of sewage is deposited into this lake which may be a cause for the high BIP values during summer the same being diluted during monsoon. It is much distributed and is exposed to extreme sunlight for a major part of the day.

It may be concluded that calculated BIP values indicate that, the season wise index shows that winter Kukkarahally lake tends to become a septic zone, probably due to less flow of water and more evaporation but during monsoon it behaves like a zone of moderate decomposition, but the consideration of the BIP values during summer indicated it has become a zone of active decomposition. The low values during summer may be due the fact that phytoplankton increases considerably. Similarly, Dalvai lake which receives city sewage also ranges from a zone of moderate decomposition to septic zone, the values being low during monsoon due to inflow of rain water and dilution effects, while the sewage oxidation pond ranges between the zone of moderate decomposition to the septic zone.

References

Canter, H.M. and Lund, J.W.J., 1969. The Parasitism of planktonic desmids by fungi. *Osterr. Bot.*, 2(116): 351–377.

Hosessi, B.B. and Patil, H.S., 1992. Enzyme activities as indicator of effluent quality. *Bioresourse Tecnhol.*, 39: 215–220.

Hosmani, S.P. and Bharathi, S.G., 1880b. Limnological studies in ponds and lakes of Dharwar comparative phytoplankton ecology of four water bodies, 19(1): 27–43.

Hosmani, S.P. and Bharathi, S.G., 1980a. Algae as indicators of organic pollution. *Phykos*, 19(1): 23–26.

Judit Pedisa, Ga bor Berics, Istva Grigorszky and Eva Soroczki Pinte, 2006. Use of phytoplankton assemblages for monitoring ecological studies of lakes within the water framework directive: The assemblage index. *Hydrobiol.*, 553: 1–14.

Liviana Marcela Mercado, 2003. A comparative analysis of the phytoplankton from six pampean lotic systems (Buenos Aires, Argentina). *Hydrobiol.*, 495: 103–117.

Mahapatra, P.K. and Mohanty, R.C., 1992. Determination of water quality of two water bodies using algal bioassay method, Utkal Univer. *Phykos*, 31(1 and 2): 77–84.

Norris, R.H. and Norris, K.R., 1995. The need for biological assessment of water quality: Australian perspective. *Australian Journala–1, Ecology*, 20(1) 1–6.

Patil, C.S. and Gounder, B.Y.M., 1985. Ecological study of freshwater zooplankton of a subtropical pond (Karnataka State), India. *Int. Rev. Ges. Hydrobiol.*, 70: 259–267.

Rajendran Nair, M.S., 1999. Seasonal variation of phytoplankton in relation to physico-chemical structures in a Village Pond at Imalia (Vidisha), India. *J. Ecotoxicoloty Environ. Monit.*, 9(3): 177–182.

Reynolds, C.S., 1998. What factors influence the species composition of phytoplankton in lakes of different trophic status? *Hydrobiol.*, 369/370: 11–26.

Stambuck-Giljanovic, N., 1999. Water quality evaluation by index in Dalmatia. *Water Res.*, 33(16): 3423–3440.

Chapter 29

Rotifer Biodiversity of Phutala Lake of Nagpur City, Maharashtra

☆ *S.R. Sitre and S.B. Zade*

ABSTRACT

Phutala lake of western Nagpur is a beautiful natural lake having recreational value since long. In order to assess the biodiversity of rotifers in this beautiful lake of Nagpur city studies were undertaken for a year span on monthly basis and qualitative as well as quantitative analysis of rotifers was done.

During the study period total 17 rotifer species belonging to 12 genera were identified and recorded during year span. The *Brachionus* was represented by 4 different species. The seasonal rotifer biodiversity showed the peak in density and diversity during summer season while lower values were observed in rainy season. Fourteen species were recorded in winter, 16 species in summer while lowest 11 species were recorded in monsoon season. The lake was slowly progressing towards enrichment stage as man made activities are burdening this beautiful lake through activities like throwing the garbage, nirmalya and religious offerings throughout the year in the lake basin.

Introduction

Rotifera is one of the fascinating group of zooplankton in the aquatic ecosystem. Rotifers occur almost universally in freshwater habitat and makes an important group of zooplankton community. The abundance of rotifers is more or less governed by the interaction of number of physical, chemical and biological processes and is related to the suitable conditions for their survival in the lake.

The zooplankton in general are important for their role in the tropho-dynamics and in energy transfer in an aquatic ecosystem. They provide food for fishes in freshwater ponds, lakes and play a major role in fish production. The occurrence and abundance of zooplankton depends on its productivity which in turn is influenced by abiotic factors and the level of nutrients in the water. The freshwater zooplankton form an important group as most of them feed upon and incorporate the primary producers into their bodies and make themselves available to higher organisms in food chain (Michael,1973). The knowledge of their abundance, diversity and distribution is important in understanding trophodynamics and trophic progression of water bodies. With the global loss of thousands of species as a result of population and habitat disturbance, assessment of species diversity and richness are needed (May, 1986).

In this context in the present investigation an qualitative as well as quantitative assessment of rotifers has been undertaken from the Phutala lake of western Nagpur situated near Satpuda botanical garden.

Materials and Methods

Water samples from 4 different spots of phutala lake were collected on monthly basis for period of one year *i.e.* from October 2006 to September 2007 for qualitative and quantitative estimation of rotifers. From each sampling spot 50 Liter of water was filtered through plankton net of bolting silk no. 2. Filtered water sample was preserved in 4 per cent formalin and few drops of glycerin were added to it and was kept aside overnight for better sedimentation. Supernatant was removed and concentrated sample was then observed under binocular microscope for qualitative and quantitative estimation by using Sedgewick Rafter Cell Method (APHA, 1998). Rotifers were identified by using keys and Monographs given by Edmondson (1959) published research paper by (Chandrashekhar and Kodarkar 1995) and other regional publications (Dhanpathi, 1974; Dhanpathi, 2000 and Kodarkar, 1992). The physico-chemical parameters of lake water were estimated (APHA, 1998) to show its relation with density and diversity of rotifers.

Results

Investigation of water sample of Phutala lake was done for a period of one year in order to study the physico-chemical parameters and density and diversity of rotifers. Table 29.1 shows the seasonal variations of various physico-chemical parameters of Phutala lake during the study period. Parameters like water temperature, total dissolved solids, conductivity, pH, total alkalinity, chlorides, total hardness, sulphate, total phosphorus and nitrates were maximum during summer while dissolved oxygen showed its peak in winter.

Table 29.1: Seasonal Variations of Physico-chemical Parameters of Lake Water

Sl.No.	*Parameter*	*Winter*	*Summer*	*Monsoon*
1.	Water Temperature (°C)	19.9 ±0.10	24.05±0.18	22.8±0.2
2.	Total Dissolved Solids (mg/l)	490 ± 0.3	760 ± 0.20	230 ± 0.28
3.	pH (mg/l)	8.20 ± 0.1	8.80± 0.16	7.45± 0.20
4.	Conductivity (μmho/cm)	118.00± 0.2	185.00± 0.29	110.00± 0.3
5.	Total Alkalinity (mg/l)	125.00± 0.26	158.00± 0.32	87.00± 0.25
6.	Dissolved Oxygen (mg/l)	8.50± 0.1	7.30± 0.12	6.50± 0.2
7.	Chlorides (mg/l)	51.6± 0.14	70.7± 0.19	38.00± 0.3
8.	Total hardness (mg/l)	85.00 ± 0.28	100.00± 0.25	65.00± 0.29
9.	Sulphate (mg/l)	0.16± 0.003	0.21+ 0.007	0.12+ 0.100
10.	Total Phosphates (mg/l)	0.17± 0.006	0.25+ 0.004	0.14+ 0.009
11.	Nitrate (mg/l)	0.21± 0.001	0.50+ 0.005	0.19+ 0.009

In the present study total 17 species of rotifera were recorded belonging to 12 different genera. The most diversified genera was *Brachionus* represented by 4 species namely *Brachionus angularis*, *Brachionus falcatus*, *Brachionus forficula* and *Brachionus calyciflorus*. The least dominant genera which were represented by a single species were *Asplanchna*, *Colurella*, *Filinia*, *Keratella*, *Notommata*, tedtudinella. (Table 29.2). Monthly population density of rotifer showed its peak during March 2007 while least density was recorded in July 2007 (Table 29.3).

Table 29.2: Diversity of Zooplankton in Phutala Lake in Different Seasons

Name of Species	*Winter*	*Summer*	*Monsoon*
Tripleuchlanis plicata	+	+	–
Trichocerca tigris	+	+	–
Testudinella patina	–	+	+
Notommata copeus	–	+	–
Lepadella ovalis	–	+	–
Cephalodella exigna	+	+	+
Colurella adriatica	+	–	+
Filinia longiseta	+	+	–
Keratella tropica	+	+	+
Lecane luna	+	+	+
Monostyla bulla	+	+	+
Lepadella ovalis	+	+	+
Asplanchna sp.	+	+	+
Brachionus angularis	+	+	+
Brachionus falcatus	+	+	+
Brachionus calyciflorus	+	+	+
Brachionus forficula	+	+	+

+: Present; –: Absent

Parameters like water temperature (24.05°C), total dissolved solids (815.50 mg/lit), conductivity (187.20 µmho/cm), pH (8.3), total alkalinity (159.0 mg/l), chlorides (975.0 mg/l), total hardness (102.60 mg/l), sulphate (0.27 mg/l), total phosphorus (0.27 mg/l) and nitrate (0.57 mg/l) were maximum during summer while dissolved oxygen (8.50 mg/l) showed its peak in winter season.

Rotifera was represented by 16 species during winter and summer while in monsoon *Filinia longiseta*, *Testudinella patina*, *Notomata* and *trichocerca tigris* were absent from the lake.

Discussion

The seasonal rotifer biodiversity study of Phutala lake of Nagpur city showed the peak in density and diversity during summer season indicating the influence of various physico-chemical factors which was supported by positive correlation between summer temperature, high pH, alkalinity nutrients and rotifer population. pH and temperature are the main factors in the appearance and abundance of different rotifers (Banik and Datta, 1991). Rotifers are chiefly freshwater forms and presence of these organisms in abundance is related to suitable conditions for their survival (Dhanpathi,

2000). Kaushik and Saxena (1995) have also reported abundance of *Brachionus* in various water bodies of central India. An abundance of *Brachionus* in tropical region has been registered and various species of this genus dominate plankton community in warmer part of peninsular India (Fernando, 1980; Sharma, 1983). Occurrence of *Keratella* with *Brachionus* is indicative of nutrient rich status of the water body (Berzins and Pejler, 1987). In Phutala lake maximum density of rotifers was noticed at pH range of 8.80 in summer season. According to Dhanpathi (2000) many species of rotifers are having preference for more alkaline water. The species like *Brachionus, Keratella, Mytilina* and *Platyias* build higher population during period when alkalinity is high. In the present investigation nutrients like sulphate, phosphate and nitrate were higher in summer due to decreased water level,evaporation and more input load due to manmade activities. Rotifers utilize nutrients more rapidly to build up their population (Saboor and Altaf, 1995). Our results are well in agreement with above findings.

Table 29.3: Population Density of Rotifers in Phutala Lake during Different Months

Sl.No.	Months	Population Density (Ind/L)
1	October	80.60±0.08
2	November	93.16±0.07
3	December	101.79±0.09
4.	January	107.07±0.10
5	February	113.83±0.12
6	March	121.01±0.15
7	April	78.11±0.05
8	May	70.02±0.06
9	June	65.10±0.05
10	July	18.39±0.02
11	August	20.42±0.03
12	September	32.45±0.04

Lower values of rotifer population density and diversity were observed during monsoon which could be due to dilution of water resulting in less nutrients or could be due to depletion of important factors such as transparency, dissolved oxygen or pH (Chandrashekhar, 1996; Kumar 2001). Similar trend was also recorded by Kedar and Patil (2002) in Rishi lake of Maharashtra, by Jeelani *et al.* (2005) in Dal lake of Kashmir. According to Dhanpathi (2000) succession rhythm of rotifer population may be disturbed by seasonal flooding of the ponds in monsoon.

Brachionus formed the dominant and diversified genus among the rotifers throughout the study period in this lake. Sunkad (2004) and Pawar and Pulley (2005) also observed the dominance of *Brachionus* in Rakaskoppa reservoir of Belgaum, North Karnataka and Pethwadaj dam of Nanded District in Maharashtra. Dhanapati (1974) reported the distribution of various *Brachionus* species from different lakes of India. The Diversified rotifer fauna of Phutala lake can be linked to favorable conditions and availability of abundant food in the form of bacteria, nanoplankton and suspended detritus in the lake water (Edmondson, 1965, Baker, 1979 and Dhanpathi, 2000).

References

APHA, 1998. *Standard Methods for Examination of Water and Wastewater*. American Public Health Association, Washington, D.C.

Banik, S. and Datta, N.C., 1991. Ecology of Sessile rotifer on artificial substrate in a freshwater lake, Calcutta. *Environment and Ecology*, 9(1): 29–32.

Baker, R.L., 1979. Specific status of *Keratella cochlearis* (gosse) and *K. earlinare*, Ahlstrom (Rotifera: Brachionidae), morphological and ecological consideration. *Can. J. Zool.*, 57(9): 1719–1722.

Berzins, S.L., 1979. Specific status of *Keratella cochlearis* (Gosse) and *K. Ahlastrar* (Rotifer: Brachionidae):

Ecological considerations. *Can. J. Zool.*, 7(9): 1719–1722.

Chandrashekhar, S.V.A. and Kodarkar, M.S., 1995. Studies on *Brachionus* from Saroornagar lake, Hyderabad. *J. Aqua. Biol.*, 19(1): 48–52.

Dumont, H.J., 1983. Biogeography of rotifers. *Hydrobiologia*, 104: 19–30.

Dhanpathi, M.V.S.S.S., 1974. Rotifers from Andhra Pradesh Part I. *Hydrobiol.*, 45(4): 357–72.

Dhanpathi, M.V.S.S.S., 2000. Taxonomic notes on the rotifers from India. IAAB, Hyderabad, p. 1–78.

Edmondson, W.T., 1959. *Freshwater Biology*, Edward and Whipple 2nd edn. John Wiley and Sons Inc., New York.

Edmondson, W.T., 1965. Reproductive rate of planktonic rotifers as related to food and temperature. *Ecol. Manoir.*, 35: 61–111.

Fernando, C.H., 1980.The freshwater zooplankton of Sri Lanka with a discussion of topical freshwater zooplankton composition. *Int. Revueges Hydrobiol.*, 65: 85–129.

Jeelani, M., Kaur, H. and Sarwar, S.G., 2005. Population dynamics of rotifers in the Anchar lake Kashmir (India). In: *Ecology of Plankton*, (Ed.) Arvind Kumar. Daya Publishing House, Delhi, p. 55–60.

Kaushik, S. and Saxena, D.N., 1995.Trophic status of rotifer fauna of certain water bodies in central India. *J. Environ. Biol.*, 16(4): 283–291.

Kedar, G.T. and Patil, G.P., 2002. Studies on the biodiversity and physico-chemical status of Rishi lake Karanja (Lad) M.S. *Ph.D. Thesis*, Amravati University, Amravati.

Kodarkar, M.S., 1992. Methodology for water analysis, physico-chemical biological and micro-biological. Indian Association of Aquatic Biologists, Hyderabad, Publ. 2, pp. 50.

Kaushik, S. and Saksena, D.N., 1995.Trophic status and rotifer fauna of certain water bodies in central India. *J. Environ. Biol.*, 16(1&2): 285–291.

Kumar, K.S., 2001. Studies on freshwater copepods and cladocera of Dharmapuri Dist. Tamil Nadu. *J. Aqua. Biol.*, 16 (1 &2): 5–10.

May, R.M., 1986. How many species are there? *Nature*, 324: 514–515.

Pawar, S.K. and Pulley, J.S., 2005. Qualitative and quantitative analysis of zooplankton in Pethwadj Dam Nanded District (Maharashtra). *J. Aqua. Biol.*, 20(2): 53–57.

Saboor, A. and Altaff, K., 1995. Qualitative and quantitative analysis of zooplankton population of a tropical pond during summer and rainy season. *J. Ecobiol.*, 7(4): 269–275.

Sharma, B.K., 1983. The Indian Species of the genus Brachionus (Eurotatoria, Monogonata, Brachionidae). *Hydrobiol.*, 104: 31–39.

Sunkad, B.N., 2004. Diversity of zooplankton in Rakaakoppa reservoir of Belgaum of North Karnataka, India. *J. Environ. and Ecoplan.*, 8(2): 399–404.

Chapter 30

Phytoplankton Diversity in Wet Soil of Wetlands from Tasgaon Tahsil, Maharashtra

☆ *S.A. Khabade, M.B. Mule and S.S. Sathe*

ABSTRACT

In present investigation, in wet soil of Siddhewadi tank 24 phytoplankton were reported. In wet soil of Pundi tank 14, in wet soil of Anjani tank 17, in wet soil of Lodhe tank 11 and in wet soil of Bastawade tank 09 phytoplankton were reported during May 2005.

The present study reveals that in the summer season, when the tanks gets dried, wet mud from the interior region protects germplasm of phytoplankton and in subsequent season these phytoplanktons flourish the phytoplanktonic population.

Introduction

Phytoplankton consists of chlorophyll bearing organisms. They mainly pertain to the groups cyanophyceae, chlorophyceae, bacillariophyceae, euglenophyceae and dinophyceae.

Phytoplankton are small plants, mostly microscopic either are weakly motile or drifted in water subject of the action of waves and currents. They form the base of the food for many animals and therefore is extremely important to the ecosystem. Phytoplanktons are the product and they belong to first trophic level (autotrophs) (Dholakia, 2004).

The study of phytoplankton from water has got applied significance as it reflects the potential of aquaculture. Apart from it the phytoplankton also serves as indicators of the quality of water (Khatri and Amardeep, 1991).

According to Srivastava (1988) other than physico-chemical parameters, the biogenic capacity of the water body or pond is also important. The biogenic capacity means its capacity to produce the natural food to feed and sustain fish. In general, water is said to be poor, average or rich depending upon its biogenic capacity. However, the biogenic capacity will not be the same for different fish species or for different conditions of water (still water, slow, moderate or fast running water).

Phytoplankton being the primary producers play a significant role in primary production of freshwater ecosystem. Phytoplankton forms a base food for zooplanktons and other higher organisms. Bhosale *et al.*(1994) reported about 18 species of phytoplanktons from waterbodies of Sangli district. Kamat (1965) reported many algal species around kolhapur. Similar type of observations is noticed during the study of Ped water body (Sathe *et al.* 2000). Goel *et al.* (1985) listed 15 phytoplankton from freshwater bodies of southwestern Maharashtra where no species was common to all water bodies.

Still today nobody has studied the phytoplanktons of wet soil *i.e.* mud of the water bodies from Tasgaon tahsil. In summer season water of the tanks reduces significantly. In such situation mud protect certain algal germplasm which restore the algae after recharging the water bodies in rainy season. Hence the phytoplankton study of wet soil *i.e.* mud from various tanks *i.e.* wetlands during summer season has been undertaken.

All the five tanks studied are situated in the eastern drought prone area of Tasgaon tahsil, between 16°43′ to 17° 15′ North latitude and 73°41′ to 74° 50′ East longitude.

Materials and Methods

The wet soil *i.e.* mud from various sampling sites from five wetlands was collected and bring into the laboratory for the study of phytoplankton during month of may 2005. Mud was dissolved and diluted using distilled water and then studied under microscope.

Results and Discussion

The role of the chemical constitution of the soil in the productive potential of a pond is generally not fully understood. There are often significant differences in the productive capacity of ponds situated in similar surroundings. The productivity of an unfertilized pond may some times be the same or more than that of fertilized ones, which can often be traced to the chemical and physical properties of the soil. The soil fertility is of special importance in the growth of benthic vegetation. While water fertility will contribute largely to the production of plankton. Soil fertility will under favorable physical conditions induce a good growth of benthic flora (Hora and Pillay, 1962).

In present study, the wet mud from the five selected water bodies during summer season (May 2005) was collected and qualitatively analyzed for the occurrence of phytoplanktons and the results are summarized in Table 30.1.

In present investigation, in wet soil of Siddhewadi tank 24 phytoplankton were reported. In wet soil of Pundi tank 14, in wet soil of Anjani tank 17, in wet soil of Lodhe tank 11 and in wet soil of Bastawade tank 09 phytoplankton were reported during May 2005.

In wet soil of Siddhewadi tank cyanophyceae members reported were 10, in wet soil of Pundi tank 02, in wet soil of Anjani tank 05, in wet soil of Lodhe tank 04 and in wet soil of Bastawade tank 02 members of cyanophyceae were reported during May 2005.

In wet soil of Siddhewadi tank chlorophyceae members reported were 07, in wet soil of Pundi tank 06, in wet soil of Anjani tank 05, in wet soil of Lodhe tank 02 and in wet soil of Bastawade tank 03 members of chlorophyceae were reported during May 2005.

Table 30.1: Phytoplankton Diversity in Wet Soils of Wetlands of Tasgaon Tahsil

Sl.No.	Phytoplankton	Name of the Tanks (Wetlands)				
		Siddhewadi	Pundi	Anjani	Lodhe	Bastawade
A)	**Cyanophyceae**					
1.	*Pleurococcus*	- -	- -	- -	- -	+ +
2.	*Schizothrix*	+ +	- -	- -	- -	- -
3.	*Scytonema*	+ +	+ +	- -	- -	- -
4.	*Gloeocapsa*	+ +	- -	- -	- -	- -
5.	*Nostoc*	+ +	- -	+ +	- -	- -
6.	*Aphanotheca*	+ +	- -	- -	- -	- -
7.	*Layngbya*	+ +	- -	+ +	- -	- -
8.	*Chroococcus*	- -	- -	+ +	+ +	+ +
9.	*Anabaena*	+ +	- -	- -	+ +	- -
10.	*Oscillatoria*	+ +	+ +	+ +	+ +	- -
11.	*Phormidium*	+ +	- -	+ +	+ +	- -
12.	*Spirulina*	+ +	- -	- -	- -	- -
		10	02	05	04	02
B)	**Chlorophyceae**					
1.	*Pediastrum*	- -	+ +	+ +	- -	- -
2.	*Tetradron*	+ +	- -	- -	- -	- -
3.	*Ankistrodesmus*	+ +	- -	- -	- -	- -
4.	*Scenedesmus*	+ +	+ +	- -	- -	+ +
5.	*Cladophora*	+ +	- -	+ +	- -	- -
6.	*Protococcus*	- -	- -	- -	+ +	+ +
7.	*Chlorella*	+ +	+ +	+ +	+ +	+ +
8.	*Spirogyra*	+ +	+ +	+ +	- -	- -
9.	*Tribonema*	+ +	+ +	+ +	- -	- -
10.	*Nitella*	- -	+ +	- -	- -	- -
		07	06	05	02	03
C)	**Bacillariophyceae**					
1.	*Pinnularia*	+ +	+ +	+ +	+ +	+ +
2.	*Nitzschia*	+ +	+ +	+ +	+ +	+ +
3.	*Surivella*	+ +	+ +	- -	+ +	- -
4.	*Fragilaria*	+ +	- -	+ +	+ +	- -
5.	*Navicula*	+ +	+ +	+ +	- -	+ +
6.	*Synedra*	+ +	- -	+ +	- -	- -
7.	*Achnanthes*	- -	+ +	+ +	- -	- -
8.	*Cyclotella*	+ +	+ +	+ +	+ +	+ +
		07	06	07	05	04
	Total	24	14	17	11	9

In wet soil of Siddhewadi tank bacillariophyceae members reported were 07, in wet soil of Pundi tank 06, in wet soil of Anjani tank 07, in wet soil of Lodhe tank 05 and in wet soil of Bastawade tank 04 members of bacillariophyceae were reported during May 2005.

The present study reveals that in the summer season, when the tanks gets dried wet mud from the interior region protects germplasm of phytoplanktons and in subsequent season these phytoplankton flourish the phytoplanktonic population.

References

Bhosale, L.J., Sabale, A.B. and Mulik, N.G., 1994. Survey and status report on some wetlands of Maharashtra. Final Report submitted to Shivaji University, Kolhapur, India, 60 p.

Dholakia, A.D., 2004. *Fisheries and Aquatic Resources of India*. Daya Publishing House, Delhi, pp. 171, 185, 192.

Goel, P.K., Trivedy, R.K. and Bhave, S.V., 1985. Studies on the limnology of a few freshwater bodies in south western Maharashtra. *Indian J. Environ. Prot.*, 5(1): 19–25.

Hora, S.L. and Pillay, T.V.R., 1962. *Handbook of Fish Culture in the Indo Pacific Region*. FAO. Fish Biol. Tech. Paper 14. pp. 204, 230.

Kamat, N.D., 1965. Ecological notes on the algae of Kolhapur. *J. Biological Sciences*, 8(2): 47–54.

Khatri, T.C. and Amardeep, 1991. Phytoplankton of Dilthaman Tank, South Andaman, Oikoassay, 1 and 2.

Sathe, S.S., Khabade, S.A. and Hujare, M.S., 2000. Studies on wetlands of Tasgaon tahsil and its importance in relation to fisheries and agricultural productivity, Project Report submitted to U.G.C. New Delhi, Western Regional office, Pune.

Srivastava, S.K., 1988. Factors affecting plankton population in a tropical pond: A statistical approach. *J. Environmental Biology*, 9(4): 401–408.

Chapter 31

The Bray-Curtis Similarity Index for Assessment of Euglenaceae in Lakes of Mysore

☆ *T.B. Mruthunjaya*

ABSTRACT

Altogether 27 species of euglenaceae were recorded in 3 water bodies of Mysore spread over a period of one year. They constituted members of Euglenaceae, *Phacus*, Lopocinoides, Trachelomonas and Astasia. They varied in number in the three lakes. In order to understand their ecological behavior the data was analyzed statistically. A similarity index was developed to group the different members of the Euglenaceae with 80 per cent value and high. *Euglena* species dominated the associations. A large cluster of 7 species was a prominent feature. Measurement of similarity and dissimilarity can help take action, plan and decision based prediction of data.

Introduction

Assessment of eutrophication of lakes is an important criterion to detect the extent of increase in concentration of inorganic nutrients have altered or are likely to alter the balance in an aquatic community. Lake communities are dominated everywhere except the margins, by planktonic organisms. The changes brought about by nutrient pollution serve to intensify this dominance together with a shift in species composition due directly to the increased nutrients or indirectly due to changes in the turbidity, dissolved oxygen, pH from higher planktonic levels. Biological organisms are necessary for monitoring the effects of organic discharges, where other aquatic populations are at risk. Relatively little attention has been devoted to determine the ranges of species tolerance to different organic pollutants. Therefore both biological and chemical parameters are essential to monitor pollution.

Multivariate techniques such as similarity indices are standard analytical tools in community studies (Bloom, 1981). Measurement of similarity can help take action plan and decision based on the structure and prediction of the data.

During a detailed ecological study in these water bodies of Mysore, a large number of euglenaceae were recorded. In order to understand the ecological behaviour in the three water bodies the data was analyzed statistically and the out put has been discussed.

Materials and Methods

Collection, preservation, identification and enumeration of Euglenaceae are as per the methods described by Hosmani and Bharathi (1980). The description of the study sites are presented in Table 31.1. the Bray Curtis similarity index was obtained using the PAST programme (Hammer *et al.*, 2001).

Table 31.1: Diversity Indices of Euglenaceae in Three Water Bodies of Mysore City

Sl.No.	*Euglenaceae*	*Dalvoi Lake*	*Kukkarahalli Lake*	*Sewage Oxidation Pond*
1.	*Euglena oxyuris* Schmarda	180	180	80
2.	*E. gracilis* Klebs	165	0	0
3.	*E. acus* Ehrenberg	180	0	0
4.	*E. enrenbergii* Klebs	165	0	0
5.	*E. limnophyla* Lemmermann	145	0	0
6.	*E. elastica* Prescott	145	180	0
7.	*E. minuta* Prescott	110	0	180
8.	*E. polymorpha* Dangeard	180	180	0
9.	*E. elongate* Schewiakoff	0	180	85
10.	*Astasia klebsii* Lemnermann	0	160	0
11.	*Lepocinclis ovum* (Her.) Lemnermann	175	180	65
12.	*L. sphagnophila* Lemn	180	0	0
13.	*L. acuta* Prescott	80	0	0
14.	*L. fusiformis* (Carter) Lemn	170	0	65
15.	*L. playfairiana* Deflandre	70	0	0
16.	*Phacus indicus* Skvortzow	20	0	180
17.	*P. anacoelus* Stokes	70	180	0
18.	*P. curvicauda* Huebner	80	0	0
19.	*P. tortus* (Lemn.) Skvortzow	120	160	80
20.	*P. acuminatus* Stokes	180	180	0
21.	*P. orbicularis* Huebner	170	180	80
22.	*P. helikoides* Pochmann	160	0	0
23.	*P. circumfilexus* Pochmann	160	0	0
24.	*Trachelomonas giradiana* (Palyf.) Deflandre	0	180	20
25.	*Tr. bulla* (Stein) Delfandre	170	0	80
26.	*Tr. volvocina* Ehrenberg	180	0	0
27.	*Tr. robusta* Swirenko	150	110	20

Results and discussion

The Bray-Curtis similarity index is presented in the Figure 31.1. it is bound between 0 and 1, where '0' means the two sites have the same composition or they show all the species, and '1' means that the two sites do not share any of the species. In ecology and biology the index is a statistics used to quantify the compositional dissimilarity between two different sites. It is equivalent to the total number of species that one unique to any one of the two sites divided by the total number of species over the two sites or it is the ratio between the turn over of species and the total species richness over the two sites (Bray-Curtis, 1957). In the analysis, clusters of high similarity will approach 0, while intermediate links become obscured. The expression of the data will be a dendrogram, one of the standard methods of rending the dendrogram is to employ the 'fixed stopping rule' or to arbitrarily select a threshold similarity (Bloom, 1981, Bloom *et al.*, 1957). If the linkage of a cluster is greater than that level the cluster is regarded as important, otherwise the cluster is ignored (Bosch, 1977).

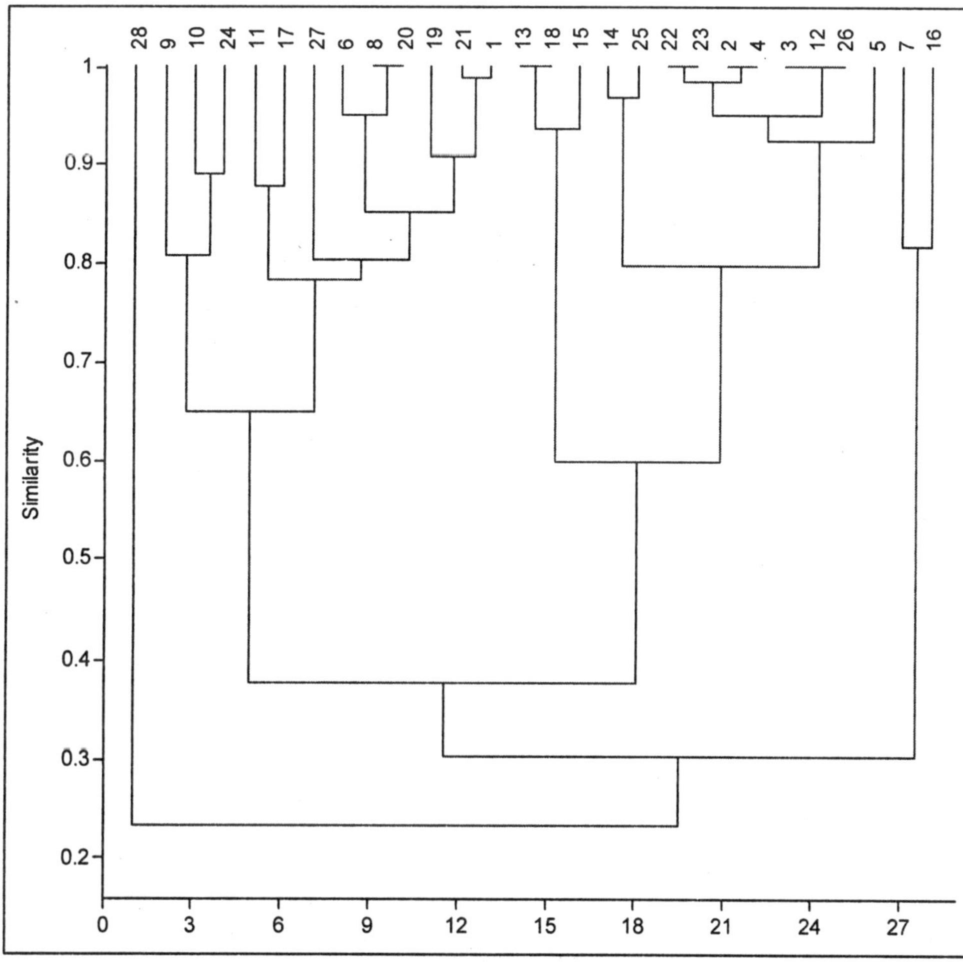

(Numbers 1-27 indicate species of Euglenaceae in the three water bodies)

Figure 31.1:The Bray-Curtis Similarity Indices for Euglenaceae

In the present study 27 species of euglenaceae (Table 1) were recorded of which 9 (*Euglena*), 5 (*Lepocinitis*), 8 (*Phacus*), 4 Tracheomonas and one species of *Astasia* were present. Maximum number occurred in Dalvai lake and least in the sewage oxidation pond. The three habitats studied are quite different, but the similarity distribution of euglenaceae was significant. High number of clusters were observed above 80 per cent level.

Close associations observed between *Euglena elongata, Astasia klebsii* and *Trachomonas girardiana*. The next high similarity association was that of *Trachomonas robusta, Euglena elastica, E. polymorpha, Phacus accuminatus, P. Tortus, P. orbicularis* and *Euglena oxyeris*. The highest similarity index was that of *Lepocuiclis fusiformis, L. sphageophila, Trachelomonas bulla, Tr. Volvocine, Phacus helipoides, P. circumflescens, Euglena clarpergi, E. aces,* and *E. Limophilia* indicating that Euglena dominated the association. A small cluster of two associations of *Lepoinclis ovum* and *Phacus analelus* and *Euglena minuta* with *Phacus* indices which did not appear in one of the lakes.

As per the distance indices it is observed that *Euglena polymorpha* and *Phacus accuminatus, Lepociclis acuta* and *Phacus curvicanda, Pacus helipoides* and *Phacus circumflesces* reach a similarity value of '1' and as per the index these do not share any other species on the other hand *Phacus helipoides, P. circumflescens, Trachomonas girarlriana, Tr. Bulla, Tr. Volvocine* attain '0' values and share the habitat with other species. Euglenoids have great adaptability to varied habitats).

Conclusion

Multivariate techniques such as similarity indices are standard analytical tools in community studies. Measurement of similarity and dissimilarity can help take action, plan and decision based on the structure and prediction of the data. The present data helps in grouping various species of Euglenaceae based on their association and their relationship with each other.

Acknowledgement

Authors are thankful to S.P.Hosmani Retired Professor and Chairman DOS in Zoology, University of Mysore Manasagangotri, Mysore.

References

Bloom, S.A., Santos, S.L. and Field, J.G., 1977. A package of computer programs for benthic community analysis. *Bull. Mar. Sci.*, 27(3): 577–580.

Bloom, S.A., 1981. Similarity indices in community studies: Potential pitfalls. *Mar. Ecol. Prog. Ser.*, 5: 125–128.

Boesh, D.F., 1977. Application of numerical classification in ecological investigations of water pollution. Special Scientific Report 77. VIMS (EPA–600/3–7703).

Bray, J.R. and Curtis, J.T., 1957. An ordination of the upland forest communities of southern Wisconsin. *Ecol. Monog.*, 27(4): 325–329.

Hammer, O., Harer, D.A.T. and Rayan, P.D., 2001. PAST palentological statistics software package for education and data analysis. *Palentological Electronica*, (1): 9.

Hosmani, S.P. and Bharathi, S.G., 1980. Limnological studies in ponds and lakes of Dharwar. Comparative phytoplankton ecology of four water bodies. *Phytos*, 19(1): 27–43.

Kumar, S. and Dutta, S.P.S., 1991. Studies on Phytoplankatonic population dynamics in Kunjwani pond, Jammu. *Hydrobiol.*, 7: 55–59.

Reynolds, C.S., 1998. What factors influence the species composition of phytoplankton in lakes of different tropic status? *Hydrobiol.*, 369, 370: 11–26.

Chapter 32

Study of Hydrophytes and Amphibious Plants Occurred in Panchanganga River, Maharashtra

☆ *V.G. Patil, S.A. Khabade and S.K. Khade*

ABSTRACT

The present study will inspire enthusiastic researchers to come forward and have local aquatic weeds identified with assistance of local taxonomists and utilize them for the national benefit. In present investigation about 9 hydrophytes were reported during study period and 06 amphibious plants were reported throughout the year during 2008-09. It is also revealed that *Salvinia species, Eichhornia crassips, Pistia, Lemna* and *Wolfia* were the floating hydrophytes while *Nitella* species, *Chara* species *Ceratophyllum* species, *Vallisneria* species were the submerged hydrophytes.

The amphibious plants reported were *Typhya species, Cyperus rotundus, Saccharum species, Dendrocalsmus strictus, polygonum plebagum* and *Asclepias curasvica*.In present study it is clear that the above mentioned hydrophytes were not originated in the river but they were deposited in river channel by the stagnant water bodies occurring nearby the Kolhapur district during rainy season, when flooded condition. But amphibious plants recorded were grown throughout the year at the bank of Panchgannga river, around Ganpati Temple.

Introduction

Aquatic plants are also called as hydrophytes or hydrophytic plants which have adapted to living or on aquatic environment, Aquatic environment is divisible brodly into freshwater, marine water and estuarine water. Freshwater bodies such as ponds, lakes, dams, rivers and freshwater marshy places. Aquatic plants abundantly occurs in eutrophic stagnant water bookies. They can only grow in water or permanently saturated soil. For living on or under water surface they shows numerous special adaptations.

The study of submerged and floating vegetation is of great importance as far as food supply to fish species. The submerged microphytes will not stress the oxygen reserves and alter the pH of the system as these are losing about 90 per cent of their dry matter throung physical or autolytic leaching and insignificant amount of organic matter is left for biological decomposition. On the contrary, floating macrophytes like *E. crassipes* lose insignificant dry mass through physical and autolytic leaching and will be putting enormous stress on the oxygen reserves of the system due to microbial respiration on its detritus and subsequent oxidation of dissolved organic constituents. Thus it may be argued that the extent of physical and autolytic losses of organic matter in the macrophytes may directly indicate the quality of water of the system. (Sing and Rai,1988).

In riverine system, during rainy season majority of hydrophytes are deposited through water currents from nearby stagnant water bodies like ponds and lakes.

Panchagnga river of Kolhapur district is a polluted river which also contains the number of hydrophytes throughout the year. These hydrophytes directly affect the physico-chemical characteristics of the river water and ultimately fauna of the river, so the present study has been carried out during June 2008 to May 2009.Varadvinayak Ganapati Temple site of Ichalkaranji is situated on the bank of Panchganga river which is selected for the study of hydrophytes.

Materials and Methods

Monthly collection of hydrophytes was done from the period of June 2008 to May 2009. The hydrophytes were collected from the study site in polythene bags and brought to laboratory. Then preserved in 4 per cent formalin and identified by using standard literature (Sardesai and Yadav, 2002).

Results and Discussion

Aquatic weeds are superficially treated rather some what neglected in the publications of plant biology. The plant community o9f floating aquatic weeds. Forming vast impermeable floating mats in natural waters, canals, lakes, rivers and other water bodies has created a nuisance, since it blocks drainage, channels and fishing almost impossible. In some parts of the world aquatic weeds are spreading at an alarming rate; obviously it has become a serious concern to the Governments of many countries and international agencies.

The menace of aquatic weeds can be overcome by their utilization in a number of ways. Aquatic plants constitute an integral component of an aquatic ecosystem. They may serve as good source of food to the mankind, a palatable feed to the water birds and animals thus forming a base for aquatic wild life conservation practices. They also serve as a potential source of energy (Majid 1986).The present study will inspire enthusiastic researchers to come forward and have local aquatic weeds identified with assistance of local taxonomists and utilize them for the national benefit.

In present investigation about 09 hydrophytes were reported during study period and 06 amphibious plants were reported throughout the year during 2008-09. It is also revealed that *Salvinia species, Eichhornia crassips, Pistia, Lemna* and *Wolfa* were the floating hydrophytes while *Nitella species, Chara species Ceratophyllum species* and *Vallisneria species* were the submerged hydrophytes.

The amphibious plants reported were *Typhya* species, *Cyperus rotundus, Saccharum* species, *Dendrocalsmus strictus, Polygonum plebagum* and *Asclepias curasvica.*

Table 32.1: The Hydrophytes and Amphibious Plants Observed in Study Area

Sl.No.	*Name of Hydrophyte and Amphibious Plant*	*Month and Year*											
		Jun 2008	*July 2008*	*August 2008*	*Sept. 2008*	*Oct. 2008*	*Nov. 2008*	*Dec. 2008*	*Jan. 2009*	*Feb. 2009*	*March 2009*	*April 2009*	*May 2009*
A)	**Hydrophyte**												
1.	*Salvinia* species	- -	- -	- -	- -	- -	- -	- -	- -	- -	+ +	+ +	- -
2.	*Nitella* species	- -	- -	- -	- -	+ +	+ +	+ +	+ +	+ +	- -	- -	- -
3.	*Echhornia crassipus*	+ +	+ +	+ +	- -	- -	- -	- -	- -	- -	- -	- -	- -
4.	*Chara*	- -	- -	- -	+ +	+ +	+ +	+ +	+ +	+ +	+ +	+ +	- -
5.	*Pistia*	+ +	+ +	+ +	+ +	+ +	+ +	+ +	+ +	+ +	+ +	- -	- -
6.	*Lemna*	+ +	+ +	+ +	+ +	+ +	+ +	+ +	+ +	+ +	+ +	- -	- -
7.	*Wolfia*	+ +	+ +	- -	- -	- -	- -	- -	- -	- -	+ +	+ +	+ +
8.	*Ceratophyllum*	- -	- -	- -	- -	+ +	+ +	+ +	+ +	+ +	+ +	- -	- -
9.	*Vallisnaria*	- -	- -	- -	- -	+ +	+ +	+ +	+ +	+ +	+ +	- -	- -
B)	**Amhibious plant**												
1.	*Typha*	+ +	+ +	+ +	+ +	+ +	+ +	+ +	+ +	+ +	+ +	+ +	+ +
2.	*Cyprus rotundus*	+ +	+ +	+ +	+ +	+ +	+ +	+ +	+ +	+ +	+ +	+ +	+ +
3.	*Saccharum* species	+ +	+ +	+ +	+ +	+ +	+ +	+ +	+ +	+ +	+ +	+ +	+ +
4.	*Dendrocalamus srictus*	+ +	+ +	+ +	+ +	+ +	+ +	+ +	+ +	+ +	+ +	+ +	+ +
5.	*Polygonum glabrum*	+ +	+ +	+ +	+ +	+ +	+ +	+ +	+ +	+ +	+ +	+ +	+ +
6.	*Asclepias curasavica*	+ +	+ +	+ +	+ +	+ +	+ +	+ +	+ +	+ +	+ +	+ +	+ +

+ +: Present: - -: Absent.

In present study it is clear that the above mentioned hydrophytes were not originated in the river but they were deposited in river channel by the stagnant water bodies occurring nearby the Kolhapur district during rainy season, flooded condition. But amphibious plants recorded were grown throughout the year at the bank of Panchgannga river, around Ganpati Temple. Similar results were reported by Yadav and Sardesai (2002). According to Bhosale *et al.* (1994) there is a network regarding the fauna of the wetlands of the Satara district. Similar networking was observed in this study, in case of the macrophytes and amphibious plants.

Acknowledgement

The authors are thankful to Principal, Dr. Ashok Karande for their encouragement and availability of laboratory facility during study.

Chapter 33

An Overview of Nile Tilapia (*Oreochromis niloticus*) and Low Cost Feed Formulation Technique for its Culture

☆ *B.R. Chavan and A. Yakupitiyage*

ABSTRACT

The review study include the tilapia biology, feeding behaviour, nutrient requirement, current feeding and feed management practice, economic feeding and current culture problems and potential solution of Tilapia and investigated on-farm feed formulation and its management strategies for enhance higher economic efficiency of tilapia culture (*Oreochromis niloticus*). The feed was prepared on the basis of feed requirement of tilapia. The prepared feed was tested by tilapia culture experiment in earthen pond. The new proposed feed formulation and feeding method was contribute significantly in reducing operational cost of tilapia farming. This is better than existing practice carried by the small scale fish farmers. The total cost of new formulated feed is 16.71 rupees kg^{-1} as wet basis (moisture 8.91 per cent) and the new formulated feed cost accounts 43-57.36 per cent in the gross revenue with 80 per cent satiation. Partial replacement of fish meal 22 per cent, soybean meal 26 per cent, rice bran 26 per cent and maize flour 22 per cent with corn oil 2 per cent, vitamin premix 1 per cent and mineral premix 1 per cent in the new feed formulation was contributed in reducing feed cost and more economic efficiency in tilapia farm operation. The formulated feed was tested by conducting 8 weeks experimental study, with a population of male Nile tilapia fry were stocked at 400 individuals in cages suspended in earthen pond and cement tank. Temperature, dissolved oxygen, pH, total phosphorus, total nitrogen, and suspended solids were monitored weekly. Growth of tilapia fingerlings during prolonged nursing with formulated feed (56 days) in cement tank, were compared with fish nursed with company feed in cement tank. The tilapia fed with formulated feed at 0 per cent feed showed no much growth (stunted growth). Mean daily weight gain of stunted tilapia were 0.012 g per fish day^{-1}, 0.06 g per fish day^{-1} for 50 per cent feed fed and 0.10 g per fish day^{-1} for 100 per cent feed fed were as 0.19 g per fish day^{-1} for 100 per cent feed fed in cement tank.

Introduction

Tilapia is one of the most important species for the 21st century aquaculture and is produced in more than 100 countries (Fitzsimmons, 2000). Tilapia, *Oreochromis niloticus* have become one of the most abundantly produced fish in aquaculture (Lovshin, 1997). Tilapia farming is now in a dynamic state of worldwide expansion to satisfy the demand from both domestic and international markets and to provide an affordable source of animal protein. Although several tilapia species are cultured worldwide, the most popular is the Nile tilapia. Nile Tilapia is extremely fast growing reaching harvest maturity of 1-2 kg in 8 to 10 months (Mary King, 2000).

Tilapia is the second largest farmed fish in the world (next to salmon) with production on the order of two billion pounds per annum. Nile tilapia is one of the most important freshwater aquaculture species with a production of 1.22 million metric tones in 2002; it is cultured in a total of 50 countries, in 19 of these at commercial scale with an annual production above 1000 metric tones (FAO 2004).

Tilapia have become a top priority in aquaculture because of their ability to efficiently use natural foods, being herbivore in nature, primarily vegetation and algae eaters and are often stocked in canals and artificial lakes for algae and vegetation control. Tilapia resistant to diseases and handling, ease of reproduction in captivity and tolerant to wide range of environmental conditions and relatively grows fast and can easily be bred (Guerrero, 1982).

Tilapia is also known as Aquatic chicken, St. Peter's fish, Nile perch and Hawaiian sunfish, is a member of the cichlids family, originated from Africa. Tilapia is very hardy and can thrive in salt, brackish or freshwater. Tilapia range in skin color from brilliant golden red, pale red, white, gray, gray-blue, dark blue to black. (Thomas and Michael, 1999).

In this chapter, brief review of the biology, feeding, behaviour, nutrient requirement, current feeding and feed management practice, economic feeding and current culture problems and potential solution of Tilapia is focus and as well new feed formulation technique to reduce the current feeding cost.

Literature Review

General Biology of Nile Tilapia (*O. niloticus*)

Taxonomic Classification and Distribution

Commercially important tilapia are currently divided into three major taxonomic groups, according to Trewavas (1983), based largely on reproductive characteristics. All are nest builders and substrate spawners, except in: Tilapia spp. guards the developing eggs and fry in the nests, *Oreochromis spp.* females incubate eggs and fry orally and *Sarotherodon spp.* males and/or females incubate eggs and fry orally.

Tilapia is a generic term used to designate a group of commercially important food fish. Tilapia belongs to the family Cichlidae which can be distinguished from other families of bony fish. Tilapia belongs to class actinopterygii, order perciformes, the family cichlidae, and genus *Oreochromis*.

2. Breeding Biology

Nile tilapia matures at about 10 to 12 months and 3/4 to 1 pound (350 to 500 grams). When growth is slow, sexual maturity in tilapia is delayed a month or two but stunted fish may spawn at a weight of less than 20 grams, under good growing conditions in ponds. (Popma and Masser, 1999).

In all *Oreochromis* species the male excavates a nest in the pond bottom *O. niloticus* build relatively small nests with diameter only about twice as greatest as the fish's length (Trewavas, 1983), (generally in water shallower than 3 feet) and mates with several females.(Popma and Masser, 1999). There is positive relationship between the sizes of males and females in pairs, the female average 60 per cent of the male's length (Schwanck, 1987). After a short mating ritual the female spawns in the nest (about two to four eggs per gram of brood female), (Popma and Masser, 1999), female tilapia has a large variation in the number of eggs they produced. The blue female tilapias are reported to lie around 9-10 eggs per g of body weight. The eggs vary in size from an average of 2-4 mm in diameter; it is depending on the species and number of spawns, almost of the spawns occurs during afternoon, with 79 per cent of these occurring between 1 p.m. and 4 p.m. (Gautier, 2000). The male fertilizes the eggs and then female holds and incubates the eggs in her mouth (buccal cavity) until they hatch. Fry remain in the female's mouth, (Popma and Masser, 1999). When the fry totally absorb their yolk and become free-swimming, the female tilapia normally releases them from her mouth and they are left by themselves to search for exogenous food. (Hewitt *et al.*, 1985). Bruton and Bolt (1975) suggested that females may breed three to four times per year in one breeding season.

Feeding Biology

Tilapia ingests a wide variety of natural food organisms, including plankton, some aquatic macrophytes, planktonic and benthic aquatic invertebrates, larval fish, detritus, and decomposing organic matter. Tilapias are often considered filter feeders because they can efficiently harvest plankton from the water. The gills of tilapia secrete a mucous that traps plankton. (Popma and Masser, 1999). Two mechanisms help tilapia digest filamentous and planktonic algae and succulent higher plants: Physical grinding of plant tissues between two pharyngeal plates of fine teeth; and a stomach pH below 2, which ruptures the cell walls of algae and bacteria. (Popma and Masser, 1999).

The individual species may have preferences between these materials and are more or less efficient depending on species and life stages in grazing on these foods. They are all somewhat opportunistic and will utilize any and all of these feeds when they are available. Feeding rates is varying with fish size and water temperature. The appropriate amount is measured as a per cent of the average body weight. As the fish weight increases, the per cent body weight fed decreases.

General Behaviour

Tilapias are generally vertical barred, but in subdued colors that blen extremely well with their background. This is assisted by a modest ability to change color by controlling the chromotophores in the skin. Therefore, tilapias can modify their overall appearance to become pale or dark and response to stressors (Ross, 2000). Tilapias are mainly lacustrine fish and are well adapted to enclosed water. They are fast growing, resistant to disease and handling, easy to reproduce in captivity and able to tolerate wide range of environmental conditions. (El-Sayed, 1999). This species is usually restricted to relatively shallow waters (Bruton and Bolt, 1975). Juveniles, appear better adapted to inhabit deeper waters than adults (Caulton and Hill, 1973).

They also have strong reputation for tolerance of low dissolved oxygen and are quite resistant to reasonable handling more than most other fish species. Their natural distribution was also pointed out the wide range of colonized habitats, particularly in temperature range, current velocities, salinity and alkalinity (Philippart and Ruwet, 1982). Tilapia occurs in freshwater and estuaries along the coast, tolerating a broad range of salinity's (Trewevas, 1983). Tilapia may be able to spawn in salinities of up to 30 ppt and survive in salinities of up to 40 ppt. Philippart and Ruwet, 1982 suggested that tilapias exploit shallower, warmer water during day time for feeding but they can be retired to deeper, cooler water at night to carry out digestion.

Tilapia does not tolerate temperatures below 10°C (Trewevas, 1983)., Bruton and Bolt (1975) reported seasonal movements to deeper waters during the cold season and to shallower waters in the warm season, with colder temperatures limiting the length of the breeding season. In tropical waters, tilapia breeds throughout the year. Tilapia generally stops feeding when water temperature falls below 15.55°C (Popma and Masser, 1999). Tilapia survives routine dawn dissolved oxygen (DO) concentrations of less than 0.3 mg L^{-1}, considerably below the tolerance limits for most other cultured fish. In aquaculture, environmental requirement of *O. niloticus* is suggested that DO should be more than 3 ppm, temperature from 28-30°C, pH is between 6-8, Secchi depth is between 30-40 cm, salinity is between 0-25 ppt, alkalinity is more than 20 ppm, ammonia union is lower 0.08 ppm, BOD is between 8-15 ppm, and COD is between 20-30 ppm (Boyd, 1997).

Nutritional Requirement

Like other animals, tilapia has specific requirements for nutrients such as amino acids from protein, fats, minerals and vitamins. Tilapia exhibit their best growth rates when they are fed a balanced diet that provides a proper mix of protein, carbohydrates, lipids, vitamins, mineral and fiber. (Jauncey and Ross, 1982). The nutritional requirements are slightly different for each species and more importantly vary with life stage. Fry and fingerling fish require a diet higher in protein, lipids, vitamins and minerals and lower in carbohydrates as they are developing muscle, internal organs and bone with rapid growth. Sub-adult fish need more calories from fat and carbohydrates for basal metabolism and a smaller percentage of protein for growth (Stickney, 1996).

Protein Requirement

Feed formulators will adjust protein sources to fit the desired pattern of amino acids through the growth cycle. Brood fish may require elevated protein and fat levels to increase reproductive efficiency (Chang *et al.*, 1988). For fry (first feeding to approximately 0.5g), dietary protein ranging from 36-50 have been shown maximum growth (El-Sayed and Teshima, 1992), for juveniles (approximately 0.5-5g), 29-40 per cent (Cruz and Laudencia, 1977) and young adult fish (up to 40 g), 27.5-35 per cent (Siddiqui *et al.*, 1988), protein is considered to be a very important component of tilapia diets (Steffens, 1981). In semi-intensive tilapia farming is 20-25 per cent protein in feed required for fertilized pond culture and 28-32 per cent protein in feed required for cage culture.

Inadequate protein in the diet results in a reduction or cessation of growth and loss of weight due to withdrawal of protein from less vital tissues to maintain the functions of more vital tissues. On the other hand if excess protein is supplied in the diet, only part of it will be used for protein synthesis and the remainder will be converted to energy (Wilson, 1989), diet containing 33.32 per cent protein, dm which resulted to the best performance of fish (Ogunji and Wirth 2000).

In term of protein digestibility, it needs to be considered also for selecting source of protein for feed formulation. Kamarudin et al. (1989) reported that apparent crude protein digestibility of rice brain, fish meal, shrimp meal, soybean meal and copra meal was at 99.9, 99.5, 99.4, 91.6 and 91.1 per cent, respectively for tilapia.

Essential Amino Acids (EAA) Requirement

Dietary essential amino acids requirements for *O. niloticus* are expressed as the essential amino acids composition of the diet with a protein content of 30-32 per cent in dry matter. This follows the concept that protein requirements are the minimum amount needed to meet amino acid requirements and ensure maximum growth. (Santiago and Lovell, 1988). It is believed that of the total requirement, not less than 60 per cent should be EAA and up to 40 per cent can be non-essential amino acids (Juancey, 2000).Tilapia require the same ten amino acids as other fish suggested by Santiago and Lovell, 1988 are Arginine-4.20 per cent, Histidine-1.72 per cent, Isoleucine-3.11 per cent, Leucine-3.39 per cent, Lysine-5.12 per cent, Methionine-2.68 per cent, Phenylanine-3.75 per cent, Threonine-3.75 per cent, Tryptophan-1.00 per cent and Valine-2.80 per cent. Tilapias actually have a requirement for sulphur containing amino acids, which can be met by either methionine alone or the proper mixture of methionine and cystine (Juancey and Ross, 1982).

Lipid Requirement

The lipid requirements for fish under two grams represent 10 per cent of the diet. This decreases to 6-8 per cent from two grams to harvest. The lipids should contain both omega 3 and omega 6 fatty acids. Each fatty acid should represent 1 per cent of the diet, although some reports suggest that fish grow better with a higher proportion of omega 6 to omega 3. The fiber component is usually the reciprocal of the lipid content (El- Sayed and Teshima, 1992).

Dietary lipids are the only source of essential fatty acids needed by the fish for normal growth and development. They are also important carriers and assist in the absorption of fat-soluble vitamin. Lipids, especially phospholipids are important for cellular structure and maintenance of membrane flexibility and permeability. A lipid also serves as precursors of steroid hormones and prostaglandins, improve the flavour of diets and affect the diets texture and fatty acid composition (Shiau, 2002). Chou and Shiau (1999) demonstrated that both *n-3* and *n-6* highly unsaturated fatty acids are essential for maximum growth of hybrid tilapia (*O. niloticus* × *O. aureus*). *Omega-6* (*n-6*) fatty acid enhanced spawning success and fry production in *O. niloticus* and than *n-3* fatty acids increased weight gain but reduced reproduction performance. The lipid, fatty acid may come from natural food in extensive tilapia farming. However there requirement is depends on fertilization regime, stocking density, biomass/ unit area and critical standing crop.

Carbohydrate Requirement

Fish in general utilize dietary carbohydrate poorly. Moreover, different types of carbohydrate may not be equally utilized by tilapia. In tilapia, a number of factors appear to be associated with carbohydrate utilization. The intestinal absorption of carbohydrate is low when diets contain fiber, regardless of source. Meal frequency affects carbohydrate utilization in tilapia. Some carbohydrate metabolic enzymes are altered due to changes in meal frequency. The ability of tilapia to utilize certain carbohydrates changes with the size or age of the fish. Complex carbohydrates such as starch or dextrin can spare some protein when the dietary protein level is low (Shi, 1997) the utilization of starch was significantly higher than that glucose by tilapia (Shiau and Lin, 1993).

An appropriate level of carbohydrates in tilapia diets is, however, required to avoid any disproportionate catabolism of proteins and lipids for the supply of energy and metabolic intermediates (Wilson, 1994). Tung and Shiau (1993) have studied the effect of daily feeding frequency on the utilization of carbohydrates by tilapia. Diets containing 44 per cent of glucose, dextrin, and starch (carbohydrate sources) were fed 6 times a day had significantly higher weight gain. The recommended inclusion level of fiber is less than 10 per cent of the diet. Biotic factors, such as physiological stages also influenced the utilization of carbohydrates by tilapia.

Energy

Energy is not a nutrient but is a property of nutrients that are released during the metabolic oxidation of proteins, carbohydrates and lipids. The energy should be first nutritional consideration in diet formulation of tilapia. Excess energy may produce fatty fish, reduced diet consumption (reduce total protein intake) and inhibit proper utilization of other feed stuffs (Shiau, 2002). The optimum dietary protein to energy (P:E) ratio for rapid and efficient gain of juvenile tilapia aurea was shown to fall with increasing size of fish. The optimum concentration of protein and energy also fell with growth (Winfree and Stickney, 1981). The digestible energy requirements for economically optimum growth are similar to those for catfish and have been estimated at 8.2 to 9.4 kcal DE (digestible energy) per gram of dietary protein (Popma and Masser, 1999).The highest digestible energy contents of feedstuffs for *O. niloticus* were determined with rice brain and groundnut cake which were 17.9 MJ/kg for both feedstuffs (Kamarudin *et al.*, 1989).

Vitamins Requirement

Vitamin supplements are often not included in the diet for tilapias stocked at moderate densities in fertilized ponds. In intensive systems where limited or no natural foods are available, supplemental vitamins must be added (Shiau and Suen, 1992). Vitamins and minerals are critical to proper nutrition in tilapia (Jauncey and Ross 1982). Water soluble vitamin requirements of tilapias that have been studied are: thiamine, riboflavin, pyridoxine, niacin, biotin, cholin, pantothenic acid and ascorbic acid. Thiamine deficiency in red hybrid tilapia causes reduced growth and diet efficiency and low haematocrit. For prevention and growth dietary thiamine level of 2.5mg kg^{-1} is required (Lim and Master, 1991).

Tilapia produced vitamin B_{12} in their gastrointestinal tract through bacterial synthesis and did not have a dietary requirement for this vitamin (Shiau and Lung, 1993). Niacin is a dietary essential for the *O. niloticus* × *O. aureus* hybrid but the level required varies depending on the source of dietary carbohydrates (Shiau and Suen, 1992). Shiau and Lo (2000) established that the optimal dietary choline requirement for *O. niloticus* × *O. aureus* was 1000 mg kg^{-1} diet. Vitamin C (ascorbic acid) is essential for *O. niloticus*. Quantitative requirement for vitamin C was found 1250 mg/kg diet for *O. niloticus*, equalized to level of 420 mg/kg body weight (Soliman *et al.*, 1994). Fat- soluble vitamins for tilapia concern only vitamin D and E. The optimum vitamin D_3 (cholecalciferol) requirement for maximum growth of *O. niloticus* × *O. aureus* is 374.8IUkg^{-1} diet (Shiau and Hwang, 1993). The dietary vitamin E requirement increased with increasing level of dietary lipid. The dietary vitamin E requirement of *O. niloticus* was reported to be 50- 100 mg kg^{-1} diet for diet containing 5 per cent lipid (Satoh *et al.*, 1987).

Minerals Requirement

Five minerals are required for tilapia namely; calcium, phosphorous< 0.9 per cent magnesium-0.059- 0.077 per cent, zinc-0.003 per cent and potassium have been quantified for their requirements in tilapia diet. Calcium and phosphorous kg^{-1} diet is required for normal growth and bone mineralization of *O. aureus* (Robinson *et al.*, 1984). Phosphorus is an essential component of both hard skeletal material and cartilage. It also plays in energy and cell metabolism and inorganic phosphates serve as buffers regulating body fluid pH (Watanabe *et al.*, 1980). Zinc is an enzyme cofactor playing a role in metabolism of protein, lipid and carbohydrate being particularly active in RNA and protein synthesis and metabolism. Zn requirement was suggested at 30 mg/kg in *O. niloticus* (Eid and Ghoneim, 1994).

Current Feeding and Feed Management Practices

The Cassava starch, Soy Bean Meal, Rice bran, Fish Meal, Fish oil/soy oil, mineral mix and vitamin premix, ingredients are using as a sources of the protein and other nutritional requirements such as lipids, carbohydrates, vitamins and minerals. In generally recognized that smaller fish consume more diet on a present weight basis than larger fish. Suggested feeding rates for different size of tilapias are as follows.

Sl.No.	Size (Grams)	Daily Feeding Rate (Per cent body weight)	Daily Feeding Frequency
1	2 days old to 1g	30-10	8
2	1-5 g	10-6	6
3	5-20 g	6-4	4
4	20-100 g	4-3	3-4
5	>100 g	3-2	2-3

Water temperature influences the metabolic rate and expenditure, thus having profound effects on feeding rate. Tilapia is consuming less diet in cold weather than in warmer weather. Feeding tilapia should be stop at water temperature below 16°C. In semi-intensive culture systems, natural food can make a significant contribution to the nutrient requirement of fish. Under these conditions, the amount of diet used should be less than that of fish grown under intensive systems (Luquet, 1991).

Currently, diets are offered to fish by hand, blower, automatic feeder or demand feeder. Hand feeding is labour intensive but has advantages over other methods of feeding in that it allows the feeder to observe feeding activity and feeding behavior and thus to regulate the amount of diet fed to prevent over feeding. However, this method of feeding is not feasible in large commercial farms. The most common method of feeding with large ponds is blowing the diet on the water surface using mechanical devices that are either mounted on or pulled behind vehicles. An automatic feeder, driven by time clock or electrical devices, allows farmers to preset the amount of diet to be fed at various time intervals. Demand feeder consists of a hopper with top opening for loading the diet and a bottom opening, which serves as a movable gate for diet delivery. Attach to the gate is a rod whose tip extends down in to the water, where it can be activated by the fish as long as the fish continue to hit the rod, diet will continue to flow out.

Economics of Current Feeding

The cost of feed for tilapia culture is the major factor for effective farming. At present, fish feed constitutes over 60 per cent of the operating costs in intensive fish farming. Identifying daily feeding rate is also important to reduce feed cost for tilapia culture farms (Nwanna, 2003).

Maximizing production efficiency can be done by supplemental feeding and fertilizer strategy. An other study shown that 50 per cent supplemental feeding and fertilizer in red tilapia culture in brackish water was the most efficient feeding rate in term of low FCR (0.93), good growth and yield performance, high economic return and potential for growing to greater size (Yi *et al.*, 2003). Net return of fertilizing ponds through the culture period and feeding Nile tilapia starting from day 80 was better (\$734.9 $ha^{-1}crop^{-1}$) than fertilizing ponds until day 80 and feeding starting from day 80 (-\$512.5 $ha^{-1}crop^{-1}$) (Thakur *et al.*,2001).

Fish oil may be replaced by plant oil because of freshwater fish species, generally, have higher capacities for the conversion of 18:2n-6 and 18:3n-3 in the body to higher C20 and C22 homologues to satisfy tissue requirements (Sarget and Tacon, 1999). The Poultry based diet produced higher carcass lipid than other diets. Thus the partial replacement of dietary fish meal (FM) with plant protein sources will may helps to minimize the cost of feed.

Current Culture Problems and Potential Solution

Problems in Hatchery Management

Fry Production

In many areas of the world, fry and fingerling production is the main technical problem yet to be overcome (Lam, 1982). In India, fry production is carried out in earthen ponds and the broodstock cross-bred with wild dwelling populations during rainy season (floods). The fry produced are supplied to the farmers and in most cases they reproduced at an early stage (2.5 to 3 months) leading to overcrowding of the ponds and resulting in small stunted fish. In order to solve the above problems, selective breeding of the tilapia will minimize these problems.

Feed

No proper feed available for the maintenance of brood fish and the rearing of fry. They fed on whatever feed is available (rice bran and wheat bran). This may have contributed to the production of poor quality fry.

Problems in Rural Tilapia Farming

Manpower

Most farmers still lack basic knowledge on proper fish pond management. In addition, there is a lack of trained extension staff and also lack of an efficient data collection system which further makes rural fish farming development slow and difficult to monitor. Most farms are rural based and dispersed over a wide area and there is a need for a regular extension support. The government provides basic extension services only and there is a need for more trained extension staff and the training of fish farmers.

Feeding

In India, tilapia is fed on various locally available feeds usually agricultural by-products which are in powder form. These are wheat bran, wheat pollard, rice bran, rice pollard, copra meal, bone and meat meal and fish meal. These feeds are mixed such that the crude protein content is 20–25 per cent. This feed quality has proven to be barely adequate and if pellets can be locally manufactured, production could be increased.

Sometimes it is difficult to obtain feed ingredients due to competition with established practices. Some feeds are available only in certain areas and the required amounts are not available to conduct proper feed studies. Due to the scattered location of the fish farms, farmers feed whatever is available and sometimes no feeding is provided. The feed millers usually sell feeds in 40–50 kg bags and this is very expensive for a small scale farmer.

Financing Initial Costs

Commercial Tilapia culture is still at an experimental stage and the financial assistance in the form of loans to other farmers may depend on the success of the government pilot project. The high cost of pond construction, water pumps, feed and rent are also a hindrance to commercial Tilapia culture.

Materials and Methods

Pearson Square Method and Linear Programming Least Cost Optimal Formulation Technique

The square method is helpful to novice feed formulators because it can get them started in diet formulation without the need to resort to trial and error. The square method can also be used to calculate the proportion of feed stuffs to mix together to achieve a desired dietary energy level as well as a crude protein level.

Linear programming is one of the types of mathematical programming models concerned with the efficient allocation of limited resources to known activities with the objective of meeting desired goals such as maximizing profit or minimizing costs. Linear programming is used here for allocating the available raw materials for feed formulation with their suitability to the existing environment (Rao, 1984).

Diet Formulation

In the first step in diet formulation is balanced the crude protein and energy levels by the square method for both crude protein level and energy level and then adjusted by solving simultaneous equations. During the initial balancing of protein and energy levels: one high in protein and high in metabolic energy (ME), one low or intermediate in protein and high in ME, and one low or intermediate in both protein and ME. Once practice makes one more proficient at diet formulation any number of feedstuffs can be used. One must remember to reserve room in the formulation for any feed additive, such as a vitamin or mineral pre-mix. In second step in diet formulation is checked the levels of indispensable amino acids in the formulation to meet the dietary level requirements of the animal to be fed. The requirements of fish for indispensable amino acids is expressed as the dietary level (as a per cent of the diet) or as a per cent of the dietary protein level. To convert an amino acid level from the per cent of diet to per cent of protein, divide the dietary level of each amino acid by the dietary protein level. Once the amino acid requirements are met, the dietary protein and energy levels are rechecked to, see if any substitution of ingredients has imbalanced the formulation. Diet mixing sheets are constructed to standardize diet formulation. In feed formulation, pellet quality and acceptability are considered in addition to nutrient levels and cost.

Feeding Experiment

The feeding experiment was conducted for 8 weeks in cages suspended in earthen pond (40m×40m×1m) and cement tank (5000L) by using a Completely Randomized Block Design (CRBD) (Plate 33.1). Pond was considered as a block and each cage and tank was treated as an experimental unit. The four earthen pond (200 m^2) located in AIT and four cement tank (3.31 m^3) located in hatchery, was used for the experiment. Male sex-reversed tilapia fry (*Oreochromis niloticus*) of initial weight 0.40-0.47 gram was used for experiment. Fry was randomly distributed into 16 groups of 400 fry each and stocked into 12 net cages (3m×1.8m×0.8m, 3mm mesh size) each 3 cages group that suspended in earthen pond and remaining 4 groups was stocked in cement tank. Wooden walkways were made to connect the cages to the pond bank. Water depth in the pond was maintained 1m throughout the experiment. Feeding rate: determined by 100 per cent satiation for treatment 1, 50 per cent satiation for treatment 2 and 0 per cent satiation for treatment 3. Every week feeding rate was determined by satiation method. Fertilizers were added in the each experimental pond weekly. Fishes were fed 2 times.day^{-1} at morning and afternoon for entire experiment.

Plate 33.1: Experimental Trials Conducted in Happa in Earthen Pond and Cement Tanks

Proximate Composition

Proximate composition of initial fish (before stocking) and harvested fish were analyzed. The fish moisture content was determined by drying the samples at 105°C to constant weight; the crude protein content was determined with the Kjeldahl method, using conversion factor of 6.25; the crude fat content was determined with the Soxhlet technique, with petroleum ether as a solvent (Folch *et al.*, 1957); the total ash content was determined by combusting the samples at 550°C and the crude fiber was also determined by using fibertec system (Foss, 2000).

Statistical Analysis

Data from individual fish were treated as independent samples. Proximate composition of the diets, growth performance and feed utilization efficiency parameters, were all subject to one-way analysis of variance (ANOVA). Analyses were performed by using statistical application. Data of water quality parameters open water were analyzed by using statistical t-test. Differences were considered significant at an alpha level of 0.05.

Results and Discussion

Water Quality

All the parameters of water quality that were monitored during the course of the study only pH were within the ranges considered suitable for the growth and survival of tilapia (*O. niloticus*). Water temperature and pH at the middle of the water column ranged 28.79-29.69 and 7.27-7.53, respectively throughout the experimental period. There was significant difference water temperature, which fluctuated. Measured DO concentrations at dawn significantly decreased during the experiment in all

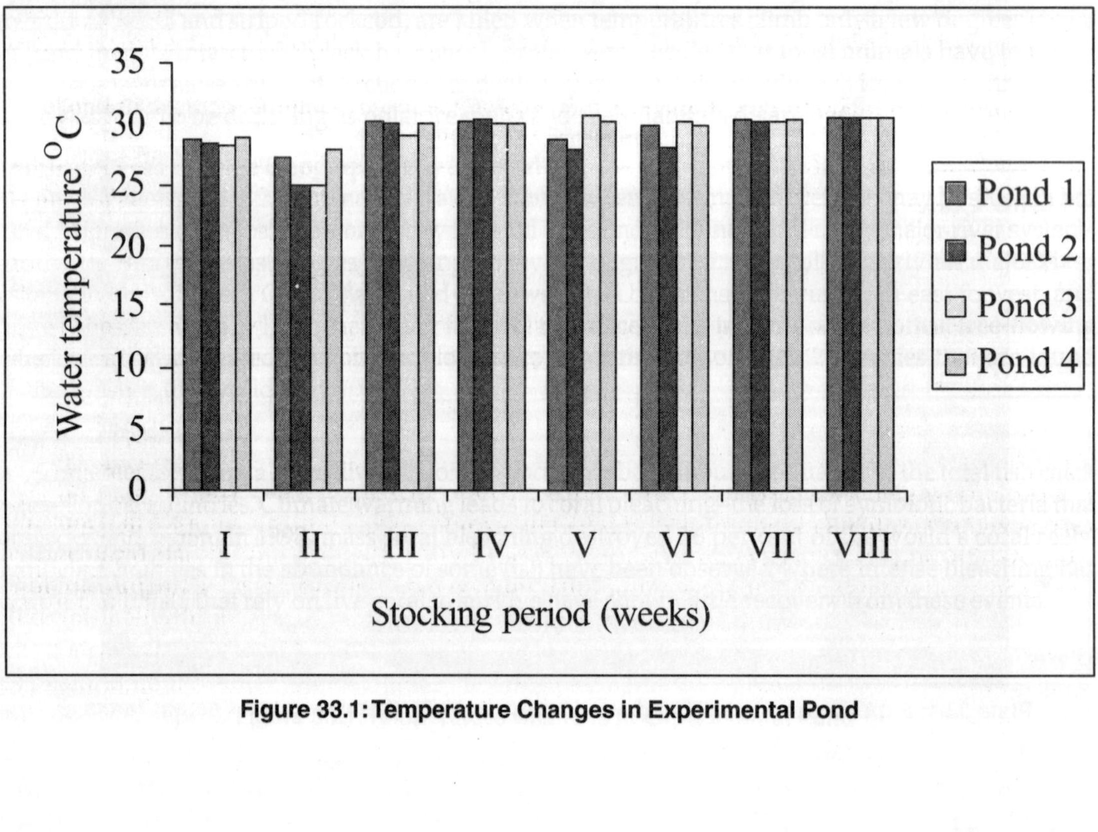

Figure 33.1: Temperature Changes in Experimental Pond

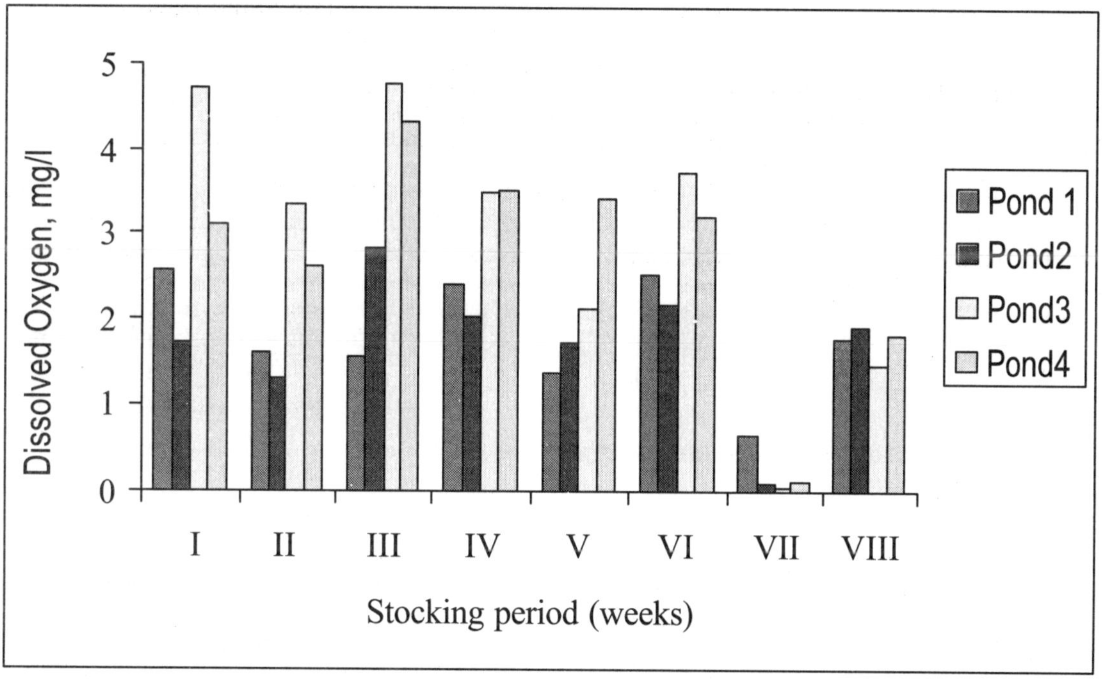

Figure 33.2: Dissolved Oxygen Changes in Experimental Pond

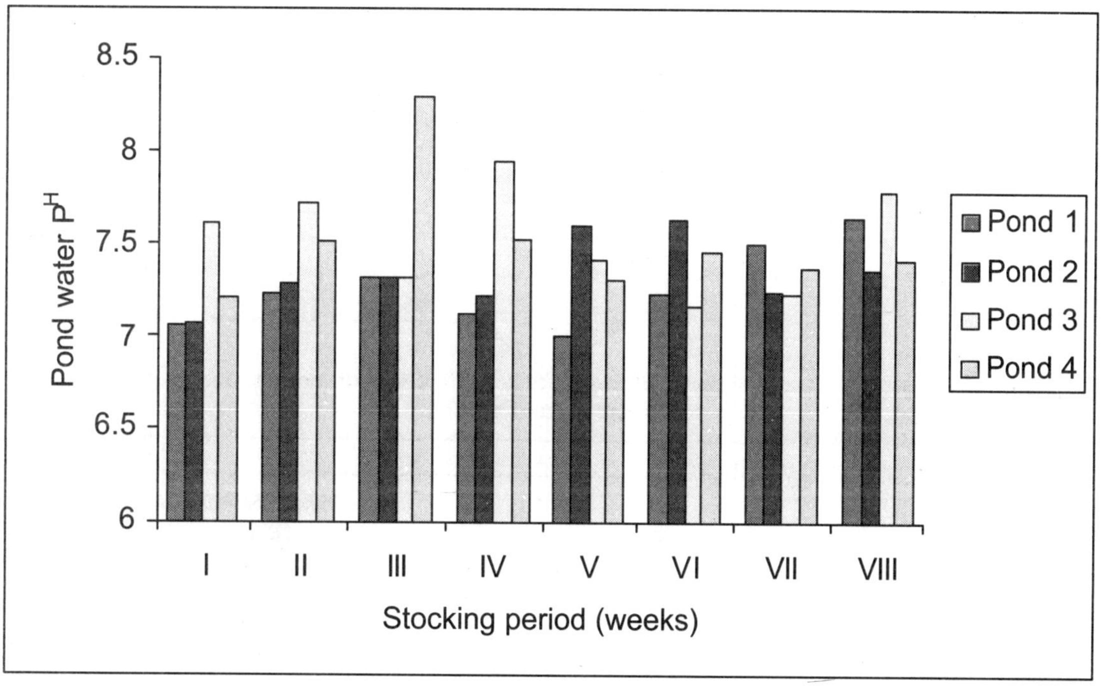

Figure 33.3: Water pH Changes in Experimental Pond

the ponds. The water temperature showed a small changed in the second week and after that there were the temperature were within the range (Figure 33.1).

The average dissolved oxygen was ranged 1.72-2.96 mg/l (Figure 33.2). The DO was very low in the seventh week which was below 0.5 $mg.l^{-1}$. Low dissolved oxygen can stress the fish, affecting their appetite, which in turn affects fish growth and can even lead to massive fish kills at very low levels (0.5$mg.l^{-1}$) (Rai, 1988). The pH value remained mostly within the range of 6.5-9.0 is suitable for aquaculture (Boyd, 1984). The pH value ranged between 7.27-7.53 throughout the experimental period (Figure 33.3).There was no significant difference in pH over time, but there were water temperature and dissolved oxygen was fluctuated.

The composition of feed is dependent on the contribution of nutrient come from natural food. Herbivorous/omnivorous fish requires relatively less per cent of macro-nutrients in their feed. In most animal diets, protein is the most expensive portion and is usually the first nutrient that is computed in diet formulation. The energy level of the diet is then adjusted to the desired level by addition of high energy supplements) which are less expensive than protein supplements. The square method is an easy way to determine the proper dietary proportions of high and low protein feedstuffs to add to a feed to meet the dietary requirement of the animal to be fed. In the new feed formulation, more than two feedstuffs are used in a feed. Rice bran having 13 per cent crude protein, Soybean meals having 45 per cent crude protein, fish meal 54 per cent crude protein and maize flour 14 per cent crude protein, they are grouped into basal feeds (CP<20 per cent) and protein supplements (CP>20 per cent), averaged within each group, and plugged into the square method. The desired protein level of the feed is 30 per cent crude protein, placed in the middle of the square. Next, the protein level of the feed is subtracted from that of the feedstuffs, placing the answer in the opposite corner from the feedstuff. Ignored positive or negative signs (Hardy, 1980).

The square method is helpful to novice feed formulators because it can get them started in diet formulation without the need to resort to trial and error. The square method can also be used to calculate the proportion of feed stuffs to mix together to achieve a desired dietary energy level as well as a crude protein level.

100- 4 (Lipid 2 per cent, Mineral 1 per cent and Vitamin 1 per cent)

Therefore, Protein diet is 96 per cent

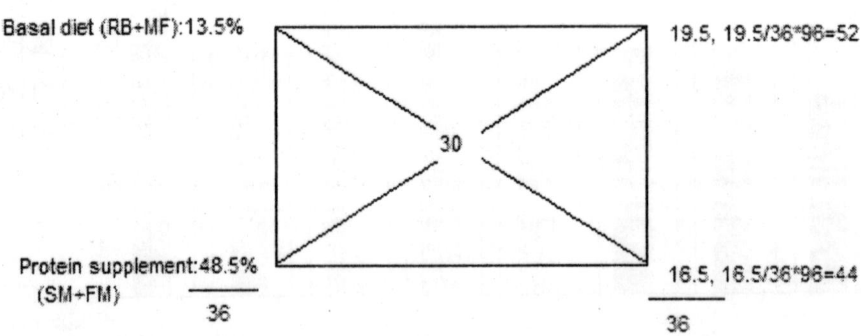

To make the 30 per cent crude protein in 100 kg of tilapia feed, we must mix:

1. Basal diet (RB+WF)=19.5/36*96=52
2. Protein supplement (SM+FM)=16.5/36*96=44

Table 33.1: New Feed Formulation and Cost per kg

Ingredients	*Rice Bran*	*RB Add*	*Maize Flour*	*MF Add*	*Mineral*	*Min Add*	*Fish Meal*	*FM Add*	*Soy-bean*	*SB Add*	*Corn Oil*	*CO Add*	*Vit. Mix.*	*Vit. Add.*	*Prox. Com*			
		26.00		22.00		1.00		22.00		26.00		2.00		1.00	100.00			
Protein	13	3.38	14.00	3.08	0.00	0.00	54.00	11.88	45.00	11.70	0.00	0.00	0.00	0.00	30.00			
Lipid	11.3	2.94	3.70	0.81	0.00	0.00	8.00	1.76	0.90	0.23	100.0	2.00	0.00	0.00	7.78			
Fibre	18.3	4.76	2.70	0.59	0.00	0.00	3.50	0.77	9.30	2.42	0.00	0.00	0.00	0.00	8.54			
Moisture%	10.10	2.92	9.00	2.18	0.00	0.00	9.50	2.31	8.00	2.26	0.00	0.00	10.00	0.11	9.78	8.91 Moisture-Final Mix		
Ash	10.2	2.65	1.40	0.31	100.0	1.00	18.00	3.96	0.60	0.16	0.00	0.00	80.00	0.80	8.88			
NFE	37.1	9.65	69.20	15.22	0.00	0.00	7.00	1.54	36.20	9.41	0.00	0.00	10.00	0.10	35.92			
Total	100.0	26.00	100.0	22.00	100.0	1.00	100.0	22.00	100.0	26.00	100.0	2.00	100.0	1.00	100.00			
Price (Rupees/kg As fed)	7	0.20	7.00	0.17	40.00	0.04	26.00	0.63	20.00	0.57	35.00	0.07	140.00	0.16	18.34	Rupees/kg Dry matter (US$ 0.40/kg dry feed)		
Ca	0.10	0.03	0.03	0.01	16.00	0.16	3.50	0.77	0.20	0.05	0.00	0.00	0.00	0.00	1.01			
Zn	0.01	0.00	0.02	0.00	14.00	0.14	0.50	0.11	0.10	0.03		0.00	0.00	0.00	0.28			
P	0.50	0.13	0.18	0.04	18.00	0.18	2.61	0.57	0.73	0.19	0.00	0.00	0.00	0.00	1.11			
Amino acid % DM																*g/100g P*	*Tilapia Req.*	*Che. Score*
Arg.	0.68	0.18	0.43	0.09	0.00	0.00	3.64	0.80	3.26	0.85	0.00	0.00	0.00	0.00	1.92	6.39	4.20	152.16
His.	0.30	0.08	0.25	0.06	0.00	0.00	0.65	0.14	1.10	0.29	0.00	0.00	0.00	0.00	0.56	1.87	1.72	108.77
Iso.	0.37	0.10	0.47	0.10	0.00	0.00	2.65	0.58	2.20	0.57	0.00	0.00	0.00	0.00	1.35	4.51	3.11	144.99
Leu.	0.72	0.19	0.87	0.19	0.00	0.00	4.96	1.09	3.30	0.86	0.00	0.00	0.00	0.00	2.33	7.75	3.39	228.58
Lys.	0.49	0.13	0.25	0.06	0.00	0.00	4.85	1.07	2.82	0.73	0.00	0.00	0.00	0.00	1.98	6.60	5.12	128.90
Met.	0.23	0.06	0.18	0.04	0.00	0.00	1.95	0.43	0.62	0.16	0.00	0.00	0.00	0.00	0.69	2.30	2.68	85.66
Cys.	0.14	0.04	0.30	0.07	0.00	0.00	0.48	0.11	0.66	0.17	0.00	0.00	0.00	0.00	0.38	1.26	0.00	
Phe.	0.47	0.12	0.60	0.13	0.00	0.00	2.54	0.56	2.16	0.56	0.00	0.00	0.00	0.00	1.37	4.58	3.75	122.02
Tyr.	0.19	0.05	0.34	0.07	0.00	0.00	2.23	0.49	1.50	0.39	0.00	0.00	0.00	0.00	1.00	3.34	0.00	
Thr.	0.42	0.11	0.33	0.07	0.00	0.00	3.16	0.70	1.72	0.45	0.00	0.00	0.00	0.00	1.32	4.41	3.75	117.55
Try.	0.10	0.03	0.10	0.02	0.00	0.00	0.56	0.12	0.62	0.16	0.00	0.00	0.00	0.00	0.33	1.11	1.00	110.65
Val.	0.59	0.15	0.50	0.11	0.00	0.00	3.16	0.70	2.24	0.58	0.00	0.00	0.00	0.00	1.54	5.13	2.80	183.21

Santiago and Lovell (1988).

US$1= 45.00 Indian Rupees.

Crude protein level of new feed:

1. Basal diet=13.5/96*52=7.31 per cent CP.
2. Protein supplement=49.5/96*44=22.69 per cent CP.

Therefore, total crude protein in new feed is 30 per cent. After balancing the crude protein in new feed. Partial replacement of fish meal with plant protein source meal due to the cost of fish meal is much higher than plant protein source meal. However, in order to get accurate feed formulation, least cost optimal formulation technique is used and showed in Table 33.1.

The Composition of new formulated feed is Rice Bran (RB) 26 per cent, Maize flour (MF) 22 per cent, Soybean meal (SM) 26 per cent and Fish meal (FM) 22 per cent. In addition, other ingredients should be added in the feed such as corn oil 2 per cent (as lipid source), vitamin 1 per cent and materials 1 per cent (Table 33.2). The lipid, vitamin and mineral may get from the natural food in the fertilized pond. The digestible energy of new formulated feed showed 13.40 and protein energy ratio was 22.41 mg protein.KJ^{-1} (Table 33.3).

Table 33.2: New feed formulation and cost per kg of wet basis

As Fed Mix	*Dry per cent*	*Moist per cent*	*Wet*	*g/kg Feed Mix*
Rice bran	26.00	10.10	289.21	266.14
Maize flour	22.00	9.00	241.76	222.48
Mineral	1.00	0.00	10.00	9.20
Fish Meal	22.00	9.50	243.09	223.71
Soybean	26.00	8.00	282.61	260.07
Corn Oil	2.00	0.00	20.00	18.40
Vit. Mix.	1.00	10.00	11.11	10.22
TOTAL	100.00	8.91	1086.67	1000.00
Dry Matter (g)			989.88	910.93
Moisture (g)			96.79	89.07
Moisture per cent			8.91	8.91
Feed Cost Rupees/kg WET (U$ 0.37/kg wet feed) 16.71				

US$ 1=45.00 Indian Rupees.

Table 33.3: Protein to Energy Ratio

	K. cal	*CF*	*KJ*	*Dig per cent*	*DE*	*Feed DE*
Carbohydrate	4.20	4.18	17.57	0.65	11.42	4.10
Lipid	9.45	4.18	39.54	0.95	37.56	2.91
Protein	5.65	4.18	23.64	0.90	21.28	6.39
Total DE						13.40
PE ratio						22.41 mg protein/KJ

Development of Feeding Methods

The success of a dry diet feeding regime is dependent not only on the formulation and manufacturing process used to produce the diet, but also on the method of presentation of the feed to the fish. Although the majority of large commercial tilapia farms in Europe normally use a fixed dietary feeding regime to administer their feed to the fish. Applying feeding rate at 3 per cent body weight per day, during starting supplemental feeding fertilizing the culture period, Fish feeding twice a day, morning and afternoon and applying 50 per cent supplemental satiation feeding level by fertilizing pond.

The hand feeding or manual feeding is best method for the tilapia feeding. The advantage of hand feeding is that it is the fish that dictates how much it wants to eat (and not the feeding technician), and by so doing allows the farmer to keep a regular check on fish feeding behavior and health, and water quality. Required quantity of food can be supplied. i. e. can avoid under feeding (slower growth) or over feeding (waste/pollution). Operator can disperse food over wide area. However, disadvantages are high labor costs especially for large farms and will increased food handling. The water temperature may affect on feeding rate of tilapia, so by manual feeding we can feed when water temperature is higher. Hand feeding to 70-80 per cent satiation may help to obtain best FCR and reduced the feeding cost. With this feeding strategy, the FCR of tilapia would be 0.75. However, this FCR is applicable in fertilized pond culture, so natural food should have satisfied in the culture pond. Therefore, tilapia farming cost may reduce significantly and more economic and environmental efficiency.

Growth Performance

At the end of the 8-week experimental period, the growth performance of fish fed different percentage of diets (0, 50, 100 per cent and tank) and feed utilization were evaluated using a number of parameters that are shown in Table 33.4.

Table 33.4: Daily Weight Gain, Specific Growth Rate, Net Yield and Survival Rate of Nile Tilapia, Fed Artificial Different Condition

Treat	*DWG (g/day)*	*SGR (per cent/day)*	*Net Yield (g)*	*Survival Rate (per cent)*
1	0.013 ± 0.003^{a}	0.921 ± 0.160^{a}	60.3 ± 83.290^{a}	68.06 ± 9.714^{NS}
2	0.060 ± 0.013^{b}	2.583 ± 0.303^{b}	890.1 ± 236.685^{b}	74.13 ± 16.379^{NS}
3	0.105 ± 0.029^{c}	3.358 ± 0.407^{c}	1763.1 ± 193.266^{c}	81.69 ± 6.745^{NS}
Tank	0.195 ± 0.026^{d}	4.382 ± 0.242^{d}	2716.7 ± 125.908^{d}	66.00 ± 8.496^{NS}

Values are mean ± s.e.m. (N=4).

Means within the same column not sharing a common superscript are significantly different ($P<0.05$)

Fish appeared healthy at the end of the trial with survival ranging from 66.00±8.49 to 81.69±6.74 per cent among all group of fish, while there were a number of mortalities unrelated to diet. Daily weight gain were of 0.012 g per fish day^{-1} in 0 per cent diet, 0.06 g per fish day^{-1} in 50 per cent, 0.10 g per fish day^{-1} in 100 per cent diet and 0.19 g per fish day^{-1} in 100 per cent diet in tank, where as specific growth rate (SGR) were in the ranges of 0.92 per cent/day in 0 per cent diet, 2.58 per cent/day in 50 per cent, 3.35 per cent/day in 100 per cent diet and 4.38 per cent/day in 100 per cent diet in tank. Net yield were of 60.3±83.29 g per fish day^{-1} in 0 per cent diet, 890.1±23.68 g per fish day^{-1} in 50 per cent, 1763.1±19.26 g per fish day^{-1} in 100 per cent diet and 2716.7±125.90 g per fish day^{-1} in 100 per cent diet in tank.

Protein intake (PI) were ranges 0.650±0.48 per cent in 50 per cent diet, 1.300±0.97 per cent in 1 diet and 1.656±0.08 per cent in 100 per cent diet in tank, Protein efficiency ratios (PER) were within the range 0.376±0.38 in 50 per cent diet, 1.35±0.10 in 100 per cent diet and 1.641±0.02 in 100 per cent diet in tank. Food conversion ratio (FCR) were 2.310±0.79 in 50 per cent diet, 2.188±0.18 in 100 per cent diet and 1.80±0.27 in 100 per cent diet in tank. Food conversion efficiency (FCE) were 0.465±0.13 in 50 per cent diet, 0.460±0.33 in 100 per cent diet and 0.555±0.01 in 100 per cent diet in tank and Apparent net protein utilization (ANPU) were 29.339±2.17 in 50 per cent diet, 14.591±0.89 in 100 per cent diet and 11.488±0.64 in 100 per cent diet in tank (Table 33.5).

Table 33.5: Protein Intake, Feed Conversion Ratio, Feed Conversion Efficiency, Protein Energy Ratio and Apparent Net Protein Utilization of Nile Tilapia, Fed Artificial Different Condition

Treat	*PI*	*PER*	*FCR*	*FCE*	*ANPU*
2	0.650±0.48^{a}	0.376±0.38	2.310±0.79	0.465±0.13	29.339±2.17^{a}
3	1.300±0.97^{b}	1.356±0.10	2.188±0.18	0.460±0.33	14.591±0.89^{b}
Tank	1.656±0.08^{c}	1.641±0.02	1.800±0.27	0.555±0.01	11.488±0.64^{c}

Values are mean ± s.e.m. (N=4).

Means within the same column not sharing a common superscript are significantly different (P<0.05)

The treatment second (50 per cent feed) having good survival rate, 74.13±16.37 per cent. Daily weight gain was 0.060 g/fish day^{-1}, specific growth rate (SGR) was 2.58 per cent/day and net yield was 890.1±23.68 g/fish day^{-1}.That indicates, the treatment second (50 per cent feed) having good growth performance and have lower FCR *i.e.* only 50 per cent feed was supplied. However the growth performance parameter of treatment 3 and 4 were lower by considering intake of daily foods.

The survival rate of treatment 4 was also very lower because, may be the lower dissolve oxygen in the tank. The FCR was also very high in the tank and treatment third as compare to other treatments.

Feeding Economics

With new feeding strategy, the FCR of tilapia is 0.75, and the amount of feed required around 750 kg for production of one tone of fish. The cost of new formulated feed is around Rs.16.71/kg as fed basis (U$ 0.37/kg fed basis, US$1=45.00 Rupees).The feed manufacturing cost is around Rs.0.50/kg of feed. So the total cost of new formulated feed is around 17.21Rupees/kg (US$ 0.38/kg fed basis) as fed basis. The farm gate price of tilapia is around Rs.30 per kg of fish (US$ 0.66 per kg of fish). Final feed operation cost is 43-57.36 per cent of farm gate price on FCR 0.75:1.The price of company formulated feed is around 19.00/kg of feed having FCR 1-1.2 in commercial farm condition and the feed operation cost is 63.33-76 per cent of farm gate price.

Therefore, with the above feeding strategy, the gross income from one tone of tilapia would be Rs. 30000.00 (US$ 666.66), while cost of feed for production of one tone tilapia is Rs. 12907.00-17210.00 (US$ 286.8-382.44). It accounts around 43-57.36 per cent of the final feed operation cost of farm gate price of tilapia farming. This new feed formulation would reduce the cost of feed and more economic efficiency for tilapia culture.

Conclusion

One of the great advantages of tilapia for aquaculture is that they feed on a low trophic level. The members of the genus *Oreochromis* are all omnivores, feeding on algae, aquatic plants, small invertebrates, detritus material and the associated bacterial films. This provides an advantage to farmers because the fish can be reared in extensive situations that depend upon the natural productivity of a water body or in intensive systems that can be operated with lower cost feeds with 80 per cent satiation. In extensive aquaculture, the fishes are able to grow by eating algae and detritus matter and the farmer can grow more fish in a given area because the fish are depending directly on the primary productivity of the body of water, primary consumers.

In new farm made fed a prepared high percentage of plant proteins, nutritional studies which substitute plant proteins supplemented with specific amino acid supplements may lower costs.

The new proposed feed formulation and feeding method would contribute significantly in reducing tilapia farming operation cost. This is better than existing practice carried by the small scale fish farmers. The total cost of new formulated feed is around 16.71Rupees/kg as wet basis (moisture 8.91 per cent) and the new formulated feed cost accounts 43-57.36 per cent in the gross revenue with 80 per cent satiation.

This could concluded that Partial replacement of fish meal 22 per cent, soybean meal 26 per cent, rice bran 26 per cent and maize flour 22 per cent with corn oil 2 per cent, vitamin premix 1 per cent and mineral premix 1 per cent in the new feed formulation will contribute in reducing feed cost and more economic efficiency in tilapia farm operation. The new formulated on farm made feed with local and cheaper available ingredient materials may reduce feed cost around 19.48 per cent in the gross revenue by manual feeding and fertilizing the pond.

References

Boyd, C.E., 1984. *Water Quality in Warm Water Fish Ponds.* Auburn University Agriculture Experimental Station, Auburn: 359.

Boyd, C.E., 1997. Pond water aeration system. *Aquaculture Engineering,* (18): 9–40.

Bruton, M.N. and Bolt, R.E., 1975. Aspects of the biology of *Tilapia mossambica* Peters (Pisces: Cichlidae) in a natural freshwater lake (Lake Sibaya, South Africa). *Journal of Fish Biology,* (7): 423–445.

Caulton, M.S. and Hill, B.J., 1973. The ability of *Tilapia mossambica* (Peters) to enter deep water. *Journal of Fish Biology,* (5): 783–788.

Chang, S.L., Huang, C.M. and Liao, I.C., 1988. Effects of various feeds on seed production by Taiwanese red tilapia. In: *Proceedings of the 2nd International Symposium on Tilapia in Aquaculture.* ICLARM, Bangkok.

Chou, B.S. and Shiau, S.Y., 1999. both n-3 and n-6 fatty acids are required for maximal growth of juveniles hybrid tilapia. *North American Journal of Aquaculture,* (61): 13–20.

Cruz, E.M. and Laudencia, I.R., 1977. Protein Requirements of tilapia. Fingerlings. *Philippines Journal of Biology,* (6): 177–182.

Eid, A.E. and Ghonim, S.I., 1994. Dietary zinc requirement of fingerlings *O. niloticus. Aquaculture,* (119): 259–264.

El-Sayed, A.F.M. and Temisha, S.I., 1992. Protein and energy requirements of Nile tilapia, *Oreochromis niloticus,* fry. *Aquaculture,* (103): 55–63.

El-Sayed, A.F.M., 1999. Alternative dietary protein sources for farmed tilapia. *Aquaculture,* (179): 149–168.

FAO, 2004. Fish stat Plus. *Aquaculture Production 1950–2002.*

Fitzsimmons, K., 2000. "Tilapia: the most important aquaculture species of the 21st century". In: *Tilapia Aquaculture in the 21st Century, Proceeding from the Fifth International Symposium on Tilapia Aquaculture,* (Eds.) K. Fitzsimmons and J. Carvalho Filho. Rio de Janeiro, Rio de Janeiro, Brazil, 3–8.

Folch, J., Lees, M. and Sloane-Stanley, G.H., 1957. A Simple method for the isolation and purification of total lipid from animal tissue.*J. Biol. Chem.*, (226): 497–509.

Foss, Tecator., 2000. M6 1020/1021 manual: The digestion of nitrogen according to Kjeldahl block digetion and steam distillation, pp.11.

Gautier, J.Y., Lefaucheux, B., Foraster, M., jalabert, B. and Baroiller, J.F., 2000. Periodicity and duration of papillary, sexual and begaveral cycle in the tilapia *Oreochromis niloticus*. In: *Tilapia Aquaculture in the 21st century. Proceedings from the Fifth International Symposium on Tilapia Aquaculture,* Brazil, Sept., Vol. 1.

Guerrero, R.D., 1982. Control of tilapia reproduction, pp. 309–316. In: *The Biology and Culture of Tilapias,* (Eds.) R.S.V. Pullin and R.H. Lowe-McConnell. ICLARM Conference Proceedings 7. International Center for Living Aquatic Resources Management, Manila, Philippines, p. 432.

Hardy, R., 1980. Fish feed formulation. Lectures presented at the FAO/UNDP Training Course in Fish Feed Technology, held at the College of Fisheries, University of Washington, Seattle, Washington. 9 October–15 November, 1978. FAO, 1980. ADCP/REP/80/11. pp. 233–240.

Hewitt, R.P., Theilacker, G.H. and Lo, C.H., 1985. Causes of mortality in young jack. Hardy, R. University of Washington Seattle, Washington, FAO Publication.

Jauncey, K. and Ross, B., 1982. *A Guide to Tilapia Feeds and Feeding.* University of Sterling, Scotland.

Jauncey, K., 2000. Nutrients requirements. In: *Tilapias: Biology and Exploitation,* (Eds.) M.C.M. Beveridge and B.J. McAndrewn. Kluwer Academic Publisheires, pp. 327–375.

Kamarudin, M.S., Kaliapan, K.M. and Siraj, S.S., 1989. The digestibility of several feedstuffs in red tilapia. In: *Fish Nutrition Research in Asia,* (Ed.) S.S. Desilva. Proceeding of The Third Asian Fish Nutrition Network Meeting, Spec. Publ. Asian Fish. Soc., pp. 118–122.

Lam, T.J., 1982. Fish culture in Southeast Asia. *Cand. J. Fish. Aquat. Sci.*, 39(1): 138–142.

Luquet, P., 1991. Tilapia. In: *Handbook Nutrient Requirement of Finfish,* (Ed.) R.P. Wilson. CRC press, Boca Raton, Florida, pp. 169–179.

Lim, C. and Master Lea, B., 1991. In: Program and abstract, *World Aquaculture Society 22nd Annual Conference and Exposition,* p.39.

Lovshin, L.L., 1997. Worldwide tilapia culture. In: *Workshop International Aquaculture,* October 15–17, Sao Paulo, Brazil, pp. 96–166.

Mary King, 2000. Program Assistant for Family and Consumer Sciences Cooperative Extension Service for Sarasota County as written for the Sarasota Herald–Tribune, J. Food Section.

Nwanna, L.C., 2003. Risk Management in Aquaculture by Controlled Feeding Regimen. © Asian Network for Scientific Information 2003. *Pakistan Journal of Nutrition,* 2(6): 324–328.

Ogunji, J.O. and Wirth, M., 2000. Effect of dietary protein content and sources on growth, food conversion and body composition of tilapia *O. niloticus* fingerling fed fish meal diet. *J. of Aquaculture in the Tropics,* 15(4): 381–389.

Philippart, J.Cl. and Ruwet, J.Cl., 1982. Ecology and distribution of the tilapia. In: *The Biology and Culture of Tilapia. ICLARM Conference Proceeding*, Vol. 7. (Eds.) R.S.V. Pullin and R.H. Lowe-McConnell (eds). ICLARM, Manila, pp. 15–59.

Popma, T. and Masser, M., 1999. *Tilapia: Life History and Biology*. Southern Regional Aquacultural Center, 283 pp.

Rai, A.K., 1988. Effects of water quality and zooplankton communities on growth of hybrid carp (bighead carp *Aristichthys nobilis* 3 silver carp *Hypophthalmichthys molitrix*) in cages. *Master's Thesis*, Auburn University, Auburn, pp. 51.

Robinson, E.H., Rawles, S.D., Yette, H.E. and Greene, L.W., 1984. An estimation of dietary calcium requirement of tilapia reared in calcium free water. *Aquaculture*, 41: 389–393.

Rao, S.S., 1984. *Optimization: Theory and Application*. 2nd edn. Wiley Estern, New York.

Ross, L.G., 2000. Environmental physiology and energies. In: *Tilapias: Biology and Exploitation*, (Eds.) M.C.M. Beveridge and B.J. McAndrewn. Kluwer Academic Publishers, pp. 81–128.

Santiago, C.B. and Lovell, R.T., 1988. Amino acid requirements for growth of Nile tilapia. *The Journal of Nutrition*, 118: 1212, 1540–1546.

Sargent, J.R. and Tacon, A.G.J., 1999. Development of farmed fish: A nutritionally necessary alternative to meat. *Pro. Nutr. Soc.*, 58: 377–383.

Satoh, S.T., Takeuchi, T. and Watanabe, T., 1987. Requirement of tilapia for α-tocopherol. *Nippon Suisan Gakkaishi* 53., 119– 124.

Shiau, S.Y. and Suen, G.S., 1992. Estimation of the niacin requirements for tilapia fed diets containing glucose or dextrin.*Journal of Nutrition*, 122: 2030–2036.

Shiau, S.Y. and Lung, C.Q., 1993. No dietary vitamin B_{12} required for juvenile tilapia, *O. niloticus* × *O. aureus*. *Comparative Biochemistry and Physiology*, 105A: 147–150.

Shiau, S.Y. and Lin, S.F., 1993. Effect of supplemental dietary chromium and vanadium on the utilization of different carbohydrates in tilapia, *O. niloticus* × *O. aureus*. *Aquaculture*, (110): 321–330.

Shiau, S.Y. and Hwang, J.Y., 1993. Vitamin D requirements of juvenile hybrid tilapia, *O. niloticus* × *O. aureus*. *Nippon Suisan Gakkaishi*, 59: 553–558.

Shiau, S.Y. and Lo, P.S., 2000. Dietary choline requirements of juvenile tilapia, *O. niloticus* × *O. aureus*. *Journal of Nutrition*, 130: 100–103.

Shiau, Shi Yen, 2002. Nutrient requirements feeding of finfish. *Aquaculture*, p. 273–291.

Shi, Y.S., 1997. Utilization of carbohydrates in warm water fish with particular reference to tilapia, *O. niloticus* × *O. aureus*. *Aquaculture*, 151(1): 79–96.

Soliman, A.K., Jauncey, K. and Roberts, R.J., 1994. Water-soluble vitamin requirements of tilapia: Ascorbic acid (vitamin C) requirement of tilapia, *Oreochromis niloticus* (L.). *Aquaculture Fish. Management*, 25(3): 269–278.

Steffens, W., 1981. Protein utilization by rainbow trout (*Salmo gairdneri*) and carp (*Cyprinus carpio*): A brief review. *Aquaculture*, (23): 337–345.

Stickney, R.R., 1996. Tilapia update, 1995. *World Aquaculture*, 27(1): 45–50.

Thakur, D.P., Yi, Y., Diana, J.S. and Lin, C. K., 2001. Effects of fertilization and feeding strategy on water quality, growth perfomance, nutrient utilization and economic return in nile tilapia (*oreochromis niloticus*) ponds.

Thomas, Popma and Michael, Masser, 1999. *Tilapia: Life History and Biology*. Southern Regional Aquacultural Center, pp. 283.

Trewavas, E., 1983. Tilapiine fishes of the genera Sarotherodon, Oreochromis and Danakilia. *British Musium* (Natural history), London, Publication No. 878.

Tung, P.H. and Shiau, S.Y., 1993. Carbohydrate utilization versus body size in tilapia, *O. niloticus* × *O. aureus. Comparative Biochemistry and Physiology*, 104A: 585–588.

Watanabe, T., Takeuchi, T., Muralami, A. and Ogino, C., 1980. The availability to *Tilapia nilotica* of phosphorous in white fish meal. *Bull. Jap. Soc. Sci. Fish*, 46(7): 897–899.

Wilson, R.P., 1989. Amino acids and proteins. In: *Fish Nutrition*, (Ed.) J.E. Halver. Academic Press London, p. 111–151.

Wilson, R.P., 1994. Utilization of carbohydrate by fish. *Aquaculture*, 124: 67–80.

Winfree, R.A. and Stickney, R.R., 1981. Effects of dietary protein and energy on growth feed conversion efficiency and body composition of *Tilapia aurea*. *Journal of Nutrition*, 1(6): 1001–1012.

Yi, Y., Diana, J.S. and Lin, C.K., 2003. *Supplemental Feeding for Red Tilapia Culture in Brackishwater*. AIT Publication, pp. 451–462.

Chapter 34

Impact of Probiotic Supplementation in Freshwater Fish *Cyprinus carpio*

☆ *R. Thamarai Selva, V. Vignesh, S. Jothi Lakshmi and R. Thirumurugan*

ABSTRACT

The infections of bacteria are thought to be a major cause of mortality in fish hatcheries. In order to overcome these infections, antibiotics and chemotherapeutics treatment are employed as control measure in the aquaculture industry. But the use of antibiotics makes microorganisms gradually drug-resistant organisms. In this study, an attempt has been made to study the role of probiotic supplementation and their antagonistic properties in freshwater fish *Cyprinus carpio*. Probiotic organisms were isolated from dairy products. The fish pathogens were obtained from infected ornamental fish. Primary and secondary screenings were performed for antimicrobial activity of probiotics against fish pathogens. In addition, the antimicrobial activity of commercial antibiotic discs against fish pathogens was assessed. The potential probiotics were used as supplements with feed in different concentrations (100, 250, 500 and 1000 µl/1g feed). It showed that the highest input of probiotics along with feed increased the weight and survival rate when compared to others.

Introduction

In the recent years, aquaculture has become a very fast growing sector in animal production. However, the number of outbreaks of bacterial diseases in cultured fish has also increased. Microbiological research has been generally focused to detect the harmful intestinal bacteria and the methods to control them. Very little research has been done to detect the beneficial bacteria, especially the lactic acid bacteria group and its applications. Beneficial bacteria in the best cases could be used to substitute the use of antibiotics as preventive agents of disease (Nikoskelainen *et al.*, 2001) and as growth promoters (Bvun *et al.*, 1997).

A growing concern about the high consumption of antibiotics in aquaculture has initiated a search for alternative methods of disease control (Gildberg *et al.*, 1997). Probiotic are living preparations of microbial cells that, when ingested in high enough concentration, beneficially affect the host's health and growth by improving the intestinal microbial balance (Fuller, 1989; Havenaar *et al.*, 1992). Selection of probiotic strains is achieved by screening procedures for several characteristics *in vitro*, such as inhibitory activities against several fish pathogens and gastric and intestinal secretions (Byun *et al.*, 1997). Once the strains have been selected, a way to supply them to the host via feeds or water needs to be developed. This is a very important aspect as great losses in viability during processing and storage are generally reported (Havenaar *et al.*, 1992).

Probiotic supplementation has been proposed as a means to increase the numbers of beneficial bacteria and to reduce the numbers of harmful bacteria (Gildberg *et al.*, 1997). Probiotics can influence physiology and health through direct or indirect effects in the gastrointestinal tract. For most of the applications, it is recommended to use probiotic strains able to survive the gastrointestinal passage, and preferentially with the capacity to adhere to the host intestinal tissues in order to prolong their health effects (Fuller, 1989; Havenaar *et al.*, 1992; Ouwehand *et al.*, 1999). Most of the probiotics, however, are completely excreted in the following days after its ingestion (Robertson *et al.*, 2000). If a probiotic does not colonize, they should be regularly administered to obtain its effects. Knowledge on their pharmokinetics is needed to answer the questions how much probiotic should be consumed, how often, for how long, and what concentration of probiotics should be present in the commercial products and also to correlate the effects with the concentrations of the probiotic at the target site (Marteau and Veda, 1998).

Materials and Methods

Isolation of Probiotics

Probiotic organisms were isolated from different milk products. The probiotic sources used in the present study were the different varieties of milk products such as Raw milk, Pasteurized milk, Milk powder, Cheese, Curd and also Lactobacillus powder and sporolac powder. 1 g or 1 ml of different probiotic sources were added to separate tubes containing 9 ml of sterile distilled water and serially diluted upto 10^{-10} dilutions. 0.1 ml suspension of the selected dilutions was spread pated on the De Man Rogosa Sharpe (MRS) medium. Then the plates were incubated in anaerobic condition for 2 days.

Pure cultures of bacterial colonies were picked as single colony by repeated streaking on the fresh MRS agar medium. The pure cultures were transferred to freshly prepared agar slants and preserved at 4°C.

Test Pathogens (Bacteria)

The selected 4 bacterial isolates were inoculated into nutrient broth and incubated for overnight at 37°C (*Escherichia coli*, *Enterobacter* sp., *Pseudomonas aeruginosa* and *Vibrio* sp.).

Screening for Antagonistic Activity

The antibacterial activity of the isolated probiotic against the test microbial isolates was done by Modified Agar Overlay method and Agar well Diffusion method.

Primary Screening

The 26 isolated probiotic strains were screened against all the test pathogens using Modified Agar Overlay method.

Modified Agar Overlay method

The modified agar overlay method was used to test for the presence of antimicrobial activities among the probiotic isolates. Isolated probiotic organisms were inoculated into Mullar Hinton medium in gridded plates by tooth pick method, incubated at 37°C for 24 hrs. 5 ml of Melted soft agar tubes are added with 5 µl of test organisms which is then added to the respective gridded plated. Then the plated are observed for the inhibition zone around the colony.

Secondary Screening

To determine the potent antimicrobial activity of the isolates, the extracts obtained from 26 probiotics were tested against all the test pathogens using Agar Well Diffusion Method.

Extraction of Bioactive Compounds

The screened probiotic bacterial isolates were inoculated into MRS broth tubes and incubated at 37°C for overnight. The supernatant was obtained by the centrifugation at 10,000 rpm for 20 minutes and then used for further analysis.

Well Diffusion Assay

It was described by Schillinger and Lucke (1989) was used for the test of antagonistic activities of probiotic isolates against indicator test organisms. Required wells each of 7 mm in diameter were made in test organism swabbed agar plates using cork borer and 100 µl probiotic extract were transferred into each well. The plates were incubated for 24 hrs at 37°C and examined for clear inhibition zone around the well. The assay was carried out in duplicate for the all the test organisms.

Antimicrobial Activity Using Standard Antibiotic Discs

The isolated pathogens were subjected to antibiotic technique to find out this resistance and susceptibility to a group of selected antibiotic discs using Kirby-Bauyer method. The susceptibility of the isolated microbes to selected eight antibiotics by Kirby-Bauyer method (cappuccino and Sherman, 1996). The following commercial discs antimicrobial substances used were: Cephaloridine (Cr^{30}), Amoxycilin (Am^{10}), Chloramophenicol (C^{30}), Doxycycline hydrochloride (Do^{30}), Rifampicin (R^{5}), Erythromycin (E^{15}), Amikacin (Ak^{30}) and Ampicilin (A^{10}).

Bacteria were inoculated on Nutrient broth and were incubated for 24 hrs at 37°C. Then sterile cotton swabs dipped into the culture were used evenly to inoculate the Muller-Hinton agar plate. After, the agar surface sets dried for about 5 minutes, the antibiotic discs were placed on it with a sterile forceps. The plates were incubated for 24 hrs at 37°C. After overnight incubation at 37°C and zone of inhibition was measured (mm) and the result were interpreted according to the instruction of the manufacturer.

Selection of Potent Probiotic

Selection of potent probiotic and most susceptible bacterial isolates are determined from the result of agar well diffusion assay (zone of inhibition in mm). After selection of potent probiotic isolates, these isolates were subjected to autoclaving at 121°C for 15 minutes. Then they were streaked on MRS medium and incubated. The isolate that grow even after autoclaving was used for feed supplementation of the experimental fishes.

Feed Preparation

Commercially available pellets from feeds were brought from the aquarium. The experimental diets such as D_2 to D_5 were supplemented with selected the potent probiotic *i.e. Lactobacillus sps* 3RM-isolated from raw milk sample at 4 different levels, 100, 250, 500 and 1000 µl per 1 gram of pellet feed respectively. The control diet D_1 was not supplemented with the potent probiotic. The supplemental feed was mixed with desired concentration of probiotics in MRS broth suspension culture. The homogenous mixture of different feed were spread over the aluminum foil and dried at 40°C in a hot air oven. These dried mixtures were provided to the fishes in the different experimental levels.

Experimental Animals and Design of Experiments

Juveniles of *Cyprinus carpio* was collected from A.M. Aqua farm, Arumpanoor pudur, Madurai, Tamil Nadu, and were transported to the laboratory in double polyethylene bags, filled with bond water and oxygenated to the saturation level. In the laboratory condition, the fishes were acclimatized in 10 liter plastic trough for days. After acclimation the health and approximately same size fish were selected and weighted individually. Five fishes were stocked in each trough. The fishes were starved for 24 hrs before starting the experiment in order to empty the gut contents.

The experimental tanks constituting the treatments were arranged according to completely randomized design with three replications per treatment. Feeding trials were conducted indoors (laboratory) under the approximate photoperiod of 12 hrs light : 12 hrs dark. An electric aerator provided continuous but slow aeration to the experimental tanks except during placement of feed and during collection of faeces and uneaten feed. One third of the water in each tank was changed daily with fresh dechlorinated tap water.

Dead animals if any, were collected and recorded. Fishes were fed at 5 per cent of the biomass once a day at 24 hrs. The faeces were siphoned out manually as soon as they appeared in the trough in order to minimize the time of contact between the water and faeces and thus prevent leaching. The excess feed that remained at the bottom of the trough was also siphoned out onto a bolting cloth before offering the next ration. The collected faecal pellets and unconsumed feed were stored separately in glass plates after drying at 60°C in hot air oven. Feed ration adjustment was done depending upon the consumption of the previous amount of feed, in such a way minimize the wastage of feed.

The weight of the animals, at every 10 days interval, was recorded with an electronic balance after removing the excess water by using blotting paper. The duration of all the feeding experiments was 30 days.

The gathered data were treated with the following formula for the evaluation of test feeds performance

$$\text{Weight gain (\%)} = \frac{\text{Final weight} - \text{Initial weight}}{\text{Initial weight}} \times 100$$

$$\text{Length gain (\%)} = \frac{\text{Final length} - \text{Initial length}}{\text{Initial length}} \times 100$$

$$\text{Specific growth rate (SGR)} = \frac{(\text{Wt} - \text{Wi})}{\text{T}} \times 100$$

where,

Wt: The average weight of fish at time 't'

Wi: Is the initial average weight of fish

T: Is the duration of feeding trial in days.

Feed Conversion Ratio (FCR)

$$\text{Feed conversion ratio (FCR)} = \frac{\text{Dry feed intake (g)}}{\text{Wet body weight gain (g)}} \times 100$$

$$\text{Feed efficiency (FE \%)} = \frac{\text{Wet weight gain (g)}}{\text{Dry feed consumed (g)}} \times 100$$

$$\text{Percentage of survival} = \frac{\text{Nf}}{\text{Ni}} \times 100$$

where,

Nf: Is the final number of fishes in the experiment

Ni: Is the inital number of fishes in the experiment.

Statistical Analysis

All the results (feeding experiment) were subjected to the standard statistical analysis using the technique of student's t-test.

Results and Discussion

Probiotic organisms were isolated from different milk products. The probiotic sources used in the present study wee the different varieties of milk products such as Raw milk, Pasteurized milk, Milk powder, Cheese, Curd and also *Lactobacillus* powder and sporolac powder. Twenty six probiotic organisms were isolated in accordance with differences in colony morphology and numbered as mention in Table 34.1. The main strategy of using probiotics is to isolate intestinal bacteria with favorable properties from mature animals and include them in the feed for immature animals of the same species (Gildberg *et al.*, 1997). The beneficial effects of probiotics in larvae rearing have been demonstrated in other fishes by several investigators (Dabrowski and Kaushik 1985).

The primary screening procedure was carried out to test the antagonistic property of the 26 isolates from different varieties of milk products. Totally 6 indicator fish pathogens including *Escherichia coli, Enterobacter, Pseudomonas aeruginosa, Vibrio cholera, Vibrio vulnificus* and *Vibrio sps* were used antimicrobial activity study (Table 34.1).

The antimicrobial activity of probiotic organism *Lactobacillus* (Probiotic extract 50 ìl were used in well diffusion method) against various fish pathogens. Results of antimicrobial activity of various isolated probiotic organisms *Lactobacillus* against *Escherichia coli, Enterobacter, Pseudomonas aeruginosa, Vibrio vulnificus* and *Vibrio* sps showed interesting observations. Although the antibacterial activity of various probiotic *Lactobacillus* only sensitive to the fish pathogen such as *Vibrio cholera, Vibrio vulnificus* and *Vibrio sps* when compared to other fish pathogens such as *Escherichia coli* and *Enterobacter*. There is no zone of inhibition record in almost of the *Lactobacillus* against *Escherichia coli, Enterobacter, Pseudomonas aeruginosa*. The selected 8 potent probiotics (3RM, 3PM, 6PM, 5C, 2S, 2CH, 2MP and 1SP) are give the higher zone of inhibition from the result of agar well diffusion assay. The potent probiotic 3RM was the maximum zone of inhibition then the remaining 7 potent probiotic (Table 34.2). The emergence of antimicrobial resistant following use of antimicrobial agents in aquaculture has been identified in fish pathogens (Midtevdt and Lingaas, 1992). For example, in several countries

A. salmonicida is frequently resistant to multiple drugs including sulphonamides, tetracycline, amoxicillin, trimethoprim-sulpo dimethyoxine and quinolones (Inglis *et al.*, 1991), antimicrobial agents which are commonly used in aquaculture the first isolation of *A. salmonicida* resistant to a specific antimicrobial agent has often been reported shortly the introduction of the agents into aquaculture. Similar correlation between antimicrobial agents used in aquaculture an antimicrobial resistance are also reported among other fish pathogens (Takashima *et al.*, 1985).

Table 34.1: Isolation of Probiotic Organisms *Lactobacillus* from Various Samples

Samples	*Labeling*	*Samples*	*Labeling*
Raw milk		**Curd**	
1	1RM	13	1C
2	2 RM	14	2 C
3	3 RM	15	3 C
4	4 RM	16	4 C
5	5 RM	17	5 C
6	6 RM		6 C
Pasteurized milk		**Soil sample for milk industry**	
7	1PM	18	1S
8	2:00 PM	19	2S
9	3:00 PM	20	3S
10	4:00 PM		
11	5:00 PM		
12	6:00 PM		

Table 34.2: Antimicrobial Activity of Probiotic *Lactobacillus* Against Various Fish Pathogens (Primary screening)

Sl.No.	*Name of the Test Organisms*	*Isolated Probiotics*
1	*Escherichia coli*	16
2	*Enterobacter*	16,17
3	*Pseudomonas aeruginosa*	11,20,24
4	*Vibrio cholerae*	1,2,3,4,9,10,11,12,13,19,21,23,24,30
5	*Vibrio vulnificus*	8,13,14,19,20,22
6	*Vibrio* sps	6,11,12,13,19,20,24,27

The antimicrobial activity was carried out in various fish pathogen tested against antimicrobial substance are Cephaloridine (Cr^{30}), Amoxycilin (Am^{10}), Chloramophenicol (C^{30}), Doxycycline hydrochloride (Do^{30}), Rifampicin (R^{5}), Erythromycin (E^{15}), Amikacin (Ak^{30}) and Ampicilin (A^{10}). The antibiotic such as Cephaloridine, Amoxycilin, Chloramohenucol, Doxycycline hydrochloride, Rifampicin, Erythromycin, Amikacin and Ampicilin showed high antibacterial activity of *Escherichia coli, Enterobacter, Pseudomonas aeruginosa,Vibrio cholera, Vibrio vulnificus* and *Vibrio* sps. However, Cephaloridine and Amoxycilin of these antibiotics were unable to inhibit the growth of almost of fish pathogens except of *Vibrio sps.* (Table 34.3). A growing concern about the high consumption of antibiotics in aquaculture has initiated a search for alternative methods of disease control (Gildberg *et al.*, 1997) and growth promotion (Byun *et al.*, 1997).

Table 34.3: Antimicrobial Activity of Probiotic *Lactobacillus* Against Various Fish Pathogens (Secondary screening–Well diffusion method–probiotic extract 50 µl)

Samples and Nos	*Labeling (Lactobacillus)*	*Fish Pathogens*					
		E. coli	*Enterobacter*	*P. aeruginosa*	*V. cholerae*	*V. vulnificus*	*V. sps*
		Zone of Inhibition (mm)					
Raw milk (6)	1RM	–	–	–	–	19	16
	2RM	–	–	–	–	17	14
	3 RM	–	–	–	30	29	28
	4 RM	–	–	–	–	15	18
	5 RM	–	–	–	12	16	–
	6 RM	–	–	–	15	12	17
Pasteurized milk (6)	1PM	–	–	–	–	–	12
	2 PM	–	–	–	17	19	–
	3 PM	–	–	–	26	25	22
	4 PM	–	–	–	19	17	18
	5 PM	–	–	16	–	17	19
	6 PM	–	–	–	26	23	22
Curd (5)	1C	–	–	15	12	19	14
	2 C	–	–	–	18	17	16
	3 C	–	–	–	–	16	12
	4 C	17	16	–	–	15	18
	5 C	–	18	–	24	26	22
Soil sample for milk industry (3)	1S	–		–	15	16	–
	2S	–	–	–	26	24	23
	3S	–	–	–	12	15	16
Cheese (4)	1CH	–	–	15	–	19	16
	2 CH	–	–	–	26	27	23
	3 CH	–	–	19	12	10	19
	4 CH	–	–	–	–	17	12
Milk powder (5)	1MP	–	–	–	–	–	–
	2 MP	–	–		–	26	20
	3 MP	–	–	20	–	–	16
	4 MP	–	–	–	–	–	–
	5 MP	–	–	–	15	–	16
Lactobacillus powder (1)	1LP	–	–	–	–	19	12
Sporolac powder (1)	1SP	–	–	–	25	23	22

Experimental diets (D_2-D_5) were supplemented with the isolated probiotic strain *lactobacillus* (3RM) at different levels (100, 250, 500 and 1000 µl/1g diet respectively). Probiotic strains show enhanced growth performance in several hosts including fish. In our studies we found significantly enhanced growth of fish. Pham *et al.* (2003) studied the effect of two probiotic strains on the weight gains of chickens for intervals of 10 days up to 40 days, and found significance after 30 days. Weight was increased by 10.7 per cent. Lara-Flores *et al.* (2003) found that the ingestion of *Lactobacillus* by tilapia larvae for 9 weeks improved growth rate and feed conversion. This statement is a remarkable support.

Weight gain response of *Cyprinus carpio* fed with D-series feeds is shown in the Table 34.4. The feed D_5 (probiotic *Lactobacillus* concentration is 1000 ìl) produced highest weight gain (79.68±5.1) and feed D_1 (control–without probiotic) (32.35±0.6) produced lowest weight increased. The feed conversion ratio (FCR) of the feed D_5 is also low (0.22±0.12) and feed D1 without probiotic supplementation produced higher FCR (1.22±0.4). Similarly specific growth rate was maximum in feed D_5 (7.05±1.3). Feed efficiency ranged from 10.2±4.3 to 29.25±2.4. The maximum survival rate (100 per cent) was also recorded in feed D_4 (probiotic *lactobacillus* concentration in 500 ìl). Feed D_5 (probiotic *lactobacillus* concentration in 1000 ìl) and control feed also and minimum survival rate in feed D_2 and $D_{3.}$ Another strain with probiotic effects is *Lactobacillus rhamnosus* (ATCC 53103) originally isolated from a human source (Nikoskelainen *et al.*, 2001). The bacterium was administered at two different doses (10^9 and 10^{12} CFI/g feed) to rainbow trout (average weight = 32.2 g) for 52 days. Sixteen days after the start of the *Lactobacillus* feeding, the fish were challenged with *Aeromonas salmonicida spp. Salmonicida,* which causes furunculosis. During the challenges trial the mortality was monitored. *L. rhamnosus* reduced the fish mortality significantly, from 52.6 per cent in the control to 18.9 per cent and 46.3 per cent in the 10^9 CFI/g feed and the 10^{12} CFI/g feed groups, respectively.

Table 34.4: Antimicrobial Activity of Commercial Product of Antibiotic Discs Against Various Fish Pathogens

Antibiotic Disc	*Fish Pathogens*					
	E. coli	*Enterobacter*	*P. aeruginosa*	*V. cholerae*	*V. vulnificus*	*V. sps*
	Zone of Inhibition (mm)					
Cr^{30}	–	–	–	–	–	10
Am^{10}	–	–	–	–	–	12
C^{30}	30	24	30	28	22	26
Do^{30}	14	12	16	10	16	24
R^5	–	–	12	8	10	8
E^{15}	–	–	8	12	12	–
Ak^{30}	18	16	22	20	21	22
A^{10}	–	8	–	–	22	22

–: No zone of inhibition.

Cr^{30}: Cephaloridine; Am^{10}: Amoxycilin; C^{30}: Chloramohenucol; Do^{30}: Doxycycline hydrochloride; R^5: Rifampicin; E^{15}: Erythromycin; Ak^{30}: Amikacin; A^{10}: Ampicilin.

Table 34.5: Growth Parameters and Survival of *Cyprinus carpio* Feed in Various Concentrations of Probiotic Organisms

Parameters	*Different Concentrations Feed Level (µ)*				
	Control (D_1)	*100 (D_2)*	*250 (D_3)*	*500 (D_4)*	*1000 (D_5)*
Initial weight (gm)	0.32±0.07	0.4±0.08^d	0.39±0.05^d	0.47±0.06^d	0.55±0.06^d
Final weight (gm)	1.04±0.07	1.62±0.08^b	1.82±0.05^c	1.99±0.18^d	2.58±0.06^c
Weight gain (per cent)	32.35±0.6	45.62±2.3^a	51.6±2.5^a	67.36±3.0^b	79.68±5.1^a
Initial length (cm)	3.66±0.17	3.54±0.2^d	3.56±0.2^d	3.6±0.2^d	3.68±0.2^d
Final length (cm)	4016±0.1	4.45±0.2^d	4.62±0.2^d	4.68±0.2^d	4.96±0.3^a
Length gain (per cent)	4039±1.9	4.95±1.9^d	5.39±2.1^d	9.14±0.6^a	10.46±0.3^b
Specific growth rate (SGR) (per cent)	1.96±0.03	3.88±0.02^c	4.53±0.3^a	5.05±0.7^d	7.05±1.3^d
Feed conversion ratio (FCR)	1.22±0.4	0.76±0.2^d	0.55±0.2^d	0.30±0.06^a	0.22±0.12^a
Feed efficiency (FE) (per cent)	10.2±403	14.23±0.9^a	19.76±1.3^a	20.78±1.5^a	29.25±2.4^a
Survival	100	93.33±6.6	93.33±6.6	100	100

Significance a: 0.05; b: 0.01; c: 0.001; d: no significance.

Rearing of freshwater fish on diet D_5 (supplemented with 1000 µl of *lactobacillus* per 1g of feed) resulted in better growth and survival of carp compared to those with control diet (diet D_1 without any microflora supplementation and other experimental diet. It is feasible that probiotics are mainly effective as growth enhancers when conditions are prevalent as was demonstrated many years ago with the use of antibiotics as growth promoters (McDonald *et al.*, 1973).

Acknowledgement

The authors are thankful to the UGC for providing financial support through Major research project (F.No: 38-293/2009 (SR)).

References

Byun, J.W., Park, S.C., Ben, Y. and Oh, T.K., 1997. Probiotic effect of *Lactobacillus* sp. Ds-12 in Flounder (*Paralichthys olivaceus*). *Journal of General and Applied Microbiology*, 43: 305–308.

Dabrowski, K.R. and Kaushik, S.J., 1985. Rearing of coregonid (*Coregouns schinzi pallea* Cuv. et V al.) larvae using dry and live food. III. Growth of fish and development characteristics related to nutrition. *Aquaculture*, 48: 123–135.

Fuller, R., 1989. Probiotics in man and animals. *Journal of Applied Bacteriology*, 66: 365–378.

Gildberg, A. Mikkelsen, H. Sandaker, E. and Ringo, E., 1997. Probiotic effect of lactic acid bacteria in the feed on growth and survival or fry of Atlantic cod (*Godus morhua*). *Hydrobiologia*, 352: 279–285.

Havenear, R. and Huis in veld, J.H.J., 1992. Probiotics: a general view. In: *The Lactic Acid Bacteria, Vol. 1: The Lactic Acid Bacteria in Health and Disease*, (Ed.) B.J. Wood. Elsevier Applied Science, London pp. 151–170.

Inglis, V. and Richards, R.H., 1991. The *in vitro* susceptibility of *Aeromonas salmonicida* and other fish pathogenic bacteria to 29 antimicrobial agents. *Journal of Fish Disease*, Elsevier Applied Science. London, 1: 151–170.

Lara-Flores, M. Olvera-Novoa, M.A. Guzman-Mendez, B.E. and Lopez-Madrid, W., 2003. Use of the bacteria *Streptococcus faecium* and *Lactobacillus acidophilus*, and the yeast *Saccharomyces cerevisiae* as growth promoters in Nile tilapia (*Oreochromis niloticus*). *Aquaculture*, 216: 193–201.

Marteau, P. and Vesa, T., 1998. Pharmacokinetics of Probiotic and biotherapeutic agents in humans. *Bioscience and Microflora*, 17: 1–6.

McDonald, P., Eduards, R.A. and Greenhalg, J.F.D., 1973. *Animal Nutrition*. Oliver and Boyd, Edinburgh, p. 479.

Midtvedt, T. and Lingaas, E., 1992. Putative public health risks of antibiotic resistance development in aquatic bacteria. In: *Cheotherapy in Aquaculture from Theory to Reality*, (Ed.) C. Michel and D. Alderman. Office International des Epizooties, Paris, p. 302–314.

Nikoskelainen, S. Ouwehand, A. Salminer, S. and Byhund, D., 2001. Protection of rainbow trust *Oncorhynchus mykiss* from furunculosis by *Lactobacillus*. *Aquaculture*, 198: 229–236.

Ouwehand, A.C., Krirjavainen, P.V., Gronland, M.M., Isolauri, E. and Salminen, S.J., 1999. Adhesion of probiotic micro-organisms of intestinal mucus. *International Dairy Journal*, 9: 623–630.

Pham Thi Ngoc Lan, Le Thanh Binh and Yoshimi Benno, 2003. Impact of two probiotic *Lactobacillus* strains feeding on fecal lactobacilli and weight gains in chicken. *The Journal of General and Applied Microbiology*, 49: 29–36

Schillinger, U. and Lucke, F.K., 1989. Antibacterial activity of *Lactobacillus* sp. sake isolated from meat. *Applied and Environmental Microbiology*, 55: 1901–1906.

Takashima, N., Aoki, T. and Kitao, T., 1985. Epidemiological surveillance of drug-resistant strains of *Pasteurella piscicida*. *Fish Pathol.*, 20: 209–217.

Chapter 35

Host-Parasite Relationship with Reference to Histochemical Changes in an Acanthocephalan Infected Fish

☆ *B. Laxma Reddy and G. Benarjee*

ABSTRACT

Pallisentis nagapurensis, an acanthocephalan worm infests the intestine of freshwater teleost, *Channa punctatus*. Histopathological, Histophysiological and Histochemical changes have been noticed in the intestine of fish due to infection with *Pallisentis nagapurensis*. Histopathological changes include drastic histo-architectural changes, severe damage to the Villi, inflammation, fibrosis associated with hyperplasia and metaplasia, vacuolation of submucous cells and necrosis of epithelial cells. The histophysiological changes have been observed through histo chemical studies. The histochemical changes observed in the intestine include change in the total carbohydrates, glycogen, total proteins, glycoproteins and lipids. They have shown an increase in the intestine after helminth infection. The present chapter deals with the Host-parasite relationship and histochemical nature of the infected intestine and uninfected intestine.

Introduction

Acanthocephalans are undoubtedly among the most injurious of heminth parasites. The proboscis hooks embedded in the intestinal wall are damaging to tissues. Adult acanthocephalans affecting fishes often cause extensive damage to the intestinal mucosa (Venard and Warfel 1953; Bskhovskaya–Pavlovskaya *et al.*, 1962; Bullock, 1963). The pathologic effects of carval acanthocephalans on fishes are unknown. Heavy acanthocephalan infections in fish with much vaguer pathogenicity. The acanthocephalan parasite, *Pallisentis nagapurensis* found in freshwater murrels (*Channa punctatus* and *Channa striatus*) in the Indian sub continent (Hilda Leiching, 1984; Hoffman, 1967; Kennedy and Lord,

1982; Taraschewski, 1988). The effect of parasites on the host causes change in the carbohydrate, protein and lipid content.

Studies concern to these changes have been made biochemically to visualize the pathological effects due to infection with parasites Hoswell (1973); Higgins *et al.* (1977), Pike and Burt (1983). However information is not available on the histochemical studies of organs of the host infected with parasites which of organs of the host infected with parasites which specially elucidate the physiological changes in the host infected with parasites. Hence an attempt was made to visualize the histochemical changes in the intestine of fish infected with *Pallisentis nagapurensis.*

Materials and Methods

For the present investigation fishes were procured from the local fish market and also collected directly from the fishermen who directly caught fish from the local freshwater bodies. To collect the parasites the fishes were sacrificed and screened after opening the alimentary canal. Since most of the parasites infect various organs of the alimentary canal, then entire alimentary tract was isolated from the fish and kept in petridish containing normal saline. Intestine was thoroughly screened for the presence of the parasites. The parasites were stained with Alum caramine. The intestine of the infected and uninfected fish were isolated and preserved in Bouins, Susa, Carnoy and Zenker's fluids (Gurr, 1962) for the histopathological and histochemical studies (Pearse, 1968; Bancroft, 1975). A battery of histochemical tests were applied on the microtome cut sections of intestine of both infected and uninfected fish to demonstrate and also to assess the histochemical changes that occur in the tissues of the infected organ if any due to infection with *Pallisentis nagapurensis.*

Results and Discussion

In the present study *Pallisentis nagapurensis* infected to the common edible teleost, *Channa punctatus. Pallisentis nagapurensis* is a common intestinal parasite of *Channa punctatus* and *Channa striatus* especially in this tropical region. It is known to exist at a particular region of alimentary canal, and they remain found in the same site throughout the period. It is a thorny headed worm, whitish to slight yellow in colour, having a wrinkled body. The anterior part through which it attaches to the host tissue is proboscis, cylindrical in shape and provided with row of spines. The association of parasite in the host body results the development of pathogenicity of the host body. The pathological effects on the host tissues are very much conspicuous. The pathological state starts with the attachment of parasite with innerwalls of the intestine. The pathogenicity of the worm is usually presumed on the basis of the depth of penetration of the proboscis or the intensity of pathogenicity may also presumed by a number of worms found in the host body. However the pathogenicity of acanthocephalans in fish is not as simple as the other worms (Boyd, 1951; Bullock, 1962). The assessment of pathogenicity of a parasite can be made by the study of morphological and histochemical, alterations of the tissues and cells involved.

The acanthocephalan interfere with digestion, absorption, and food intake that caused metabolic disorders. Various secretions and metabolites of the parasite brings toxic conditions in the intestine of fish. These abnormal conditions influences on the secretory and digestive nature of the intestine. Bullock (1963) reported that there is proliferation of the connective tissue of the lamina propria in the area where the proboscis of the Worm is attached. The lamina propria is increased as the result of the development of collagenous fibrous tissue, and there is an increase in the number of host cells–Primarily macrophages, fibroblasts, lymphocytes, polymorphonuclear leukocytes, and granular cells. There is also damage to the lining epithelium. The cells at the point of attachment are completely destroyed, and the cells in adjacent areas are compressed. On the other hand the increase in the thickness of the muscular layer may be considered as an adaptation to the presence of parasite. Further, the destruction of important regions of the host suggests that, there will be less supply of nutrients to the infected tissue, from other organs of the body, as such a deteriorated condition of fish exists. The damage of epithelial cells, particularly absorptive and goblet cells indicate the loss of absorptive power of intestine in the infected fish. Thus the food entering the intestinal villi is utilized by the parasite. A similar observation has been made by Bullock (1963) on the intestinal histology of some salmonoid fishes with particular reference to the acanthocephalan infections.

Table 35.1: Histochemical Tests for Intestine

Histochemical Tests Applied	*Results*	
	Un-infected	*Infected*
Periodic Acid/Schiff (PAS)	+	+++
PAS/Saliva	+	+++
Schiff's without oxidation	+	+++
Acetylation/PAS	–	–
Deacetylation/PAS	+	+++
Alcian blue 1.0 pH	++	++
Alcian blue 2.5 pH	++	++
Alcian blue 1.0 pH/PAS	++	++
Alcian blue 2.5 pH/PAS	++	++
Alcian blue/safranin	++	++
Alcian blue/Aldehyde fuchsin	++	++
Mercuric Bromophenol blue	+	+++
Ninhydrin/Schiff	+	++
Ferric ferricyanide	+	++
Congored	+	+++
p-DMAB nitrite	+	++
$KMnO_4$/Alcian Blue	++	+++
Millon's reaction	+	++
Copperpthalocyanin	++	+++
Sudan Black 'B'	+	+++
Orcein/Van Gieson	+	+++
Orcein	+	+++

+++: Strongly positive; ++: Moderately positive; +: Positive; –: Negative.

The histochemical nature of various organs reveals their physiological functional state and the pathological conditions state and the pathological conditions certainly reflects on a change in the histochemical nature which certainly influences by and large the functional status of various organs. In the present study the histochemical studies conducted on various tissue of infected fish gave a different picture when compared with the uninfected. Periodic acid/Schiff's (PAS) technique yielded an intense positively in all the layers of intestine of the uninfected fish, which suggested the occurrence of Vinyl hydroxyl group in the tissues. Further this reaction also suggests the occurrence of mucopolysaccharides and other related carbohydrate groups. The distribution of various groups of carbohydrates in the intestine of fishes was studied by Bucke (1971); Sinha and Chakravarthy, (1984); Laxma Reddy and Benarjee (2007). The distribution of carbohydrates in the intestine of infected fish with acanthocephan, Pallisentis nagapurensis revealed a different picture about the chemical nature. The changes in the distribution of carbohydrates therefore shown relationship between the host and the parasite. Even the glycogen content of the intestine showed a difference from the uninfected fish, this was evidenced when the section were subjected to PAS after diastase digestion. In the infected fish the carbohydrates and glycogen content increased significantly. This increase of carbohydrates may be the protection interruption developed by the fish to come back with the parasitic effect. When the sections of intestine of uninfected fish showed a positive response to Alcian blue and 1.0 pH and 2.5 pH. However an increase of intensity of the colour of alcian blue both at 1.0 pH and 2.5 pH in the infected fish reveals the presence of more quantity of mucopolysaccharides. This increase may be attributed to increase of mucous secretion in the intestine for the invade of the parasite. The mucous secretion is a protection interaction of the host.

The carbohydrate histochemistry of oesophagus, stomach and intestine of various teleostean fishes was reported by Reifel and Travil, 1977; Reifel, 1978, 79. They demonstrated the presence of sulfomucins, sialomucins, and neutral mucosubstances. Others who have also contributed to the knowledge of histo chemical nature of alimentary canal regions are Spicer *et al.* (1967); Yamada (1975); Hirji (1983), Chakravarthy *et al.* (1983), Woodword and Bergeron (1984); Chakravarthy and Sinha (1986).

The parasite live in the intestine of the fishes also interfere with the protein metabolism of the host due to which the host tissue may show a decrease or increase in the level of the protein content. So the quantity of protein depends on the metabolic activity of the body and cellular integrity. The strong positive reaction for Bromophenol blue gave a strong positivity indicating the presence of proteins. The intensity of the reaction is more in the infected fish compared with the uninfected fish. The increase in the protein content in the infected fish may be due to more supply of nutrients to the effected organs. However the quantum of increase depends upon the degree of infection, the degree of pathoenicity and the degree of resistance. Where as the other proteins which were found positive to histo chemical tests are tyrosine, tryptophan, glycoproteins, protein bound amino groups and disulphides. The intensity of the reaction however varied which shows the degree of pathological condition prevailed in the fish due to parasitic infection.

The lipid profiles of various organs and their distribution are influenced by the parasiticinfection. Since the parasiticinfection brings dietary change and metabolic process a change can be observed even in the various lipid contents of the organ. A similar condition has been noticed in the present study where in the infected fish an increase in the various groups of lipid including total lipids and phospholipids. The distribution of total lipids was observed by sudan black 'B' method and phospholipids was by copperpthalocyanin these staining reactions suggested the involvement of *Pallisentis nagapurensis* and the elaboration of lipid content. These results are presented in Table 35.1.

Acknowledgements

The authors thank the Head, Department of Zoology, Kakatiya University, for providing Laboratory facilities. Authors also thank Prof. Y. Prameela Devi for the critical evaluation of manuscript for as improvement.

References

Bancroft, J.D., 1975. *Histochemical Techniques*. Butterworths, London and Boston.

Boyd, E.M., 1951. A survey of parasitism of the *Sturnus vulgaris* L. in North America. *J. Parasitol.*, 37: 56–84.

Bucke, D., 1971. The anatomy and histology of the alimentary tract of the Carnivorous fish the pike *Esox lucius*. *J. Parasit.*, 48: 442–451.

Bullock, M.R., 1963. Intestinal histology of some salmonoid fishes with particular reference to the histopathology of acanthocephalan infections. *Journal of Morphology*, 112: 23–44.

Bullock, W.L., 1962. A new species of Acanthocephalus from New England fish, with observations on variability. *J. Parasit.*, 48: 442–451.

Bykhovskaya-Pavlov Skaya, I.E., Dubinina, M.N., Gusev, A.V., Izycnova, N.A., Nagabina, L.F., Rajkova, E.V., Shtein, G.A., Shulmam, S.S. and Sthlkov, T.S., 1962. Key to parasites of freshwater fish of the USSR Zool. *Inst. Acad. Sci.* CSSR (English transl. TT 64–11010. NTIS, Dept. Commerce, Springfield, Va, 919 p).

Chakravarthy, P. and Sinha, G.M., 1986. Characterization and distribution of acid mucopolysaccharides in the mucous cells of alimentary canal in an Indian freshwater carp, *Catla catla* (Ham.). A fluorescence microscopic study. In: *Proc. 2nd National Convention of Young Scientists*, Meerut Univ., Ind. Symp. Vol., pp. 212–219.

Chakravarthy, P., Mukopadyay, S. and Sinha, G.M., 1983. Sulfomucins and Sialomucins in the mucous cells of the alimentary canal in an Indian freshwater major carp, *Labeo rohita*: A fluorescence microscopic study. Folia Histochem. *Cytochem.*, 21(3/4): 181–86.

Gurr, E., 1962. *Staining Animal Tissue: Practical Theoretical*. Leonard Hill (Books) Ltd., London, p. 631.

Higgins, J.C., Wright, D.E. and Mathew, R.A., 1977. The Ultrastructure and histochemistry of the cyst wall of *Bucephalus haimeanus* Laeaze Duthiers (1854). *Parasitology*, 75: 207–214.

Hilda Lei Ching, 1984. Description of *Neechinorhynchus salmonis* sp. N. (Acanthocephala; Neoechinorhynchidae) from freshwater fishes of British Columbia. *J. Parasitol.*, 70(1): 286–291.

Hirji, K.N., 1983. Observations on the histology and histochemistry of Oesophagus of the perch, *Perca flaviatilis*. *J. Fish Biol.*, 22(2): 145–152.

Hoffman, G.L., 1967. *Parasites of North American Freshwater Fishes*. University of California Press, Berkele, California, 486p.

Hoswell, M.S., 1973. The resistance of cyst of Stictodora Lari (Trematoda; Heterophyidae) to encapsulation of cells of the fish host. *Int. J. Parasitol.*, 3: 653–659.

Kennedy, C.R. and Lord, D., 1982. Habitat specificity of the acanthocephalan Acanthocephalus Clavula in eels, *Anguilla anguilla*. *J. Helminthol.*, 56(2): 121–129.

Laxma Reddy, B. and Benarjee, G., 2007. Histochemical changes in the intestine of freshwater murrel, *Channa striatus* due to Acanthocephalan, *Pallisentis nagapurensis*. *J. Mendel.*, 24(3–4): 113–114.

Pearse, A.G.E., 1968. *Histochemistry: Theoretical and Applied*, 2nd edn. Little Brown and Company, Boston, MSS.

Pike, A.W. and Burt, M.D.B., 1983. The tissue response of yellow perch. *Perca flavescens*, Mitchill to infections with the metacercarial cyst of Adophallus brevis Ransom 1920. *Parasitology*, 87: 393–404.

Reifel Conrad, W., 1978. Structure and Carbohydrate histochemistry of the stomach in eighteen species of teleosts. *J. Morpho.*, 158: 155–168.

Reifel Conrad, W., 1979. Structure and Carbohydrate histochemistry of the intestine in ten teleostean species. *Ibid*, 160: 343–360.

Reifel Conrad, W. and Travill, A.A., 1977. Structure and Carbohydrate histochemistry of the oesophagus in ten teleostean species. *J. Morphol.*, 152: 303–314.

Sinha, G.M. and Chakravarthy, P., 1984. Mechanisms of mucous release in the alimentary canal of a freshwater major carp, *Labeo rohita* (Ham). A light and Scanning electron microscopic study. *Proc. Ind. Nat. Sci. Acad.*, 50b: 27–31.

Spicer, S.S. Horn, R.G. and Leppi, T.J., 1967. Histochemistry of connective tissue mucopolysaccharides. In: *Symposium on Connective Tissue Research Methods*. The William and Wilkins Co., Baltimore, 17: 251–302.

Taraschewski, H., 1988. Host-parasite interface of fish acanthocephalans. 1. Acanthocephalus anguillae (Palaeacanthocephala) in naturally infected fishes. LM and TEM investigations. *Diseases of Aquatic Organisms*, 4: 109–119.

Venard, C.E. and Warfel, J.H., 1953. Some effects of two species of acanthocephala on the alimentary canal of the large moulth bass. *Journal of Parasitolosy*, 39: 187–190.

Wood Word, B. and Bergeron, T., 1984. Protein histochemistry of the granule cells in the small intestine of the rainbow trout, *Salmo gardneri*.

Yamada, K., 1975. Morphohistochemical analysis of mucous substances in some epithelial tissues of the eel (*Anguilla japonica*). *Histochemistry*, 43: 161–172.

Chapter 36

Seasonal Occurrence of *Genarchopsis goppo* in Freshwater Teleost, *Clarias batrachus*

☆ *B. Laxma Reddy, P. Gowri and G. Benarjee*

ABSTRACT

The trematode parasite, *Genarchopsis goppo* infecting the catfish, *Clarias batrachus* has been studied for 2 consecutive years (2005-2007) to analyse the incidence of parasitic infection, intensity of infection, index of infection density of infection and their seasonal variation. The infection was more during July to September. The parasitic infection of affects the general metabolic state of the host and affects the growth of fish.

Introduction

The trematode, *Genarchopsis goppo* is found infected to the intestine of freshwater cat fish, *Clarias batrachus*. The occurrence of trematodes was reported in the piscine hosts by Bose and Sinha (1979), Barbara (1980), Lester (1980), Chung yui-tan (1981), Gupta and Agarwal (1984), Zarina (1990), Laxma Reddy *et al.* (2006), Laxma Reddy and Benarjee (2006), Benarjee *et al.* (2006), Benarjee and Laxma Reddy (2006), Benarjee and Laxma Reddy (2008), Pardeshi and Hiware (2010). In the present study, an analysis has been made on the percentage of infection, parasite density and seasonal variation on the occurrence of *Genarchopsis goppo* in *Clarias balrachus*.

Materials and Methods

The fish, *Clarias batrachus* were procured from freshwater bodies such as rivers, tanks, lakes and ponds, located in the Warangal district of Andhra Pradesh. To identify the infection and also to collect the parasites, the fish were dissected and various visceral organs such as digestive tract, gall bladder, liver were examined carefully after placing them in a petridish containing normal saline. The parasites collected from the infected fish are enumerated and permanent slides were prepared by preserving the fresh parasites in 10 per cent formaldehyde for 48 hours later stained with Alum carmine (Pearse, 1968; Bancroft, 1975).

The study has been conducted for two consecutive years *i.e.*, starting with July and ended with June of the next year. The observation has been done on incidence of infection, intensity of infection and the density of infection of the parasite to the host. The seasonal variation of the infection and density of parasites in the host has been recorded.

The following equations have been adopted to calculate various biostatistical parameters of the parasite.

(*a*) $$\text{Incidence of infection } (X_1) = \frac{\text{No. of hosts infected (b)}}{\text{No. of hosts examined (a)}}$$

(*b*) $$\text{Incidence of infection } (X_2) = \frac{\text{No. of parasites collected (c)}}{\text{No. of hosts infected (b)}}$$

(*c*) $$\text{Density of infection } (Y_1) = \frac{\text{No. of parasites collected (c)}}{\text{No. of hosts examined (a)}}$$

(*d*) $$\text{Index of infection } (Y_2) = \frac{\text{No. of hosts infected (b)} \times \text{No. of parasites collected (c)}}{\text{No. of hosts examined } (a^2)}$$

Results and Discussion

The data obtained on various biostatistical parameters *viz.*, the incidence of infection, intensity of infection, density of infection and index of infection during the 2 consecutive years reveal that the incidence of infection is more during July, August and September. Subsequently, the rate of infection decreases up to May then slowly starts increasing. Kisielwska (1970) reported that the seasonal fluctuations play a significant role in the helminth infection. Water temperature seems to have no clear cut impact as the occurrence pattern of helminth parasites. Jha *et al.* (1992) reported that the water temperature does not play a role in the seasonal occurrence of helminth parasites. However, the ecological factors have been held widely responsible for the occurrence of adult digenetic trematodes as suggested by Chubb (1979) and Madhavi (1978). Choice of food items consumed by fish has also been held responsible for the variation in the occurrence of nematode infestation (Chubb, 1980). The degree of infection of various parasites in their hosts shows considerable seasonal variation as noticed by Amin (1975a). Population dynamics of endohelminths of *Channa punctatus* was also reported (Gupta *et al.*, 1984).

The data (Table 36.1) obtained in the present study also suggests that the incidence, intensity, density and index of infection of the parasite, *Genarchopsis goppo* is high during July, August, September and infection decreases up to May then slowly starts increasing. Shoromendra *et al.* (2005) reported the similar results in *Clarias batrachus* due to the digenetic trematode, *Astiotrema reniferum*. In which incidence of infection is high during July–August months. The reasons for this are that the infecting agents (intermediate hosts) which are commonly snails and insects are prevalent during colder months

Table 36.1: Showing Biostatistical Indices of *Genarchopsis goppo* in the Population of Fish *Clarias batrachus* during the Years 2005-06 to 2006-07

Year	Month	No. of Hosts Examined (a)	No. of Hosts Infected (b)	No. of Hosts Uninfected	No. of Parasites Collected (c)	Incidence of Infection (b/a x 100)	Intensity of Infection (c/b)	Density of Infection (c/a)	Index of Infection ($b \times c/a^2$)
2005	JUL	30	15	15	29	50.00	1.93	0.97	0.48
2005	AUG	29	15	14	30	51.72	2.00	1.03	0.54
2005	SEP	31	17	14	33	54.84	1.94	1.06	0.58
2005	OCT	33	17	16	27	51.52	1.59	0.82	0.42
2005	NOV	32	16	16	25	50.00	1.56	0.78	0.39
2005	DEC	33	16	17	24	48.48	1.50	0.73	0.35
2006	JAN	34	16	18	23	47.06	1.44	0.68	0.32
2006	FEB	35	16	19	22	45.71	1.38	0.63	0.29
2006	MAR	29	13	16	17	44.83	1.31	0.59	0.26
2006	APR	33	14	19	16	42.42	1.14	0.48	0.21
2006	MAY	30	14	16	22	46.67	1.57	0.73	0.34
2006	JUN	31	15	16	26	48.39	1.73	0.84	0.41
TOTAL		380	184	196	294	48.42	1.60	0.77	0.37
2006	JUL	34	16	18	32	47.06	2.00	0.94	0.44
2006	AUG	34	17	17	35	50.00	2.06	1.03	0.51
2006	SEP	34	18	16	38	52.94	2.11	1.12	0.59
2006	OCT	33	16	17	28	48.48	1.75	0.85	0.41
2006	NOV	32	15	17	25	46.88	1.67	0.78	0.37
2006	DEC	29	14	15	23	48.28	1.64	0.79	0.38
2007	JAN	30	13	17	20	43.33	1.54	0.67	0.29
2007	FEB	36	15	21	22	41.67	1.47	0.61	0.25
2007	MAR	37	15	22	21	40.54	1.40	0.57	0.23
2007	APR	31	12	19	16	38.71	1.33	0.52	0.20
2007	MAY	29	13	16	21	44.83	1.62	0.72	0.32
2007	JUN	33	15	18	27	45.45	1.80	0.82	0.37
TOTAL		392	179	213	308	45.66	1.72	0.79	0.36

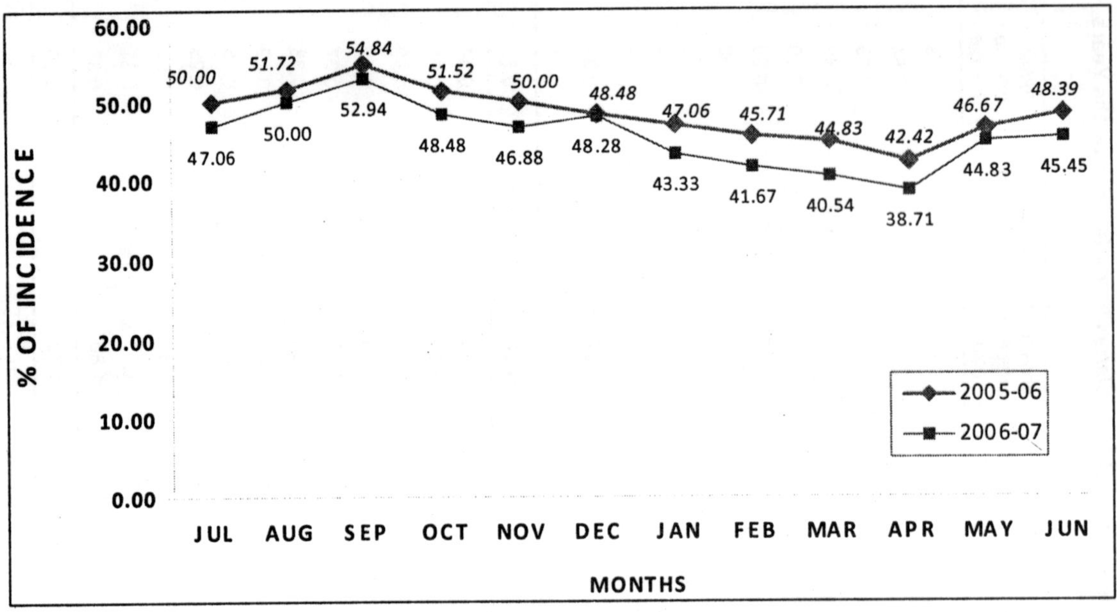

Figure 36.1: The Incidence of Infection of *Genarchopsis goppo* in the Population of *Clarias batrachus* during the Years 2005-06 to 2006-07

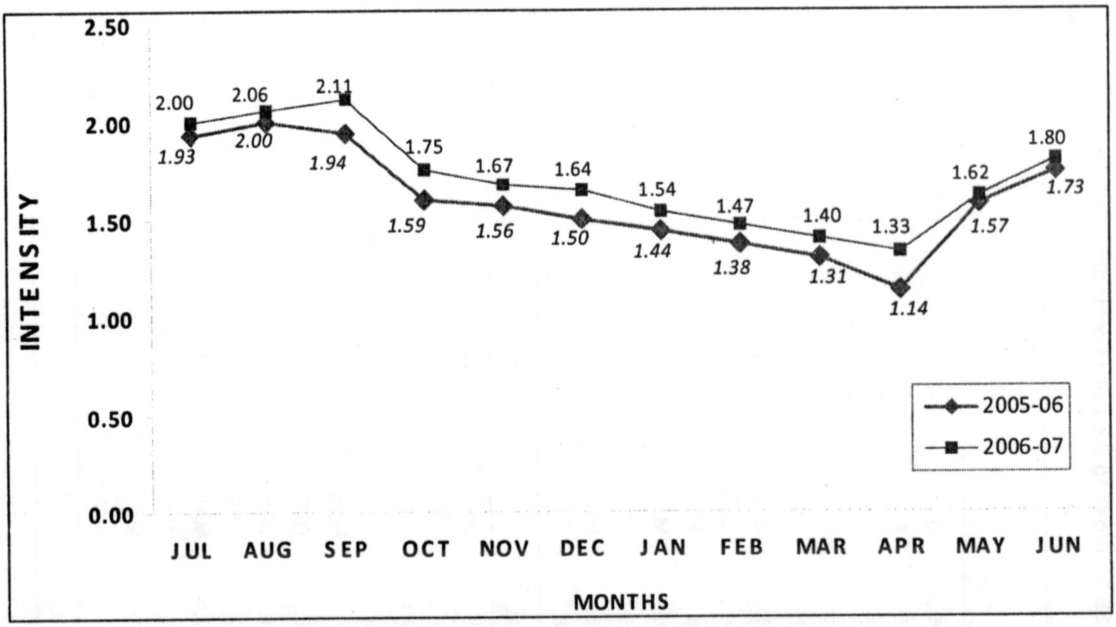

Figure 36.2: The Intensity of Infection of *Genarchopsis goppo* in the Population of *Clarias batrachus* during the Years 2005-06 to 2006-07

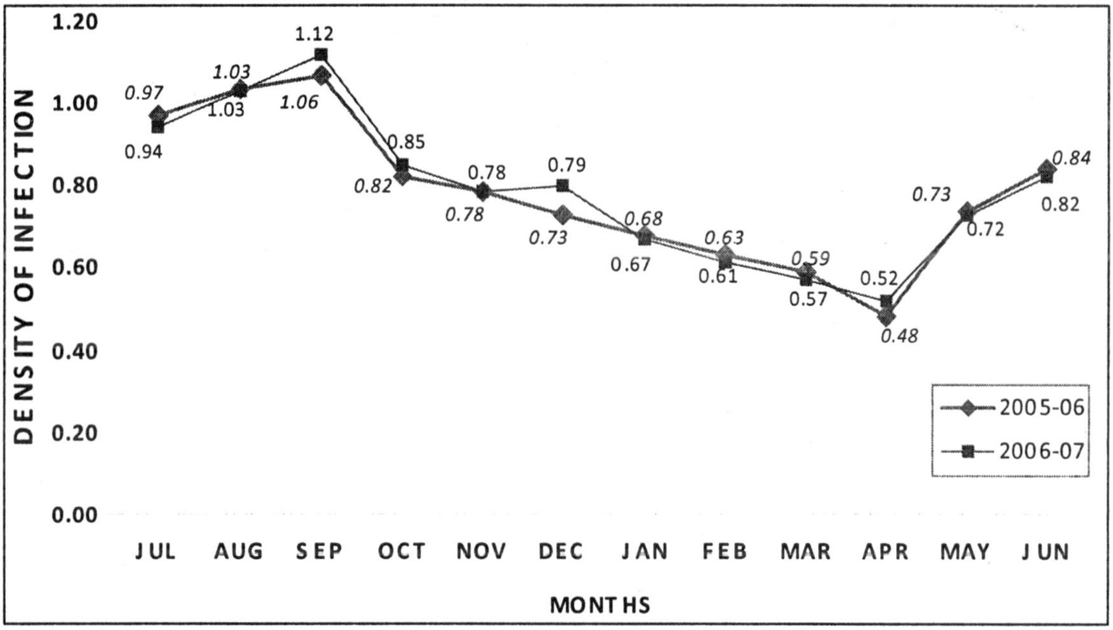

Figure 36.3: The Density of Infection of *Genarchopsis goppo* in the Population of *Clarias batrachus* during the Years 2005-06 to 2006-07

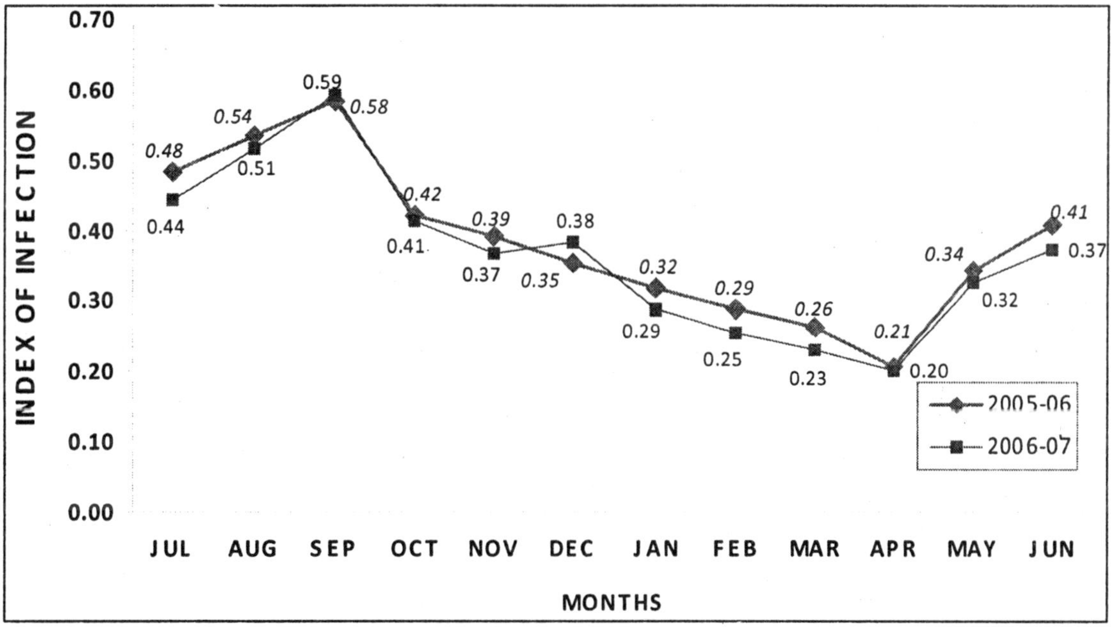

Figure 36.4: The Index of Infection of *Genarchopsis goppo* in the Population of *Clarias batrachus* during the Years 2005-06 to 2006-07

of the year. Due to this the rate of contact between the intermediate and primary hosts is more leading to an increase in the rate of incidence.

It is clear from the results that the infection of this parasite is more in cooler months than hotter months which may be due to the availability of the intermediate hosts, which play a significant role in the spread of worms. Moreover, the habitat of the hosts *i.e.*, ponds get dried up or the quantity of water is less. Due to this condition, a sort of inactivity prevails, there by the host and the vector contact may not be established. Even the number of intermediate hosts may also be less in this season, resulting in the lesser incidence of infection. However, with the advent of the rainy season, ponds and lakes will be filled up with much water and hosts become quite active. At the same time, large number of intermediate hosts also become available. Due to this, the host and vector contact is reestablished leading to the occurrence of the infection. This increases throughout the rainy and winter seasons and starts to decrease with the increase in the temperature. Therefore, in summer months, the incidence of infection will be lesser.

Acknowledgements

The authors are thankful to the Head, Department of Zoology, Kakatiya University for providing the laboratory facilities.

References

Amin, O.M., 1975a. *Acanthocephalus parksidei* sp. *Acanthocephala*. Echinorhynchidae from Wisconsin fishes. *J. Parasitol.*, 61: 301–306.

Bancroft, J.D., 1975. *Histochemical Techniques*. Butterworths, London and Bostan.

Barbara B., 1980. Pathological Changes in Cyprinid fry infected by *Bucephalus polymorphus* and *Rhipidocotyle illensis metacercariae* (Trematoda, Bucephalidae). *Acta Parasitol. Pol.*, 27(15–18): 241–246.

Benarjee G. and Laxma Reddy, B., 2008. Pathobiological and histochemical studies on liver of the freshwater murrel infected with frematode parasite *J. Ecotoxicol. Environ. Monit.*, 18(6): 565–572.

Benarjee, G. and Laxma Reddy B., 2006. Histopatological and histochemical changes in the liver of *Clarias batrachus* due to trematode, *Euclinostomum heterostomum*. *J. Natcon.*, 18(2): 251–259.

Benarjee, G., Laxma Reddy, B. and Bikshapathi, V., 2006. Histopathology and histochemistry of the intestine of *Clarias batrachus* due to trematode, *Genarchopsis goppo* (Ozaki, 1925). *J. Aqua Biol.*, 21(2): 257–262.

Bose, K.C. and Sinha, A.K., 1979. The histopathology of the stomach wall of the fish *Channa gachua* (Ham) Channidae attributable to the Digenitic trematode, *Genarchopsis goppo* (ozaki). *Current Science*, 48 (16): 747–748.

Chubb, J.C., 1979. Seasonal Occurrence of helminths parasites in fishes Part II. Trematoda. In: *Advances in Parasitology*, Academic Press, London and New York, 17: 171–313.

Chubb, J.C., 1980. Seasonal Occurrence of helminth parasites in fishes. Part III. Larval cestoda and Nematoda. In: *Advances in Parasitolosy*, Academic Press. London and New York, 18: 1–120.

Chung, Yui–Tan, 1981. A study on the histopathology in the wolfian ducts of *Hypentelium nigricans* (Osteichthyes: Catostomidae) caused by *Phyllodistomum superbum* (Trematoda: Gorgoderidae). Q.J. Talwanmus (Taipeli), 34 (3/4): 237–240.

Gupta, A.K. and Agarwal, S.M., 1984. Host-parasite relationships in *Channa punctatus* and *Euclinostomum heterostomum* III. Transaminase and total proteins and free amino acids. *Current Science*, 53: 710–711.

Gupta, A.K., Niyogi, A. and Naik, M.C., 1984. Population dynamics of endohelminths of *Channa punctatus* at Raipur, India. *Jpn. J. Parasitol.*, 33(2): 105–118.

Jha, A.N., Sinha, P. and Mishra, T.N., 1992. Seasonal Occurrence of helminth parasites in fishes of Sikandarpur reservoir, Muzaffarpur (Bihar). *Indian J. Helminth.*, 44(1): 1–8.

Kisielwska, K., 1970. Ecological Organization of intestinal helminth groupings in *Clethriomomys glareolus* (shreb) (Rodentia). I. Structure and seasonal dynamics of helminth grouping in a host population in the Bialowieza National Park. *Acta Parasitologica Polonica*, 18(13): 121–147.

Laxma Reddy, B. and Benarjee, G., 2006. Histopathological changes in the intestine of *Clarias batrachus* due to trematode, *Orientocreadium batrachoides*. *Aqua Cult.*, 7(2): 251–257.

Laxma Reddy, B., Benarjee, G., Rajender, G., Bikshapathi, V., Swamy, M. and Rama Rao, N.J., 2006. Histopathological and histochemical abnormalities induced by *Euclinostomum heterostomum* in the liver of freshwater fish, *Channa punctatus*. *J. Aqua. Biol.*, 21(2): 263–267.

Lester, R.J.C., 1980. Host parasite relations in some didymozoid trematodes. *J. Parasitol.*, 66(3): 527–531.

Madhavi, R., 1978. Life history of *Genarchopsis goppo* Ozaki. 1925 (Trematoda: Hemiuridae) from freshwater fish *Channa punctatus*. *J. Helminth.*, 52: 251–259.

Pardeshi, P.R. and Hiware, C.J., 2010. Histochemical studies on digenetic trematode parasite, *Orientocreadium striatusae* N.Sp. from *Channa striatus* (Bloch, 1793). *Recent Research in Science and Technology*, 2(8): 114–117.

Pearse, A.G.E., 1968. *Histochemistry: Theoretical and Applied*, 2nd edn. Little Brown and Company, Boston, MSS.

Shoromendra, M., Jha, A.N. and Kumar, Pankaj, 2005. Seasonal Occurrence of helminth parasites in fishes of Loktak Lake, Manipur. *U.P. J. Zool.*, 25(1): 23–27.

Zarina Z., 1990. Infection of a digenean trematode *Orientocreadium batrochii* in two species of genus *Clarias* collected from Kedah state of Malasia. Dhaka Uni. Stud. Part. E. *Biol. Sci.*, 5(2): 105–114.

Chapter 37

Modified Chinese Circular Hatchery of Marathwada, India

☆ *S.D. Niture and S.P. Chavan*

Introduction

In India for inland fisheries, development specially for stocking the fish seed into the reservoirs there is a large need of fish seed of Indian Major Carps. The most important pre-requisites for the development of fish culture are establishment of dependable fish seed resources of commonly cultivable species and the establishment of suitable fish farms. A hatchery complex not only provide the facility for spawning the fish brooders and hatching of their laid eggs but also provides the facilities for brood stock rearing and rearing of the fish seed to produce stock material or grow out ponds and other fish production units.

The chinese model of fish seed hatchery has proved an easiest, suitable, and economical method to get the fish seed of desirable fish species in large number. Therefore, the Covernment of India has constructed and established 47 commercial fish seed farms and hatcheries under both the National Programme for Fish Seed Development and the Inland Fisheries Project with World Bank assistance. The fish seed production, which was 1048 million fry in 1979–80, increased to 14429 million fry by 1993 because of the extra ordinary success of chinese type of eco-hatcheries. In India, major carp seed production in hatcheries were 420 with 342,918-lakh spawn production capacity. Magur Hatcheries were 04 with 01 lakh spawn production capacity. Trout hatcheries were 12 with 37.30 lakh spawn production capacity. One mirror carp hatchery with one lakh spawn production capacity, 41 freshwater prawn hatcheries with 2991 lakh spawn production capacity, 26 dual purpose tiger/freshwater prawn hatcheries with 3452 lakh spawn production capacity and 226 shrimp hatcheries with 105828 lakh spawn production capacity(state-wise inventory of Aqua hatcheries.To construct a good fishpond it is essential to have enough water of desirable quality, heavy impervious soil of suitable type and land of satisfactory topographical characteristics. The principal consideration for constructing a fish pond are the choice of a suitable site, the preparation of the designs and the layout of construction of dykes, the water inlet and drainage arrangement. Recently in development due to ovaprim and ovatide hormonal stimulating injections the hatchery is being compared with a machine, just you put 50 kg Male and females and put the button you will get 70.00 Lakh fertilized eggs in the incubation without any manual labors (Hussaini,1997).

Design and Construction of Ideal Model of Chinese Circular Hatchery Unit

Site Selection

The selection of a suitable site for establishment of a carp hatchery complex would mainly depend upon the water supply, the soil condition, the topography and the economical aspect of the hatchery.

A carp hatchery should be located by the side of a perennial source of good quality water. There should be adequate water to fill the ponds and maintain water level, which does not fluctuate more than 0.6 meters. Equally important is the need of avoiding excess water. The common water sources are rivers, streams, springs, canals, surface run-off from rainfall and the ground water sources.

The soil must have enough clay to make the ponds watertight. The types of soil best suited for fishpond construction are in general impervious. These are clays, salty clays, clay loams etc. The chemical composition of the soil should also be favorable for fish production.

The topography of the area proposed for construction of the hatchery complex should generally be flat or gently sloping towards the outlet. The site should be so selected that the earth available by excavation as far as possible balances with the earth required for raising the dykes.

Design Consideration

A carp hatchery complex would generally have the following units.

1. The brood stock unit for rearing and management of carp brood stock.
2. The main hatchery unit for spawn production.
3. The nursery unit for rising fry from stocked spawn (4.5 mm to 25-30 mm)
4. The rearing unit for raising fry up to fingerlings (50 mm and above)
5. The packing and marketing unit.

The Brood Stock Unit

The brood stock unit would generally have the ponds, the size of which may vary from 0.2 ha to 1.0 ha; but the preferable size would be of 0.3 ha to 0.5 ha The depth of these ponds would be 2.0 m to 2.5 m for better management of brood stock; the ponds should have the facilities of water supply and drainage, so that the desired water level in the pond can be accomplished, if needed. The bottom of the ponds should be gently sloping towards the outlet. The slopes of the dykes may generally be 2:1 (2 horizontal: 1 vertical).

The Main Hatchery Unit

The main hatchery unit for spawn production comprises the following essential sub units:

1. Water Supply
2. Brood spawning

3. Egg incubation
4. Egg and spawn collection.

Water Supply Unit

The most important requirement of an eco-hatchery (circular hatchery) is the water supply which can be accomplished either form a surface water source or from a ground water source.

The water contains fish and other organisms, which may attack eggs and spawn in the hatchery. Some surface water carries suspended solids, which may coat egg and clog the gills of fry. Filtration of surface water can be accomplished by sand filters, which should filter out particles larger than 0.025 mm size. This will remove most suspended soils including eggs and parasites. After filtering of water at low elevation for use in hatchery by gravity, it must be lifted to storage reservoir/over head tanks. The reservoir should have a capacity large enough to serve all hatchery needs for several hours. The reservoir can also be equipped with an auto water level guard switch to automatically turn electric pump on and off at pre selected levels.

Ground water (water from open wells/bore wells) is usually free from fish pathogens and suspended solids and does not require filtering before being pumped in the reservoir. However, most well water is having low dissolved oxygen (DO) and may contain ferrous salts. Therefore, for ground water, aeration may be required and the DO content would be at least 5 mg/liter. If water contain ferrous salts, it would be necessary to have two reservoirs One can be in use, while other acts as a settling tank. Water take off point for the hatchery should be well above the bottom of removing precipitates. The iron content of water should not exceed 0.2 mg/l. The water pH should preferably to 7.4-8.4 and there should not be any pollution at any source. The preferable range of temperature is 28°C to 30°C. Since all components parts of the hatchery are required to be provided with controllable water supply, a very careful attention should be given for designing the water supply installations.

Spawning/Breeding Unit for Brood Fishes

The breeding pool is a circular smooth cistern of bricks, RCC or FRP with provision of showers and water current in the pool. Diameter of the tank may vary from 6 to 8 m and depth 1 to 1.5 m for standard commercial carp hatchery. Water flows at the inlet pipe and the central outlet (egg delivery) pipe are so adjusted that the water current in circular pool is created at a speed of 0.2 to 0.5 Inch/sec.

Egg Incubation/Hatchery Unit

The hatching unit is a double circular smooth cistern made of brick, RCC or FRP material. For a commercial production unit the outer diameter ranges from 3 to 6 m and the inner chamber diameter ranges from 1 to 1.5 m with a depth of the pool from 1 to 1.5m based on the running water principal. Water enters into cistern by a series of duck mouth inlets and drains of through a screen encircled on the inner chamber. The central outlet pipe being vertically erected maintains water depth in the incubation chamber. The egg-receiving pipe from the breeding pool is generally fixed about 0.3 m above the base of the incubation chamber to avail a water cushion against mechanical injury of eggs.

The circulation of water in the pools is the most important consideration for the success of eco-hatchery. Proper circulation prevents settling of egg cells, dead animals etc. on the bottom of the tank. If they settle at the bottom, the decomposition will be anaerobic and will form harmful metabolites causing death or encouraging growth of diseases and parasites. If the particles are kept moving in water in presence of oxygen, the decomposition is aerobic and the by-products are not harmful.

For eco-hatchery the capacity of the hatching tank is about 5 cubic meters and about 500 liter (60-70 lakh) eggs can be put for hatching at a time. Initially a flow rate of 2.5 I/sec is maintained in the hatchery with circular motion in tanks. After the embryo are hatched the water flow rate is increased to 3.5 I/sec for keeping the hatchlings in floating condition as their buoyancy reduces and their swimming ability is weak.

Egg and Spawn Collection Unit

A rectangular cistern adjacent or in between the spawning and incubation pool can serve both for egg collection and spawn collection purposes. The camber provides water cushion and collects spawn by gravity flow along with water. Over flow pipe maintains water depth and adequate screen arrangement prevents escape of spawn.

The Nursery Unit

The nursery unit of the hatchery complex is for raising fry form stocked spawn (4-5 mm to 25-30 mm size). Nursery ponds should preferably be small and shallow. Ponds of 0.02 ha to o.06 ha in size and 1 m to 1.5m in depth are generally suitable. The slopes of the ponds dykes may generally be (1.5 horizontal:1 vertical) and the dykes should be right enough to have a free board of at least 0.3m. The top width of the dykes should be minimum 1.0m, but a preferable top width is considered as 2.0m. Since this culture is for a very short duration and for about 3 weeks only, the ponds should have good water supply and drainage arrangement, with controlled inlets and outlets so that they can be emptied dry and refilled at will. This will effected easy and better management and there would not be any requirement for poisoning the ponds.

The Rearing Unit

The unit is for raising fry up to fingerlings (50 mm and above). The rearing pond should be slightly larger but need not to be proportionately deep. They should preferably be 0.06 ha to 0.10 ha the ponds should have controlled inlets and outlets with independent water supply and drainage arrangement. Drainpipes with controlling valves may be the best and easiest for ensuing the desired entry and discharge of water from the ponds.

The Packing and Marketing Unit

In a hatchery complex a unit for packing and marketing is required for its layout, to facilitate the seed supply scientifically after due acclimatization and proper oxygen packing in the polythene bags. This will reduce the transportation and handling mortality leaving a good name for the hatchery to its customers

In Maharashtra, 28 chinese circular types of hatcheries are present (Year, 2010), these hatcheries use ovaprim or ovatide hormones for the induced breeding of Indian major carps. Most of the hatcheries were set up between 1989 and 1995 with the exception of Arey Hatchery Farm; it was set up near Mumbai quite some years back (Fishing Chimes, 2000. State wise inventory of Aqua hatcheries. Vol.19. No.10&11. P 164-214). In Parbhani and Hingoli district there are 3 government fish seed hatcheries under the control of state fisheries department under the administrative and working control though District Fisheries Development Officer (D.F.D.O.) recently from 2008 renamed as Assistant Commissioner of Fisheries, Parbhani. At every hatchery unit structure and working of a Chinese hatchery system is similar.

Apart from all these hatcheries, there is a private hatchery, named 'Sudarshan Fish Seed Hatchery, located at village Kanha, Post. Charthana, Tal. Jintur, District Parbhani.This hatchery is quite different from the standard Circular Chinese hatchery units. Sudarshan hatchery unit for the production of Indian Major Cartp Fish seed is the only of its kind in the Maharashtra State, due to maney modifications invented by the owner himself in this hatchery. The modifications incorporated in this hatchery are need based and to overcome the draught conditions and poor availability of water. This modified unit is cost effective, need little manpower, less budget to operate and has similar or good efficiency compared to the regular Circular Chinese hatchery. The location, structural details, fish seed production data, modifications and other details of this hatchery is explained as below.

Sagar Fish Seed Hatchery

Sagar fish seed hatchery centre is located at Kanha, Post Charthana Tq. Jintur Dist.Parbhani, Maharashtra State, India. It is a Private fish seed hatchery, under the control and ownership of Mr. Shivhar Bhaskarrao Jogwadkar, age 38. This hatchery is about 25 km. from Jintur town of Dist. Parbhani. This centre was established in 2001-2002. The silent features of this unit are given in Table 37.1.

This is the only private and modified type of chinese fish seed hatchery and successful in its business. How it is modified and how the ecological conditions forced the hatchery owner to change the design of this hatchery is explained in detail as below.

Table 37.1: Salient Features of Sagar Fish Seed Hatchery

Name	*Sagar Fish Seed Hatchery*
Location	At. Kanha, Post. Charthana Tq. Jintur Dist.Parbhani, Maharashtra State, INDIA
Total area	04.20 ha.
Water spread area	01.90 ha.
Construction started	Year 2000-2001
Construction completed	Year 2001-2002
Construction cost	16 lakh (160 thousands Indian Rupees)
Ponds	Total = Working + Dead
Nursery pond	11 = 11
Rearing pond	03 = 03
Stocking pond	02 = 02
Fish spawn production capacity	200 lakh (2 Crores)
Fish spawn stocking capacity	100 lakh (I Crore)
Fish fingerling stocking capacity	01 lakh (100 thousands)
Overhead tank cum Breeding unit	01(88,000 litre water stocking capacity) 8 m. diameter and 4 m. depth
Incubation pool (two)	3 m. diameter and 1.5 m depth
Water supply through	Bore well

Units: ha. = Hacters, m. = Meters.

Source: Mr. Jogwadkar Shivhar, Owener of Sagar Fish Seed Hatchery At. Kanha Post. Charthana Tq. Jintur Dist.Parbhani.

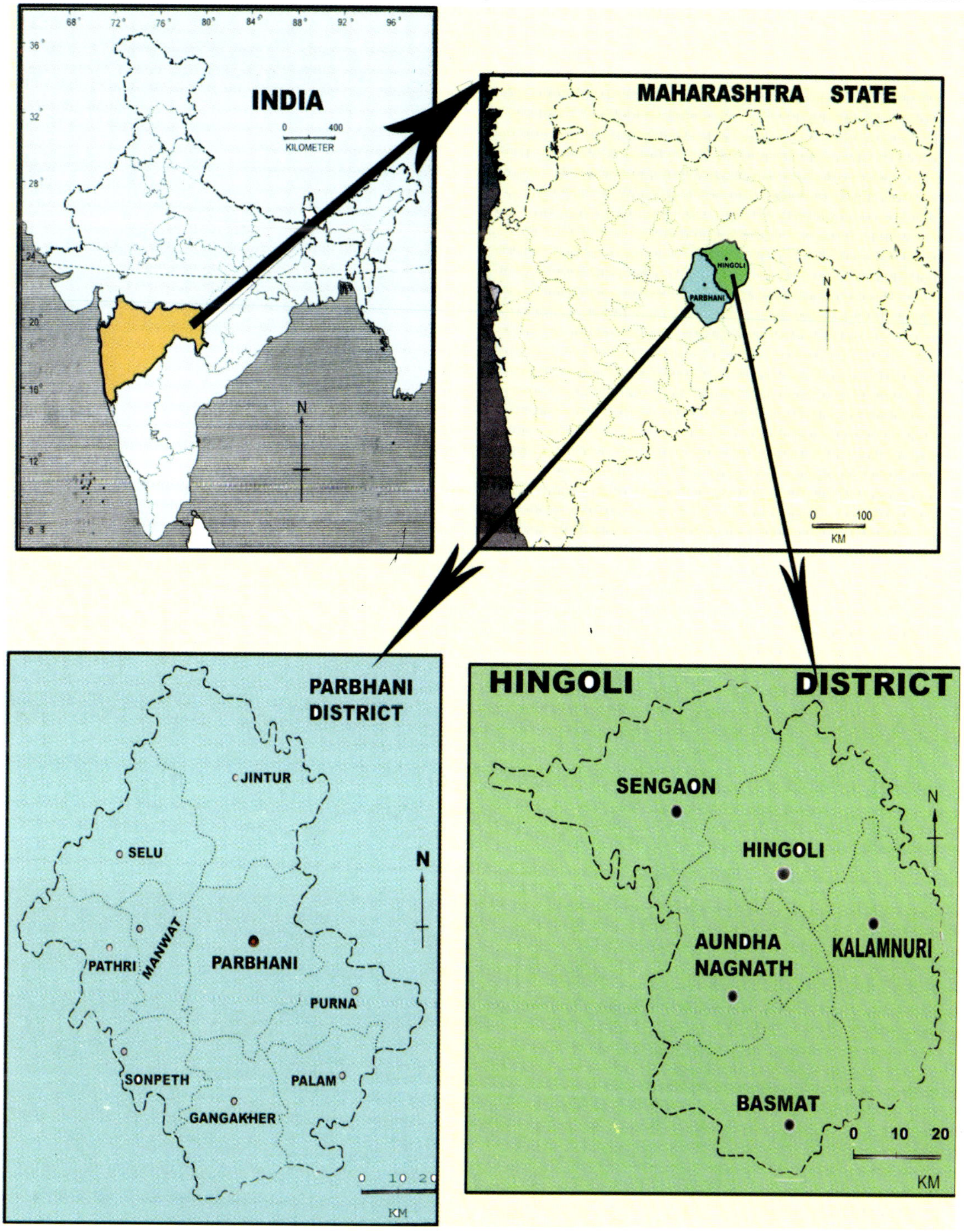

Figure 37.1: Location Map of the Study Area, Map of India, Maharashtra, Parbhani District Showing Sagar Fish Seed Hatchery (Modified Circular Chinese Fish Seed Hatchery)

View of Sagar Fish Hatchery (Indigenous New Model) at Kanha Tq. Jintur

Arrangement of Pipeline, Valves and Storage Tank of Sagar Fish Seed Hatchery

Visit of Guide Dr. S.P. Chavan to Observe the Modified Chinese Hatchery

Mr. Shivhar Jogwadkar, Owner of Sagar Hatchery Providing Information

View of Dry Nursery Ponds at Sagar Fish Seed Hatchery

Figure 37.2: Modified Circular Chinese Hatchery–' Sagar Fish Seed Hatchery of Marathwada'

Infrastructure Available at Sagar Fish Seed Hatchery

Office of Sagar Fish Seed Hatchery

A small house of Mr. Jogwadkar is being used as a office for this hatchery unit. Thermometer, pH meter, Mosquito nets for netting, potassium permagnate, oxygen cylinders, diesel engine are the important components available in the Office-cum-Laboratory of this hatchery unit, but water analysis kit, transportation vehicle for seed transportation etc were not available.

Hatchery Unit and Ponds

Hatchery unit includes one dug well used as underground water storage tank where water from bore well is piped and kept for desilting. One overhead tank cum breeding unit constructed on the underground tank. Water is lifted from underground storage tank to overhead breeding tank by motor pump or diesel pump. The over head tank cum breeding unit having inlet pipes and central outlet pipe. There is one incubation unit where eggs are incubated up to spawn stage then transferred to the nursaries, this is the speciality of this private hatchery. which was different from the well known Chinese Circular hatchery for Indian major carp seed production.

Sagar Fish Seed Hatchery unit has 11 nursery ponds, 03 rearing ponds and 02 stocking ponds. All the ponds were in working condition. The present status of types and number of ponds and their size with water storage capacity is given in the following Table 37.2.

Table 37.2: The Present Status of Number of Ponds and their Size with Water Area, Working Ponds Numbers and their Water Area at Sagar Fish Seed Hatchery (2005-08)

Sl.No	*Types of Pond*	*No. of Ponds*	*Water Area of Ponds in ha.*	*Size of Ponds*	*No. of Working Ponds in ha.*
1.	Stocking pond	02	01.19	189 X 55 X 2m 24 X 62 X 2m	02
2.	Rearing pond	03	0.50	50 X 40 X 1 m 40 x 30 x1 m 60 x 30 x 1 m	03
3.	Nursery pond	11	0.22	10 X 20 X 1 m(each)	11

Source: Mr. Jogwadkar Shivhar, Owner of Sagar Fish Seed Hatchery At. Kanha, Post. Charthana Tq. Jintur Dist. Parbhani.

The two stocking ponds of 1.19 ha.capacity were used to maintain the brood stock of fish species *Catla catla, Labeo rohita, Cirrhina mrigal, Cyprinus carpio, Heteropneustes fossilis* and *Ctenopharengedon idella*. Regular feeding to brood stock at proportion of 2 per cent of their body weight was done for the maintenance of health of brooders. The data of brooder fish species maintained at Sagar Fish Seed Hatchery during July 2005 to June 2008 is given in Table 37.3.

Induced Breeding Experiments at Sagar Fish Seed Hatchery

During 2005- 2008 it was observed that, induced breeding experiments of *Catla catla, Labeo rohita,* and *Cirrhina mrigal* were carried out in month June to August every year, where as *Cyprinus carpio* breeding experiments were not carried out. Following points of hatchery management of this hatchery unit are different from the Government Circular Chinese hatchery unit.

Table 37.3: Fish Broodstock at Sagar Fish Seed Hatchery during 2005 to 2008

Sl.No.	Duration and Fish Proportion		Fish Species											
			Catla-catla		Labeo rohita		Cirrhina mrigal		Cyprinus carpio		Silver Carp		Grass Carp	
			Male	Female	Male	Female	Male	Female	Male	Female	Male	Female	Male	Female
1.	June	No	35	34	20	20	10	10	No	No	No	No	No	No
	2005	Wt.	160	155	72	70	38	36	No	No	No	No	No	No
2.	Sept	No	33	31	19	18	08	08	No	No	No	No	No	No
	2005	Wt.	150	140	69	64	32	30	No	No	No	No	No	No
3.	Dec.	No	33	31	19	18	08	08	No	No	No	No	No	No
	2005	Wt.	150	140	69	64	32	30	No	No	No	No	No	No
4.	Mar	No	33	31	19	18	08	08	No	No	No	No	No	No
	2006	Wt.	150	140	69	64	32	30	No	No	No	No	No	No
5.	June	No	33	31	19	18	08	08	No	No	No	No	No	No
	2006	Wt.	150	140	69	64	32	30	No	No	No	No	No	No
6.	Sept	No	31	28	17	16	06	06	No	No	No	No	No	No
	2006	Wt.	139	125	61	55	22	20	No	No	No	No	No	No
7.	Dec	No	31	28	17	16	06	06	No	No	No	No	No	No
	2006	Wt.	139	125	61	55	22	20	No	No	No	No	No	No
8.	Mar	No	31	28	17	16	06	06	No	No	No	No	No	No
	2007	Wt.	139	125	61	55	22	20	No	No	No	No	No	No
9.	June	No	31	28	17	16	06	06	No	No	04	04	No	No
	2007	Wt.	139	125	61	55	22	20	No	No	44	48	No	No
10.	Sept	No	28	24	15	13	04	04	No	No	04	04	No	No
	2007	Wt.	123	105	53	42	14	13	No	No	44	48	No	No
11.	Dec	No	56	58	55	57	60	65	102	107	04	04	10	10
	2007	Wt.	212	220	200	210	190	200	280	290	44	48	20	20
12.	Mar	No	56	58	55	57	60	65	102	107	04	04	10	10
	2008	Wt.	212	220	200	210	190	200	280	290	44	48	20	20

1. Preparation and management of nursery and rearing ponds.
2. Repair and maintenance of inlet and outlet pipes of hatchery unit, diesel Engine etc.
3. Examination of brood stocks.
4. Availability of hormonal injections.
5. Examination of working hatchery unit.
6. Necessary actions on the starting day of each breeding experiment and during the incubation period.

Preparation and Management of Nursery and Rearing Ponds

At Sagar Fish Seed modified Chinese circular hatchery 11 nursery and 3 rearing ponds of average water area 0.72 ha are used to stock the spawn to rear upto fry, semi-fingerling and fingerling stage. In summer, the nursery ponds and rearing ponds remained always dried due to drought condition. It was observed that, the water was allowed to fill in nursery and rearing ponds by natural rain. One day before the stocking of spawn of Indian major carps in to the nursery and rearing ponds, they were manured by using paste made up of 5 to 6 kg fresh cow dung, 1 kg Super-phosphate and 5 kg oil seed cake for the production and growth of plankton.

Repair and Maintenance of Inlet and Outlet Pipes of Hatchery Unit, Diesel Engine

Before actual start of breeding experiments in the month of June every year, all inlet and outlet pipes and valves of breeding tank and incubation tank were checked for normal working, if necessary they were repaired. The diesel engine and electrical motor is used to fill the overhead tank cum breeding tank.

Examination of Brood Stocks

8 to 10 days before the start of breeding experiments, the brood stock was examined by observing and pressing their abdomen for eggs and milt. Drag net was used to haul the brooders. Brood fish species are kept in the same stocking pond.

Availability of Hormonal Injection

The role of hormone in induced breeding experiments is very significant. In Sagar Fish Seed Hatchery during 2005-2008, Inj. Ovaprim hormone (Syndel Laboratories, Canada) was used. Mr. Jogwadkar, the owner of this hatchery unit explained that, in year 2004-05 comparative 15 per cent more Fish brooder mortality was caused due to the use of Ovatide hormonal injections (Hemo Pharma, Mumbai) in breeding experiments resulted into adverse effect on the expected target of fish seed production. The owner of this hatchery unit also maintain the data of comparative success of the indigenous hormone and exotic hormone as a part of research to judge the the success of the only two hormones in this field of fish culture, he was a agriculture graduate from Marathwada Agriculture University, Parbhani, Maharashtra.

Examination of Working Hatchery Unit

4 to 5 days before actual breeding experiments, a demo of working condition of hatchery unit was examined by supply of water from Bore-well to Well (underground water tank) by pipeline. Upliftment of water to breeding tank cum overhead tank by electric motor or diesel engine and supply of water from overhead tank cum breeding tank to incubation tank, then from incubation tank to spawn collection unit to check the valves of inlet and outlet pipes of various tanks (Figure 37.2).

Induced Breeding of Fishes

On the first day of each breeding experiment all the necessary work related to water supply, electric motor and diesel engine, brood stock sorting, hormonal injection and syringes, potassium permanganate ($KMnO_4$) solution preparation for disinfection were personally done by Mr. Jogwadkar and the team of 3 persons remain alert for 24 hours.

On the first day of each breeding experiment, the brood stock from the stocking pond is collected, examined and some brooders are selected for the breeding experiments. The selected brood fish species are weighed and kept into mosquito bag net and then transferred to water filled overhead tank cum breeding tank of this hatchery unit.

On the day of breeding experiment between 6 to 7 p.m. with the help of 02 workers, Mr. Jogwadkar inject the ovaprim hormone to brood fish species according to their body weight *i.e.* 0.5ml/kg to female fish species and 0.4ml/kg male fish species of carps ventral to the lateral line in the abdominal musculature to every fish. After injection, all the brooders are shifted into the overhead tank cum breeding tank for breeding. The water sprinklers from circular inlet pipe fixed on the breeding tank was started and water current in circular breeding tank is created by adjusting the flow speed of inlet and central outlet pipe.For the breeding of IMC species, 6-8 hours are required after the hormonal injections to the brood fishes in this hatchery unit.

Induced Breeding of Indian Major Carps

In 2005-06 Indian Major carp (IMC) breeding experiments were carried out during 05.07.2005 to 26.07.2005 by utilizing available brood stock in the hatchery unit (Table 37.3). 31 males of IMC having total body weight 110 kg and 31 females of IMC having total body weight 108 kg was used in four sets of breeding experiments. It includes 20 (57.14 per cent) Catla males having total body weight 75 (46.87 per cent) kg and 20 (58.82 per cent) Catla females having total body weight 73 (47.09 per cent) kg, 09 (45.00 per cent) Rohu males having total body weight 29 (40.27 per cent) kg and 09 (45.00 per cent) Rohu females having total body weight 29 (41.42 per cent) kg, 02 (20.00 per cent) Mrigal males having total body weight 06 (15.78 per cent) kg and 02 (20.00 per cent) Mrigal females having body weight 06 (16.66 per cent) kg. In these four (04) sets of breeding experiments all females of IMC having total body weight 108 kg were succeed to release eggs from which 116 lakh spawn were obtained. The Percentage of spawn to per kg body weight of successful female carps is 01.07 lakh spawn to per kg body weight of successful female carps.

In 2006-07 Indian Major carp breeding experiments were carried out during 26.06.2006 to 17.08.2006 by utilizing available brood stock in the hatchery unit (Table 37.3). 25 males of IMC having total body weight 112 kg and 25 females of IMC having total body weight 107.50 kg was used in six sets of breeding experiments which includes 13 (39.39 per cent) Catla males having total body weight 63 (42.00 per cent) kg, 13(41.93 per cent) Catla females having total body weight 60.50 (43.21 per cent) kg, 06 (31.57 per cent) Rohu males having total body weight 24 (34.78 per cent) kg and 06 (33.33 per cent) Rohu females having total body weight 22 (34.37 per cent) kg 06 (75.00 per cent) Mrigal males having total body weight 25 (78.12 per cent) kg and 06 (75.00 per cent) Mrigal Females having body weight 25(83.33 per cent) kg. In these six(06) sets of breeding experiments 25 females of IMC having total body weight 107.50 kg were succeed to release eggs from which 124 lakh spawn were obtained. The Percentage of spawn to per kg body weight of successful female carps is 01.15 lakh spawn to per kg body weight of successful female carps.

In 2007-08 Indian Major carp (IMC) and Silver carp breeding experiments were carried out during 05.07.2007 to 25.07.2007 by utilizing available brood stock in the hatchery unit (Table 37.3). 40 males of IMC and Silver carp having total body weight 184 kg and 40 females of IMC and Silver carp having total body weight 165 kg was used in five (05) sets of breeding experiments. It includes 20 (64.51 per cent) Catla males having total body weight 106 (82.17 per cent) kg, 20 (71.42 per cent) Catla females having total body weight 101 (80.80 per cent) kg, 12 (66.66 per cent) Rohu males having total body weight 24 (68.85 per cent) kg, and 12 (75.00 per cent) Rohu females having total body weight 24 (72.72 per cent) kg, 06 (100 per cent) Mrigal males having total body weight 22 (100 per cent) kg and 06 (100 per cent) Mrigal Females having body weight 20 (100 per cent) kg, 02 (50.00 per cent) Silver carp males having total body weight 24 (54.54 per cent) kg and 02 (50.00 per cent) Silver carp Females having body weight 24 (50.00 per cent) kg. In these five (05) sets of breeding experiments 40 females of IMC and Silver carp having total body weight 185 kg were succeed to release eggs from which 166 lakh spawn were obtained. The Percentage of spawn to per kg body weight of successful female carps is 0.89 lakh spawn to per kg body weight of successful female carps.

Spawn Stocking and Fish Seed (Fry, Semi-fingerling and Fingerling) Production and Marketing

During breeding experiments spawn produced was either directly marketed at spawn stage from incubation tank to the private parties or the non-marketed spawn was stocked into the nursery and rearing ponds to grow in to fry stage, advanced fry, semi-fingerling and fingerling stage. The fish seed stages were marketed to fish co-operative societies and lease owners of the Zila Parishad ponds and State Irrigation Department ponds. It was observed that, in 2005–2006 out 116 Lakh spawn produced, 57 Lakh spawn *i.e.* 49.13 per cent was marketed. From the stock of 59 lakh spawn, 15 lakh (25.42 per cent) fry was produced 0f which 06 lakh fry was marketed; from stocking 09 lakh fry, 2.50 lakh advanced fry, 0.50 lakh semi-fingerling and 01.25 lakh fingerling was obtained and marketed. In 2006–07 out of 124 Lakh produced spawn 74 Lakh spawn *i.e.* 59.67 per cent stock was marketed and 50-lakh spawn was stocked into nursery and rearing ponds to grow in to fry, semifingerling and fingerling stage. But, due to heavy rain in July 2006 the embankment of those ponds were damaged and in the flood all the stocked spawn was lost and there was a huge loss to the hatchery unit. In 2007-08 out of 166 Lakh produced spawn 166 Lakh spawn *i.e.* 100 per cent spawn were marketed directly from incubation tank.It was observed that the owner of the hatchery unit had arranged the breeding experiments on the fish seed demand of different fishermen of this area by taking advance payment as the repair work of the ponds were in progress.

Table 37.4: The Rates of Fish Seed Sale during (2007-2008) at Sagar Fish Seed Hatchery

Sl.No.	Fish Seed Stages for Marketing		Rate of Fish Seed Including Packaging Charges	Rate of Fish Seed/Lakh Including Packaging Charges
1	Spawn	Catla	Rs. 1540/lakh	Rs. 1540
		Silver carp	Rs. 1540/lakh	Rs. 1540
		IMC(Mixed)	Rs. 1240/lakh	Rs. 1240
2	Fry	Catla	Rs. 160/1000	Rs. 16000
		Silver carp	Rs. 160/1000	Rs. 16000
		IMC(Mixed)	Rs. 130/1000	Rs. 13000
3	Advanced fry	Catla	Rs. 160/500	Rs. 32000
		Silver carp	Rs. 160/500	Rs. 32000
		IMC(Mixed)	Rs. 130/500	Rs. 26000
4	Semi-Fingerling	Catla	Rs. 160/250	Rs. 64000
		Silver carp	Rs. 160/250	Rs. 64000
		IMC(Mixed)	Rs. 130/250	Rs. 52000
5	Fingerling	Catla	Rs. 160/125	Rs. 128000
		Silver carp	Rs. 160/125	Rs. 128000
		IMC(Mixed)	Rs. 130/125	Rs. 104000

Source: Mr. Jogwadkar Shivhar, Owner of Sagar Fish Seed Hatchery At. Kanha Post. Charthana Tq. Jintur Dist.Parbhani.

For packing of fish seed stages in oxygen filled polythene bags Rs 10/packing was charged to the customers. In each pack about 25000 spawn or 1000 fry or 500 advanced fry or 250 semi fingerlings or 125 fingerlings were packed.

Table 37.5: Spawn Production, Spawn Sale, Spawn Stocking and Fish Seed Occurrence from Stocked Ponds and their Percentage and Income to the Centre by Fish Seed Sale during (2007-2008) (Figures in Lakh)

Sl.No	Year	Spawn Production IMC	Sale	Spawn Stalked	SALE				Total Fish Seed Produced	% of Seed Obtained	Income
					Fry	Advanced Fry	Semi Fingerling	Fingerling			
1.	2005-2006	116	57	59	06	2.50	0.50	1.25	10.25	17.37%	4.47
2.	2006-2007	124	74	50	No	No	No	No	No	No	1.11
3.	2007-2008	166	166	No	No	No	No	No	No	No	2.45

Source: Mr Jogwadkar Shivhar, Owner of Sagar Fish Seed Hatchery At Kanha Post Charthana Tq. Jintur Dist.Parbhani.

Table 37.6: Summary of Indian Major Carps and Silver Carps Breeding Experiments at Sagar Fish Seed Hatchery during (2007-2008)

Sl.No.	Details of Breeding Experiment		2005-2006	2006-2007	2007-2008
1.	Duration of breeding experiment		05.07.2005 to 26.07.2005	26.06.2006 to 17.08.2006	05.07.2007 to 25.07.2007
2.	Total number of breeding experiment sets		04	06	05
3.	Total number of successful breeding Expt. sets		04	06	05
4.	Total number of IMC males used in breeding experiment sets	No weight	31 100	25 112	40 184
5.	Total number of IMC females used in breeding experiment sets	No weight	31 108	25 107.50	40 165
6.	Total number of IMC successful females used in breeding Expt. Sets	No weight	31 108	25 107.50	40 165
7.	Hormones used in IMC breeding experiment sets	Males Females	50 ml 50 ml	56 ml 54 ml	90 ml 80 ml
8.	Total number of eggs obtained from IMC successful female brooders		N.A	N.A	N.A
9.	Per cent of eggs obtained from IMC successful female brooders to per kg of body weight		N.A	N.A	N.A
10.	Total number of spawn obtained from IMC successful female brooders		116	124	166
11.	Total number of spawn obtained from IMC successful female brooders to per lakh eggs		1.07	1.15	0.89
12.	Spawn supply out of centre		57	74	166
13.	Spawn stocked in the centre		50	50	No
14.	Fry obtained from spawn		06	No	No
15.	Semi fingerling obtained		03	No	No
16.	Fingerling obtained		1.25	No	No

Source: Mr. Jogwadkar Shivhar, Owner of Sagar Fish Seed Hatchery At Kanha Post Charthana Tq. Jintur Dist. Parbhani

Table 37.7: Mortality of Fish Brood Stock during Induced Breeding Experiment at Sagar Fish Seed Hatchery during (2005-2008)

Sl.No.	Duration		*Catla-catla*		*Labeo rohita*		*Cirrhina mrigal*		*Total*	
			Male	*Female*	*Male*	*Female*	*Male*	*Female*	*Male*	*Female*
1.	Aug	No	02	03	01	02	02	02	05	07
	2005	Wt.	10	15	03	06	06	06	19	27
2.	Aug	No	02	03	02	02	02	02	06	07
	2006	Wt.	11	15	06	09	10	10	27	34
3.	Aug	No	03	04	02	03	02	02	07	09
	2007	Wt.	16	20	08	13	08	07	32	40

Source: Prop. Mr Jogwadkar Shivhar, Owner of Sagar Fish Seed Hatchery At Kanha Post Charthana Tq. Jintur Dist. Parbhani

Modifications in the Sagar Fish Seed Hatchery Unit

According to Gupta *et al.*, (2000) the fish seed hatchery unit can be judiciously designed keeping in view business, resource and production environments for obtaining the sustainable fish seed production and profit.

1. This is a rainwater dependant hatchery; source of water for this hatchery unit is only one borewell which fulfills the demand of this hatchery. The use of rainwater is precisely and perfectly being used by the hatchery unit, the same concept can be used for the working of Government Hatchery units in this region and other draught prone regions of India. With very little improvement, this could be an ideal model of Hatchery to fulfil the fish seed demand of Indian major carps in draught prone freshwater pond culture fishery.
2. This hatchery is located in the drought prone area, sufficient water is not available to stock in the nursery and rearing pond throughout year, hence the owner of this hatchery is able to mange single stocking pond to rear the brooders.
3. A circular tank of 88,000 liter capacity is used as overhead tank-cum breeding tank by changing the water supply pipeline and arrangement of the valves.
4. The owner of this hatchery unit supply the fish seed of only one kind of species, generally *Catla catla* hence this hatchery unit is very famous to supply the fish seed of one specific species in this region. Mixed seed was also supplied from this hatchery. Getting the fish seed of only one specific fish species, this hatchery is the only reliable and authentic source for the fish culturists in this region, the farmers have more belief on this hatchery unit as compared to the Government Hatchery units. The most demanded fish seed from this hatchery is of *Catla catla* species.
5. Scarcity of water and drought conditions are the important factors for changing the design of this hatchery it is the unique and only modified Chinese Circular Hatchery Unit of Maharashtra.
6. The experience of the fish culturists in this region is, this hatchery is also famous for supply of accurately measured fish seed of advanced fry stage of Indian Major Carps as compared to the other hatchery units in the region. Due to accuracy of getting the fish seed in nearly

perfect number, the fish culturists don't need to thinning the stock or to reintroduce the stock, this save the money and time of the fish culturists and improve the fish production.

Status of Private and Government Chinese Circular Fish Seed Hatcheries in Parbhani District

There are 03 Government and 02 Private fish seed hatcheries in District Parbhani and Hingoli. Government Fish Seed hatchery in District Parbhani is Masoli Fish Seed Hatchery at Masoli reservoir and in District Hingoli are Siddheshwar Fish Seed Hatchery and Bhategaon Fish Seed Hatchery having spawn production capacity 500 lakh, 500 lakh and 300 lakh respectively. Spawn stocking capacity in the ponds of these hatcheries are 40 lakh, 180 lakh and 44 lakh respectively and fry stocking capacity is 16 lakh, 36 lakh and 11 lakh respectively. It means these three Government Fish Seed Hatcheries stocking capacity for the development of fry into fingerling stage which is very less than the estimated fingerlings stocking (174.575 lakh) in different reservoirs present in district Parbhani and Hingoli. Hence, it is recommended to construct new Government fish seed hatcheries in District Parbhani and Hingoli. From all government fish seed hatcheries, the fish seed marketed is always at the spawn stage or fry stage. Brood fish identification and selection are very important for breeding management, success of induced breeding depend on proper selection of brooders (Musa A.S.M and Abdus Salam, 2007). It seems that, infrastructure facilities were adequate at these fish seed hatcheries, but the maintenance of stoking ponds, rearing ponds and nursery ponds was not found good. Fish brooder size and weight was also less in all these fish seed hatcheries. Remarkably it was found that, the conversion ratio of spawn to fry and from fry to fingerlings at all hatcheries was less; therefore the hatcheries could not fulfill the fingerling demand to stock in to the reservoirs of this region. From these hatcheries there was marketing of spawn, fry and fingerlings to stock in reservoirs of other districts of Maharashtra like Beed, Osmanabad, Nanded and Washim. The reason behind the scarcity of brooders is, there was mortality of brooders annually during breeding experiments. The probable cause behind the death of brooders was the impact of hormonal injection and improper dose of hormonal injections during induced breeding experiments. Marimathu *et al.*, (2000) successfully produced *H.fossilis* seed by using Ovatide hormone in induced breeding and mentioned the need of culture of economically important, highly prized air-breathing catfishes like *Heteropneustes fossilis*, but such type of breeding experiments were not found in any hatchery in this region. Reddy (2000) mentioned the role of 'D' series of carp hatchery, as it provides controlled environment and continous flow of fresh clear water and maintenance of optimum dissolved oxygen *i.e.* 7-9 ml/l and increased survival rate *i.e.* 90-95 per cent from fertilized eggs to spawn. This commercially viable 'D' series hatcheries systems were widely adopted by the government organizations, entrepreneurs and fish farmers in AndhraPradesh, Madya Pradesh, Maharashtra, Uttar Pradesh, Haryana, Tamil Nadu and Rajasthan states of India, but not in use in this region. It was also found that, the number of fish brooders decreased due to mortality of brooders during breeding experiments was not replaced by the addition of new brooders in any government fish seed hatchery. It was found that, when dogs consumed hormonal injected dead brooders after breeding experiments, then the dog become weak and finally died within eight days at Masoli Fish Seed Hatchery. Therefore, the consumption of injected dead brood fishes is not safe. It can be suggested that, there is a need of recruitment of new brooders captured from the reservoirs nearby every year in to these hatcheries to compensate the loss of brooders due to the mortality of brooders during induced breeding.

The main drawback in seed supply from all these hatcheries was that, these hatcheries do not supply the proportionate mixer of IMC or pure seed of any one species of IMC, therefore two private hatcheries in the District were established. First Private hatchery was established at Jamb (Jamb Reservoir) near Parbhani City named Sudershan Fish Seed Hatchery and other at Village Kanha near Charthana Tq. Jintur Dist. Parbhani named Sagar Fish Seed Hatchery for the production and supply of proportionate quantity of Indian major carp fish seed. The private Hatcheries also market the fish seed of any one fish species of Indian major carps to private pond owner and reservoir lease owners.

References

Gupta, S.D., Rath, S.C. and Ayyapan, S., 2000. Designing and management of eco-hatchery complex for carp seed production. *Fishing Chimes*, 19(10 and 11): 27–33.

Hussaini, S.S., 1997. F.F.D.A.Akola, Information Bulletin on Hatcheries and their Management and Breeding of Major Carp.

Sinha, Manirajan, 2000. Carp seed production systems in West Bengal. *Fishing Chimes*, 19(10 and 11): 43–45.

Reddy, A.K., 2000. Development of carp hatcheries–'D' series at CIFE, Mumbai. *Fishing Chimes*, 19(10 and 11): 37–42.

Chapter 38

Variations in Growth of *Perionyx excavatus* Fed *Ad libitum* on Different Feed during Different Seasons from Hatchlings to Post Reproductive Statges

☆ *Manjunathr, Krishna Ram H. and Shivabasavaiah*

ABSTRACT

Growth in *Perionyx excavatus* fed *ad libitum* on different feed during different seasons from hatchlings to post reproductive stages was revealed. Life cycle of *Perionyx excavatus* has small immature, large immature, adult I and adult II stages. The total growth increased as function of life stages. The total growth during rainy season increased from 112.5 mg wet weight during small immature stage to as much 1010.0 during the last adult stage II with T_1 feed. This values were comparable the worms with T_3 and T_4 feed; while this growth decreased to 995.6 mg with T_2 feed during adult II stage. During winter season the total growth was maximum with feed T_3 at adult III stage which amounted to 1056.0 mg wet weigh. Next maximum growth of 1030.3 mg was found with T_2 feed. Similarly the growth differed with the change in the season. During summer period the maximum growth was found with T_4 feed which amounted to 1015.0 mg during adult II stage. Lowest growth was found during summer with all other feed T_1, T_2 and T_3. Perhaps the lower growth during summer may be due to higher temperature. The reduced growth affected production of cocoons which was the least during summer.

Introduction

Study of earthworm bioenergetics leads to understand how environmental factors and feed habitats affect growth through influences on utilization of food energy and material in their body. This being much more than the study of growth, it permits us to evaluate the cost of life under different environmental conditions. The growth rates of earthworms increase by 148 per cent to 119 per cent by anaerobic biogas slurry (Balasubramanian and Bai, 1995). When more resource is allocated to production of an organism, it is likely to increase its capability for survival under unfavorable circumstances, which is likely to increase its competitive ability (Patnaik, *et al.*, 2004). Consumption, digestion, egestion, excretion, metabolism, respiration, production and growth aspects in different animal phyla pertaining to energy allocations, are available in "Animal Energetics" (Pandian and Vernberg, 1987; Payal Garg, *et al.*, 2005).

Growth can be known by studying the food consumption by earthworm. Growth forms the outcome of the integrated activities of the whole individual; which is dependent on the metabolic state of the individual and on energy expended in maintenance and behavior (Alexander, 1999) as well as on the quantity and quality of the organic waste consumed. Rates of feeding and conversion efficiency estimates are better parameters of rates of metabolism (Paloheimo and Dickie, 1966). *Perionyx excavatus* feeds on dead plant material, compost and organic manure. Several works have found Cattle dung as natural diet forearth worm (Holter and Hendriksen, 1988. Edwards *et al.*, 1985; Hendriksen, 1991a and 1991b; 1995; 1997; Reinecke and Viljoen, 1990; Viljoen and Reinecke, 1990). A favorable environment needs to be created to permit the balance such studies need to be effectively pursued in this laboratory, to show sexual touchstones in nature for biologist with a bioenergetics point. food converted into body substance, conversion rate, conversion efficiency, gross conversion efficiency (K_1), net conversion efficiency (K_2) and metabolism in insects (Delvi, 1972; Delvi and Pandian 1972; Radhakrishna and Delvi, 1987 and 1992).

Growth involves conversion of food material into tissue through the process of feeding digestion, assimilation and synthesis. These process are common for both invertebrates and vertebrates and require expenditure of energy for maintenance and metabolism. The balance of assimilated energy, available after completing these expenditures is allocated to growth in *Perionyx excavatus* as in other earthworms. It includes somatic/segmental and reproductive. Since earthworms are hermaphrodite, developing male accessories like testes, testes sacs, spermathecae and seminal vesicles; and female organs like ovary, ovisacs and oviduct funnels and other accessories like prostate gland and clitellum that help in copulation and chitinous cocoon production, require additional energy.

The present investigations deals with the bioenergetic studies of the tropical epigeic, oriental compost earthworm *Perionyx excavatus* from hatching to the post reproductive (active) stages. This study has been carried out using agro-industrial refuse-sugar factory pressmud, Cattle dung and Banana waste; for maximum utilization of these species for the safer conversion at industrial level through the studies on rates of feeding, assimilation and conversion, efficiencies of assimilation and conversions functions of growth and the patterns of energy partitioning.

Materials and Methods

Different organic wastes Cattle dung, Press mud, Banana plant waste were decomposed and stabilized by keeping it moist for few days or till the warmth subsides, before being used for the study. The epigeic earthworm *Perionyx excavatus* were cultured under laboratory conditions. Fed *ad libitum* on cattle dung (T_1 - control feed), Press mud (T_2 - experimental feed), Banana waste (T_3 - experimental feed) and T_4 - experimental feed with combination of T_1 T_2 and T_3 in equal proportions (1:1:1), during different seasons (rainy, winter and summer season) with 75 per cent to 80 per cent moisture throughout the study period.

Sixty adult earthworms (having approximately equal age and body weight) were selected from the culture bed and were left into labeled plastic tub containing decomposed and stabilized cattle dung. After four to five days sixty cocoons were selected and their weight were recorded and placed in respective labeled plastic tub provided with filter paper for incubation. As soon as juvenile earthworms were hatched, the body weights were recorded and forty juvenile earthworms were isolated and left into four labeled aerated plastic boxes (10 juvenile worms/replicate) containing respective feed materials. One replicate with control feed T_1 (cattle dung) and three replicates with treatment feeds T_2 (Press mud), T_3 (Banana waste) and T_4 mixed feed. Body weight, feed input and feed output were recorded once in three to four days till the juvenile earthworm's attained adult stage and the above observations were subjected for different season.

Results and Discussion

The bioenergetics data, collected over a period of time, of earthworm *Perionyx excavatus* reared under laboratory conditions is helpful to understand its contribution at different climatological conditions with different food components. This leads to study of the role played by the earthworm *Perionyx excavatus* in solid waste management and also leads to study of metabolic rate for conversion of solid waste into vermicompost (earthworm faeces) and utilization of proteins produced as body tissue. Somatic growth of the earthworm *Perionyx excavatus* in the present study is considered in live weight as a function of age group. Somatic growth does not include the amount of body substance spent on coelomic fluid, nephrediel excretion and cocoon production and hence is different from conversion that is represented in dry weight. The amount of food converted into body substance is expressed in mg dry weight. Net growth refers to the amount of dry matter present in the body at death due to senescence.

Growth during Rainy Season

The growth was recorded with 20 days interval at different age groups. *Perionyx excavatus* displayed preference for growth during rainy season than during summer season. The maximum growth was during the adult II stage with all the four tested feeds. However, the total growth was the lowest of 995.6 mg during this stage with Press mud T_2 feed during rainy season. It can be seen that during rainy season the maximum growth was 1025.0 mg during adult II stage with T_3 (Banana waste). This is followed by slight decrease in growth with T_4 (Mixed feed) of 1005.0 mg. Further it decreased to 1010.0 mg with T_1 Cattle dung. *Perionyx excavatus* seems to prefer T_3, Banana waste, over other feeds.

Growth during Winter Season

The growth was recorded with 20 days interval at different age groups. *Perionyx excavatus* displayed preference for growth during rainy season than during winter season. The differences depended not only on season but also the type of feed. The growth with T_1, Cattle dung and T_4, Mixed feed, was slow and are comparable to each other. The worms took more time and were active till 100 days of the growth period. During this period the worms grew to 1006.0 mg with T_1 Cattle dung and 1008.0 mg with T_4, Mixed feed. Next the growth was comparable with T_2 Press mud where the worms grew to 1030.3 mg during adult III stage. However, *Perionyx excavatus* grew fastest attaining a body weight of 1056.0 mg with T_3 Banana waste during winter season at adult III stage.

Table 38.1: Variations in Growth of *Perionyx excavatus* Fed *Ad libitum* on Different Feed during Different Seasons from Hatchlings to Post Reproductive Stages

Season	*Age Group*	T_1	T_2	T_3	T_4
Rainy Season	Small Immature	110.6 ± 9.1	62.9 ± 14.8	67.8 ± 7.4	120.1 ± 4.6
	Large Immature	675.9 ± 17.2	109.8 ± 16.8	605.5 ± 17.5	715.2 ± 21.6
	$Adult_{(1)}$	917.0 ± 29.5	505.2 ± 52.3	990.0 ± 24.4	985.5 ± 13.1
	$Adult_{(2)}$	1010.00 ± 22.4	995.6 ± 47.8	1025.0 ± 16.7	1005.0 ± 38.6
Winter Season	Small Immature	61.0 ± 8.8	79.3 ± 9.4	79.0 ± 7.2	61.4 ± 3.6
	Large Immature	269.1 ± 24.1	307.3 ± 29.0	315.9 ± 31.9	149.2 ± 15.4
	$Adult_{(1)}$	683.5 ± 35.1	709.0 ± 15.8	896.2 ± 28.9	395.1 ± 18.6
	$Adult_{(2)}$	992.1 ± 24.1	999.6 ± 39.7	1000.0 ± 17.2	770.8 ± 30.4
	$Adult_{(3)}$	1006.0 ± 16.6	1030.0 ± 57.0	1056.0 ± 47.7	1008. ± 51.8
Summer Season	Small Immature	62.8 ± 5.4	100.0 ± 7.3	105.3 ± 13.7	60.7 ± 5.5
	Large Immature	200.7 ± 20.1	182.8 ± 13.1	214.1 ± 35.7	278.8 ± 13.5
	$Adult_{(1)}$	650.8 ± 53.9	680.5 ± 59.2	607.9 ± 25.2	557.2 ± 35.6
	$Adult_{(2)}$	1001.0 ± 28.7	1000.0 ± 16.1	1000.6 ± 71.3	1015.0 ± 103.9

T_1 Cattle dung (control); T_2–Press mud (experimental); T_3–Banana waste (experimental) and T_4–Mixed feed [Experimental (T_1:T_2:T_3–1:1:1)].

Table 38.2: Correlation Coefficient of Growth in *Perionyx excavatus* Fed *Ad libitum* on Different Feed during Different Seasons from Hatchlings to Post Reproductive Stages 'r'

Season	*Age Group*	$T_1 : T_2$	$T_1 : T_3$	$T_1 : T_4$
Rainy Season	Small Immature	0.01285	–0.17656	–0.05150
	Large Immature	–0.71225	–0.21878	–0.48708
	$Adult_{(1)}$	0.465049	–0.54229	0.033427
	$Adult_{(2)}$	0.685764	–0.63796	0.976295
Winter Season	Small Immature	–0.67142	–0.04437	0.20113
	Large Immature	0.46473	0.040011	–0.498578
	$Adult_{(1)}$	0.164864	–0.355595	–0.282896
	$Adult_{(2)}$	–0.76537	–0.674686	–0.439376
	$Adult_{(3)}$	–0.507034	–0.634988	0.703338
Summer Season	Small Immature	–0.005611	0.435636	0.288143
	Large Immature	–0.822204	0.869198	0.701335
	$Adult_{(1)}$	0.86069	0.278463	0.588203
	$Adult_{(2)}$	–0.040525	–0.056192	0.271407

Values are given as compared to growth at control (T_1)

T_1 Cattle dung (control); T_2–Press mud (experimental); T_3–Banana waste (experimental) and T_4–Mixed feed [Experimental (T_1:T_2:T_3–1:1:1)].

Growth during Summer Season

The growth was recorded with 20 days interval at different age groups. *Perionyx excavatus* displayed preference for growth during rainy season than during summer season, the worms attained different body weight with the least weight of 1000.0 mg during adult II stage with T_2, Press mud feed. Next, comperatively the growth was 1000.6 mg with T_3, banana feed and 1001.0 mg with T_1 Cattle dung feed. However, during summer the maximum growth was found with T_4, Mixed feed where the growth was 1015.0 mg during adult II stage.

There exist a relationship between the growth and the type of feed to the worms. In an attempt to compare the relative difference between the control T_1 Cattle dung and the other three types of feed T_2 press mud, T_3 Banana waste and T_4 Mixed feed, statistical analyses were carried out and coefficient of co-relation ('r' value) was obtained. The analyses were subjected separately for each season for each age group and T_1- Cattle dung fed worms were compared to T_2 - Press mud, T_1 to T_3 - Banana waste and T_1 to T_4 Mixed feed. The results so obtained have been tabulated. It can be seen from Table 38.2, that the coefficient of co-relation different considerably and either positive or negative depending on the season and/or feed. The 'r' values significantly change during each season and at various feed verities. No generalization can be drawn as the relationships change from negative to positive at each level of experiment and type of feed. The coefficient of correlations between T_1 and other feeds were also positive during summer season except in small immature worms the relation was negative with -0.005611 value for T_1:T_2. Similarly in large immature worms also the value was -0.822204 (Table 38.2). However, during adult II stage the negative values were found in both with T_1:T_2 and T_1:T_3. The overall study period during summer season revealed that the worms though fed on the different experimental feed and attained growth, their growth responses were lower compared to other two seasons. Maximum growth was found in T_4, Mixed feed, during summer season. Thus attributable to preference of feed depending on the season and life stage.

References

Alexander, R. Mc Neill, 1999. *Energy for Animal Life*. Oxford University Press, Great Clarendon.

Balasubramanian, P.R. and Bai, R.K., 1995. Recycling of cattle dung, biogas plant-effluent and water hyacinth in vermiculture. *Bioresource Technology*, 52(1): 85–87.

Delvi, M.R. and Pandian, T.J., 1972. Rates of feeding and assimilation in the grasshopper *Poecilocerus pictus*. *J. Insect Physiol.*, 18: 1829–1843.

Delvi, M.R., 1987. Food consumption and utilization in different races of *B. mori*. In: *15th Nat. Sci. Cong.*

Edwards, C.A., Burrows, I., Fletcher, K.E., and James, B.A., 1985. The use of earthworms for composting food wastes. In: *Composting of Agricultural and other Wastes*, (Ed.) J.K.R. Gasser. Elsevier, Amsterdam, pp. 229–242.

Garg, Payal, Gupta, Asha and Satya, Santhosh, 2005. Vermidecomposting of different types of waste usings *Eisenia fetida:* A comparative study. *Bioresource Technology*.

Haynie Donald, T, 2001. *Biological Thermodynamics*. Cambridge University Press, Cambridge, U.K., pp. 369.

Hendriksen, N.B., 1991c. The effects of earthworms on the disappearance of particles from cattle dung pats during decay. *Pedobiologia*, 35: 139–146.

Hendriksen, N.B., 1995. Effects of detritivore earthworms on dispersal and survival of the bacterium *Aeromonas hydrophila*. *Acta Zool. Fennica*, 196: 115–119.

Holter, P. and Hendriksen, N.B., 1988. Respiratory loss and bulk export of org. mat. from cattle dung pats: a field study. *Holistic Ecology*, 11: 81–86.

Paloheimo, J.E. and Dickie, L.M., 1966. Food and growth of fishes. II. Effect of food and temperature on the relation between metabolism and body weight. *J. Fish. Res. Board. Can.*, 23: 868–908.

Pandian, T.J. and Vernberg, F.J., 1987. *Animal Energetics, Vol. 1: Protozoa through Insecta*. Academic Press Inc.

Patnaik, H.P., Panda and Samaol, M., 2004. Preliminary studies on the earthworm species of Kendujhar (Orissa). *J. Appl. Zoo. Res.*, 15: 91–95.

Radhakrishna, P.G., and Delvi, M.R., 1992. Effect of organophosphorus insecticides on food utilization in different races of *B. mori*. (Lepidoptera: Bombycidae). *Sericologia*, 32(1): 71–79.

Reinecke, A.J. and Viljoen, S.A., 1990.The influence of feeding patterns on growth and reproduction of the vermicomposting earthworm *Eisenia fetida* (Oligochaeta). *Biology and Fertility of Soils*, 10(3): 184–187.

Previous Volumes

— Volume 1 —

2007, xvi+194p., figs., tabls., ind., 25 cm Rs. 950

ISBN 81-7035-483-8

— Volume 2 —

2008, xvi+143p., col. plts., figs., tabls., ind., 25 cm Rs. 750

ISBN 81-7035-559-5

— Volume 3 —

2010, xiv+176p., col. plts., figs., tabls., ind., 25 cm Rs. 800

ISBN 978-81-7035-633-2

— Volume 4 —

2010, xvii+182p., figs., tabls., ind., 25 cm Rs. 750

ISBN 978-81-7035-657-8

— Volume 5 —

2011, xviii+231p., col. plts., tabls., figs., ind., 25 cm **Rs. 1200**

ISBN 978-81-7035-697-4

Index